Anticancer Plants: Clinical Trials and Nanotechnology

Mohd Sayeed Akhtar
Mallappa Kumara Swamy
Editors

Anticancer Plants: Clinical Trials and Nanotechnology

Volume 3

Springer

Editors
Mohd Sayeed Akhtar
Department of Botany
Gandhi Faiz-e-Aam College
Shahjahanpur, Uttar Pradesh, India

Mallappa Kumara Swamy
Department of Crop Science,
Faculty of Agriculture
Universiti Putra Malaysia
Serdang, Selangor, Malaysia

ISBN 978-981-10-8215-3 ISBN 978-981-10-8216-0 (eBook)
https://doi.org/10.1007/978-981-10-8216-0

Library of Congress Control Number: 2018935273

Printed on acid-free paper

This Springer imprint is published by the registered company Springer Nature Singapore Pte Ltd.
The registered company address is: 152 Beach Road, #21-01/04 Gateway East, Singapore 189721, Singapore

This book is dedicated to

A. P. J. Abdul Kalam
(1931–2015)
A great personality, scientist, missile man,
and former president of India

Foreword

Cancer is one of the major life-threatening diseases of the world's population. It is characterized by the formation of abnormal cells that proliferate uninhibitedly and often spread all over the body. About 100 different types of cancers have been categorized on the basis of the affected tissue or organ of the human body. As reported by the World Health Organization, cancer is the second leading cause of human mortality around the globe. Nearly 8.8 million deaths have been reported due to cancer in 2015, and more than 200,000 individuals are being diagnosed with various types of cancers every year. Some of the major causes of cancer include smoking, pollution, dietary imbalances, and chronic infections. Several billions of US dollars are being spent annually toward cancer research in developing effective treatments. Though presently available cancer treatments, such as surgery, chemotherapy, radiotherapy, immunotherapy, and targeted therapies, reduce the growth rate of cancer cells, however they are incapable of complete cure. Many chemodrugs used against this disease cause toxicity and cause serious side effects. Further, the development of resistance to the existing chemotherapeutic drugs has compelled adoption of improved alternative treatment approaches with novel therapeutic agents including nanoproducts. In this regard, plant-based natural anticancer drugs have proven to be more effective with negligible side effects against various cancers. More than 50% of the anticancer drug molecules in clinical use are mainly derived from nature. Plant metabolites such as alkaloids, terpenoids, flavonoids, carotenoids, etc., are reported to have anticancer properties. Major plant-based compounds such as vincristine, vinblastine, camptothecin, paclitaxel, irinotecan, topotecan, and etoposide are widely used in treating cancers. Many other potential plant secondary metabolites have been experimentally proven with anticancer properties and are under clinical or preclinical trial phase. More recently, nanoparticles biosynthesized from plant extracts and their compounds have emerged as promising agents in cancer therapy. Nanotechnology as a herbal therapeutic approach comprising novel nano-based drug delivery systems has been found to be effective in treating cancers. Nano-based drug delivery systems not only increase the therapeutic value but also the bioavailability of the herbal medicine. If the guidelines of toxicology and clinical pharmacology are followed, more phytocompounds can be made available as an effective anticancer agent.

This book *Anticancer Plants: Clinical Trials and Nanotechnology (Volume 3)* to be published by Springer includes 11 chapters contributed by the potential researchers from various parts of the world. In Chapter 1, Indian authors highlight about the currently practiced cancer therapies, their management, and safety aspects. Chapters 2 and 3 by Indian researchers described in-depth nano-natural products as anticancer agents and natural anticancer compounds and their derivatives under clinical trials. Chapter 4 by Portugal researchers summarized the recent medicinal advancements of natural anticancer compounds and their analogues. Chapters 5 and 6 described the designing of natural cancerous drugs and their delivery system and the applications of nano-based novel drug delivery systems in herbal medicine-mediated cancer therapy, respectively. Similarly, Chapter 7 by collaborative works of Malaysian and Indian authors has discussed on the toxicological and pharmacological use of anticancer compounds, while the recent advancements in the clinical trial of plant-derived anticancer compounds were discussed in Chapter 8. Chapter 9 highlights the past and present studies on organosulfur compounds for exploring their potential as an adjunct in cancer chemotherapeutics and research. Chapter 10 discusses the relevance of traditional Unani system of medicine in cancer treatment. However, Chapter 11 entails the actions and reactions of anticancer plants and their signaling pathways in the cancer therapy. Overall, this book volume gives an overview of various plant-based anticancer compounds in use or under clinical trials and provides a detailed information of the applications of nanotechnology for cancer cure. Also, it discusses on several aspects of toxicology and pharmacology of different anticancer compounds and improved drug delivery systems involving nanostructures. This book volume will be very useful in developing novel therapies against cancers as it provides a wide range of data on the heterogeneity of cancers and chemodrugs and their adverse side effects and in improving the efficacy of cancer treatment approaches by exploring more potential anticancer phytocompounds and nanotechnological applications. I personally congratulate both editors, Dr. Mohd Sayeed Akhtar and Mallappa Kumara Swamy, for their noble academic efforts in bringing out this volume.

Department of Crop Science, Faculty of Agriculture
Universiti Putra Malaysia,
Serdang, Selangor, Malaysia

Uma Rani Sinniah

Preface

Cancer is one of the leading death causes of human population increasingly seen in recent times. Plants have been used for medicinal purposes since immemorial times. Though several synthetic medicines are useful in treating cancer, they are inefficient and unsafe. However, plants have proved to be useful in cancer cure. Moreover, natural compounds from plants and their derivatives are safe and effective in treatment and management of several cancer types.

The anticancer plants such as *Catharanthus roseus, Podophyllum peltatum, Taxus brevifolia, Camptotheca acuminata, Andrographis paniculata*, *Crataeva nurvala*, *Croton tonkinensis*, *Oplopanax horridus*, etc., are important source of chemotherapeutic compounds. These plants have proven their significance in the treatment of cancer and various other infectious diseases. Nowadays, several well-known anticancer compounds such as taxol, podophyllotoxins, camptothecin, vinblastine, vincristine, homoharringtonine, etc., have been isolated and purified from these medicinal plants. Many of them are used effectively to combat cancer and other related diseases. The herbal medicine and their products are the most suitable and safe to be used as an alternative medicine. Based on their traditional uses and experimental evidences, the anticancer products or compounds are isolated or extracted from the medicinally important plants. Many of these anticancer plants have become endangered due to ruthless harvesting in nature. Hence, there is a need to conserve these species and to propagate them in large scale using plant tissue culture. Alternatively, plant cell tissue and organ culture biotechnology can be adopted to produce these anticancer compounds without cultivation. The proper knowledge and exploration of these isolated molecules or products could provide an alternative source to reduce cancer risk, anti-tumorigenic properties, and suppression of carcinogen activities.

Anticancer Plants: Clinical Trials and Nanotechnology (Volume 3) provides a timely reviewed effort in this direction. This volume with 11 chapters from various contributors of the globe focuses on the concepts and experimental data in the application of anticancer plants in clinical trials and the use of nanotechnology in cancer therapy as chemotherapeutic agents and formulation of nanoproducts and its relevance in boosting immune system. This book provides an informative data packages on anticancer compounds and their use as nanomedicine to understand and solve the problems related to various cancers. We hope that this book is helpful for researchers, academicians, clinicians, oncologists, and scholars who are functioning in the fields of oncology, health care, nanotechnology, phytochemistry, toxicology, pharmacology, and herbal research.

We are highly grateful to all our contributors for readily accepting our invitation for not only sharing their knowledge and research but for venerably integrating their expertise in dispersed information from diverse field in composing the chapters enduring editorial suggestions to finally produce this venture. We greatly appreciate their commitment. We are also thankful to Professor Uma Rani Sinniah for his suggestions and writing the foreword for this volume. We also thank the team of Springer Nature, especially Dr. Mamta Kapila, for their generous cooperation at every stage of the book publication.

Shahjahanpur, Uttar Pradesh, India — Mohd Sayeed Akhtar
Serdang, Selangor, Malaysia — Mallappa Kumara Swamy

Contents

About the Editors

Dr. Mohd Sayeed Akhtar is an assistant professor at Gandhi Faiz-e-Aam College, Shahjahanpur, India. He received his PhD degree from Aligarh Muslim University (AMU), India in 2008, prior to conducting postdoctoral research at the Botanical Institute, University of Basel (BIB), Switzerland (2008–2010), and Chonbuk National University (CBNU), Republic of Korea, in 2011. He was an assistant professor at Jimma University, Ethiopia (2011–2014), and a fellow researcher at the Institute of Tropical Agriculture, Universiti Putra Malaysia (UPM) (2014–2015). Dr. Akhtar has 14 years of research and 8 years of teaching experience in soil microbiology, applied microbiology, environmental microbiology, molecular biology, plant pathology, and plant nanobiotechnology.

He is the author or coauthor of over a hundred articles in peer-reviewed journals, conference proceedings, and book chapters, and has edited five books with international publishers. His current research focuses on rhizospheric plant-microbe interactions and their molecular biotechnology, bioremediation, biomineralization, nanofertilizers, and nanobiotechnology.

Dr. Mallappa Kumara Swamy is a postdoctoral researcher at the Department of Crop Science, Faculty of Agriculture, Universiti Putra Malaysia (UPM). He received his PhD (Biotechnology) from Acharya Nagarjuna University, Guntur, India, in 2013. He has more than 14 years of teaching and research experience in the fields of plant biotechnology, secondary metabolite production, phytochemistry, and bioactive studies. Dr. Swamy has authored more than 50 research publications in peer-reviewed journals and is an editorial board member and reviewer for several high-impact international journals. His current focus is on cell and tissue culture technology for bioactive compound production and on nanotechnology for medical applications.

Marketing text: Cancer is one of the leading causes of death in human beings. Though several synthetic medicines are used to treat cancer, they are largely inefficient and unsafe. In contrast, plants, which have been used for medicinal purposes since time immemorial, have proved to be useful in fighting cancer, with natural compounds from plants and their derivatives offering safe and effective treatment and management for several types of cancer.

Plants such as *Catharanthus roseus*, *Podophyllum peltatum*, *Taxus brevifolia*, *Camptotheca acuminate*, *Andrographis paniculata*, *Crateva nurvala*, *Croton tonkinensis*, and *Oplopanax horridus* are important source of chemotherapeutic compounds. These plants have proven their value in the treatment of cancer and various other infectious diseases, and several common anticancer compounds such as taxol, podophyllotoxins, camptothecin, vinblastine, vincristine, and homoharringtonine have been isolated and purified from these medicinal plants.

Unfortunately, many of these anticancer plants have become endangered due to ruthless and irresponsible harvesting practices. Hence, there is a need to conserve these species and to propagate them on a large scale using plant tissue culture. Alternatively, plant cell tissue and organ culture biotechnology could be adopted to produce these anticancer compounds without the need for cultivation. A better grasp and continuing exploration of these isolated molecules and products could provide a powerful alternative means of reducing cancer risk.

Anticancer Plants: Volume 3, Clinical Trials and Nanotechnology provides a timely review of concepts and experimental data on the application of anticancer plants and their compounds in clinical trials, and on the use of nanotechnology in cancer therapy.

Chapter 1
Cancer Therapies: Current Scenario, Management, and Safety Aspects

Shivaswamy Santosh, Manasa Deepa Rajagopalan, Bangalore Acharlu Pallavi, Gudepalya Renukaiah Rudramurthy, Valluru Rajashekar, Katta Annaiah Sridhar, and Mallappa Kumara Swamy

Abstract Cancer is a disease state in which abnormal cells proliferate uninhibitedly and might spread all over the body. There are more than 100 types of cancers which are categorized based on the affected tissue or organ of the human body. At present, several methods of cancer treatments including surgery, chemotherapy, radiotherapy, immunotherapy, and targeted therapies are available. These therapies aim to stop or slower the growth rate of cancer cells, however incapable to completely cure cancers. Chemotherapy employs cytotoxic drugs that prevent, weaken, or reduce the reappearance of cancer cells and thereby control the development and progression of cancer. These anticancer compounds work in various ways such as impairing mitosis; targeting the cancer cells' energy source, enzymes, and hormones; preventing the angiogenesis process; and triggering apoptosis activity. Based on the nature of cancer, anticancer compounds are administered either single or in combination through oral, parenteral (intraperitoneal injection, intravenous injection), and other topical routes. These chemo-drugs exert the beneficial pharmacological action on cancer cell, but they also affect the healthy cells and lead to toxicity and clinical side effects in patients. These adverse effects vary depending on the type/group of drugs administered. Some of the common undesirable properties of anticancer compounds include weakness, xeroderma (dry skin), erythema (skin redness), alopecia (hair loss), loss of appetite, weight loss, and

S. Santosh (✉)
Department of Head and Neck Cancer, Coloumbia Asia Hospital, Bengaluru, India

M. D. Rajagopalan (✉) · V. Rajashekar · K. A. Sridhar
Department of Pharmaceutics, East West College of Pharmacy, Bengaluru, India

B. A. Pallavi
Department of Ophthalmology, M.V.J. Medical College, Bengaluru, India

G. R. Rudramurthy
Department of Biotechnology, East West First Grade College, Bengaluru, India

M. K. Swamy
Department of Crop Science, Faculty of Agriculture, Universiti Putra Malaysia, Serdang, Selangor, Malaysia

M. S. Akhtar, M. K. Swamy (eds.), *Anticancer Plants: Clinical Trials and Nanotechnology*, https://doi.org/10.1007/978-981-10-8216-0_1

numbness. This chapter provides an idea about various toxicological side effects and management of these side effects in different types of cancer therapies.

Keywords Drug delivery · Toxicity · Cancer therapy · Permeability retention · Photodynamic therapy · Secondary malignancy · Side effects

1.1 Introduction

Cancer is a disease, associated with a group of more than 100 distinct disorders. Several carcinogenic factors such as genetic, certain microbial infections, pollutants, environmental factors, etc. are involved in the development of cancer. The environmental pollutants and some industrial hazardous chemicals are also well-known causes of cancers. The most common causes of cancer are related to environmental (Sharma and O'Keefe 2007), food (Mathew et al. 2004), lifestyle, and behavioral exposures (Grant et al. 2008). Moreover, other contributing factors of cancer include chemicals (Nasca 2007), some microbial infection (bacteria, virus, and parasites) (Liao 2006), diet (Mathew et al. 2004), radiation (X-rays, γ-rays), hormones (estrogen and progesterone) (Jaiyesimi et al. 1995) and genetic or heredity factors (Kinzler and Vogelstein 1996). The carcinogenic factors induce a definite damage at the genetic and cellular level. However, the exact cause of some cancers is still unknown.

Several endogenous factors promote the release of reactive oxygen species (ROS) such as hydroxyl, superoxide, and peroxyl radicals. Reactive Oxygen Species results in an extensive oxidative cell damages and causes many health-related conditions including atherosclerosis, neurological diseases, and cancers (Halliwell 2009). The absence of apoptosis may lead to uncontrolled proliferation of cancerous cells, which would further allow the cancerous cells to invade nearby tissues (metastasis) through the formation of new blood vessels known as angiogenesis (Weinberg 2007; Garraway and Lander 2013). Mutations at the cellular and molecular level lead to uncontrolled proliferation of normal cells by altering the cell cycle, leading to the development of mass of abnormal cells known as tumor. Carcinogens induce cancer through two distinguishing steps such as initiation and promotion (Caldas 2012). Initiation is a primary rapid phase involved in the formation of cancerous cell; during this step DNA gets damaged due to metabolic activation of genotoxic carcinogen. This initial change in cells may persist for an extensive period, possibly the life span of an individual. Subsequently, initiated cells enter into promotion stage where expansion of cells (clone expansion) takes place through promoters. Changes involved in the epidermal homeostasis provide a favorable tissue environment for the clone expansion. Many agents are known to induce normal cell division. However, promoters induce abnormal/uncontrolled growth leading to the formation of tumor (Volker 2001). Anticancer agents mainly exhibit a preventative or healing role in an impaired system. Furthermore, an anticancer agent aids to activate the apoptosis (programmed cell death) signaling system to control the cell proliferation process (Dewey et al. 1995; Elmore 2007).

Cancer is due to the uncharacteristic proliferation of different types of body cells and differs markedly. According to cancer pathology, cancer can be either malignant or benign. Malignant forms of tumors possess the capability to invade their nearby normal tissues and organs and spread across the body through the lymphatic or circulatory systems. This is more often called as metastatic state of cancerous cells. While, benign tumors remain localized in their original site of occurrence and do not spread to surrounding regions of the body (Lodish et al. 2000). Cancers are classified on the basis of the type of cells, tissues, or organs that are affected. For instance, carcinomas represent the malignant forms of the epithelial cells, and it is one of the most frequent forms of human cancers (approximately 90%). Sarcomas are tumors of the connective tissues (bone, muscle, cartilage). Lymphomas and leukemias represents the cancers of the lymphatic system and blood-forming cells, respectively. Likewise, tumors are also categorized based on the type of cells involved or origin of tissues, for example, breast or lung malignancies (Lodish et al. 2000; Lahat et al. 2008).

Cancers are treated by using both traditional therapeutic methods (surgery, chemotherapy, and radiation therapy) and complementary and alternative therapeutic approaches. However, current forms of cancer treatments cause side effects and mainly affect the healthy and normal tissues or organs. In addition, side effects may differ from one person to another, even between those patients who received the same therapy. Also, a therapy may show lesser side effects in some patients, but in others, the effects may be many. The type of tumors, progress stages, amount of drugs, treatment frequencies, age, and health conditions of a patient are some of the critical and deciding factors for choosing the right cancer treatment method (NIH-NCI 2017).

The anticancer compounds involved in chemotherapy prevent cancer progress in various ways such as impairing mitosis; targeting the cancer cells' energy source, enzymes, and hormones; preventing the angiogenesis process; and triggering apoptosis activity (Croce 2008). The chemo-drugs are either single or in combination through oral, parenteral (intraperitoneal injection, intravenous injection), and other topical routes. However, the use of chemotherapy is mainly based on the nature of a tumor. Though chemo-drugs effectively inhibit cancerous cells, they also induce toxicity and other adverse health problems in the treated patients (Kroschinsky et al. 2017). Moreover, adverse effects vary with the type/group of drugs administered. Some of the common undesirable effects of anticancer compounds include weakness, xeroderma (dry skin), erythema (skin redness), alopecia (hair loss), loss of appetite, weight loss, and numbness (Shewach and Kuchta 2009; Kroschinsky et al. 2017).

1.2 Cancer Therapy: Side Effects and Management

The choice of cancer treatment depends on origin of primary cancer and stage of illness. Several different strategies are available for the treatment of cancer; the selection and application of the therapy depends mainly on the type of cancer, site of the tumor, stage of the cancerous tissue, and many other factors. The advancement in the field of medical technology has led to the development of innovative and advanced cancer therapies. These advanced therapies are known to have better advantages over earlier strategy which include improvement in the lifestyle of affected individuals, minimizing pain, and many more. The most commonly employed cancer treatment includes surgery, chemotherapy, radiotherapy, immunotherapy, etc. Some of the major anticancer drug types and their clinical effects (NIH-NCI 2017) are mentioned in Table 1.1.

1.2.1 Surgery

The main objective of surgery in cancer treatment includes diagnosis by biopsy, prevention of tumor progression, palliation, and in few cases as a treatment option. The surgical methods may vary based on the tumor and its location (Faries and Morton 2007). Surgical therapy is normally carried out to find the type and stage of the disease. Surgical treatment plays a vital role in soft tissue and bone cancers. However in hematological malignancies, surgery has minimal role. Surgical therapy may be of different types depending upon the modalities and use of therapeutic agents, such as vaporization of tumor using light known as laser therapy (Disaia and Creasman 2002), freezing cancerous cells in liquid nitrogen probe called cryosurgery (Mouraviev and Polascik 2006), laparoscopic surgery (Apgar and Brotzman 2002), electrosurgery (Sheridan and Dawber 2000), thoracoscopic surgery, and micrographic surgery (Grossman and Leffell 2009).

The most common side effects of post-surgery are pain, infections, loss of blood, blood clotting, and impairment of neighboring tissues and/or organs and many more. Apart from these common side effects, some of the long-term effects are also reported after the surgery. For instance, after prostatectomy, male patients may lose to control urine (Hart et al. 2008), loss of limb function following resection of bony tumors, loss of hearing following removal of acoustic tumor, loss of vision after removal of orbital tumors, and many more. Pain and infections are managed with the aid of pain killers such as aspirin, acetaminophen, etc. (Kimberlin et al. 2004) and antibiotics given through oral or intravenous route (Gilbert et al. 2008). Surgeries are more beneficial in preventive, in palliative, and in diagnosis and in certain cases used as a treatment. However, surgery in combination with additional therapies like radiation therapy, immunotherapy, and chemotherapy acts as a powerful tool to cure some of the cancers (Amant et al. 2005; Kantoff et al. 2010).

Table 1.1 Major anticancer drug types and their clinical effects

Anticancer drug types	Example	Mechanism of action	Clinical side effects	Uses	References
Antimetabolites	5-FU	Inhibition of thymidylate synthase and incorporation of its metabolites into RNA and DNA	Photophobia taste changes, metallic taste in the mouth during infusion, low blood counts	In several cancers that includes cancer of colon and rectal, breast, head, and neck, neuroendocrine, cervix, bladder, hepatobiliary, and thymus	Longley et al. (2003)
	Methotrexate	Inhibits the dihydrofolate reductase and thus affects nucleoside metabolism	Kidney toxicity Skin photosensitivity	Acute lymphoblastic leukemia, choriocarcinoma, cancer of the breast, neck, head, lungs, cervix	Chorawala et al. (2012)
	Mercaptopurine	Inhibits the synthetic steps during S-phase of replication	Blood in urine or stool Swelling of lower limbs Decreased appetite	Acute lymphocytic leukemia	Chorawala et al. (2012)
Alkylating agents	Chlorambucil	Interferes with DNA replication and induces cellular apoptosis via the accumulation of cytosolic p53 and subsequent activation of Bax, an apoptosis promoter	Seizure Rapid heart rate Trouble concentrating	Leukemia, lymphoma	Peterman and Braunstein (1986)
	Cyclophos-phamide	Inhibits protein synthesis by introducing alkyl groups into DNA and creates cross-linking between two DNA strands	Shortness of breath Fast heart beat Joint pain	Hodgkin lymphoma Acute lymphoblastic leukemia Retinoblastoma Ovarian cancer Nephritic syndrome	Travis et al. (1995)
	Lomustine	Alkylates and crosslinks DNA and also carbamoylates DNA and proteins, thereby inhibiting DNA and RNA synthesis and disruption of RNA processing	Troubled breathing with exertion Black, tarry stools Bleeding gums	Brain tumors Breast cancer Colon cancer Melanoma Lung cancer	Rahman et al. (2014)

(continued)

Table 1.1 (continued)

Anticancer drug types	Example	Mechanism of action	Clinical side effects	Uses	References
Plant products	Vincristine	Arrest the cell division in metaphase by binding to tubulin	Numbness in fingers and toes	Lungs and breast cancer	Tu et al. (2013)
			Pain in testicles	Neuroblastoma	
			Hallucination	Rhabdomyosarcoma	
				Testicular carcinoma	
				Kaposi's sarcoma,	
	Paclitaxel	Stabilize polymerization of tubulins and inhibit the disassembly of microtubules	Shortness of breath	Cancer of the head and neck, breast, lungs, ovary	Kampan et al. (2015)
			Increased heart beat		
Monoclonal	Trastuzumab	Targets against human epidermal growth factor receptor protein-2 (HER-2) and induce the apoptosis in breast cancer	Arthralgia	B-cell lymphoma	Lavaud and Andre (2014)
Antibodies			UTI	Chronic lymphocytic leukemia,	
			Tachycardia		
			Ear congestion	T-cell lymphoma and breast cancer.	
			Edema		
Antibiotics	Doxorubicin HCl	Intercalation into DNA, leading to inhibition of the DNA synthesis or poisoning of topoisomerase II	Low blood count	Bone sarcoma	Dhanaraj et al. (2014)
			Mouth sores	Breast, kidney, gastric, liver, uterine, endometrial cancer	
				Head and neck cancer	
				Hodgkin lymphoma	
				Non-Hodgkin lymphoma	
				Wilms' tumor	

Vaccine	Sipuleucel-T	An autologous cellular immunological agent is thought to work through APCs to stimulate T-cell immune response targeted against prostatic acid phosphatase (PAP)	Skin rashes, infection, secondary cancer	Hormone-refractory prostate cancer	Goldman and De Francesco (2009), Kantoff et al. (2010)
				Breast cancer	
	Gardasil	The proteins in Gardasil are structural, virus-like proteins (VLP) that resemble the HPV virus. The proteins can activate the immune system but cannot give rise to replicating virus	Skin rashes, infection, secondary cancer	Cervical cancer	Kane (2012)
				Anal cancer	

1.2.2 Chemotherapy

The term "chemo" is described as the chemically induced compound, i.e., the drug which flows through the bloodstream to almost each part of the human system. Chemotherapy is one of the method of treatments followed before/after surgery and/or radiations to cure and/or control/prevent certain cancers. The choice of drug (alone or in combination), and doses (lower to higher) completely depends on cause and category of the cancer and biological factors such as age of patient, past health history, and other treatments taken before the cancer treatment etc. (Hausheer et al. 2006; Kampan et al. 2015). By considering the above facts, several different chemotherapeutic drugs have been studied and formulated for the treatment of various cancers. All these chemotherapeutic drugs are distinct from one another based on their active constituents and their use in cancer therapy. The chemotherapeutic drug is largely administered through different modes such as oral, intravenous, and/or intra-lesional with the aid of special catheters known central venous catheters. Chemotherapeutic agents known as cell-cycle-specific drugs normally act on specific and/or all phases of cancer cell cycle (G_1-, G_2-, M-, S-phase), for example, methotrexate acts during S-phase and vinca alkaloids acts on M phase. However, certain drugs (non-cell cycle specific) such as alkylating agents do not act on the cell division cycle (Nies and Speilber 2006). Hence, an appropriate knowledge of tumor biology and cellular kinetics would be useful in the development of innovative and advanced drugs for cancer therapy (Kroschinsky et al. 2017). Mechanism of action of different chemotherapeutic agents (Payne and Miles 2008) against various types of cancer is listed in Table 1.1.

Earlier studies have reported that the therapeutic efficacy of chemotherapeutic drugs depends on various pathophysiological conditions of cancerous cells/tumors. Conditions such as structural impairment in blood vessels due to angiogenesis, destructive lymphatic drainage system (Lila et al. 2010), suppression of cell signaling, high interstitial fluid pressure in tumor, and large production of phosphoglycoprotein may resist the uptake of drug/s by the cancerous cells (Mahesh et al. 2006). Moreover, previous study on cancer stem cells (CSCs) also reveals that CSC metastasis helps in the development of resistance against anticancer drugs (Visvader and Lindeman 2008; Housman et al. 2014).

Chemotherapeutic drugs used in cancer treatment have major drawback, i.e., they are too toxic and damage the healthy cells surrounding the cancerous cells leading to various side effects. The side effects may remain for short period, perhaps few months to years or may be lifetime, and these side effects vary from drug to drug. The common side effects are gastritis, bleeding, pain, loss of appetite due to damaged cells in the gastrointestinal tract (Honea et al. 2007; Joshi and Ehrenberger 2001); suppression of bone marrow formation due to lowered RBC level; (Balducci et al. 2007; Shayne et al. 2009); difficulty in swallowing due to sore throat and oral ulcers (Bensinger et al. 2008); hair loss and microbial infections due to suppressed immunity (Rolston 2005). However, some of these effects may be reversed after the completion of chemotherapy. In some cases, the comparable side effects may worsen due to the interactions with other drug. For example, high-dose vitamin consumption may reduce the therapeutic activity of chemotherapeutic drugs; aspirin along with chemotherapeutic drug may reduce the platelet counts and increase risk of severe bleeding (Mora et al. 2008; Mamede et al. 2011; Kampan et al. 2015). Long time side effects may also cause or result in the formation of secondary cancer. Most frequently, after receiving chemotherapy it may cause different forms of leukemia such as acute lymphocytic leukemia and acute myelogenous leukemia. Among the chemotherapeutic agents, anthracycline compounds and topoisomerase inhibitors II are safe enough than the alkylating agent and platinum-based chemodrugs such as cisplatin and paclitaxel (Shayne et al. 2009; Kampan et al. 2015).

1.2.3 Immunotherapy

Immunotherapeutic agents have complex structures comprised of many special cells and work on human immune systems. These types of immunotherapeutic agents known as immunomodulators include vaccines, monoclonal antibodies, cytokines, etc. which induce immune cells to act against cancerous cells or control the cancer cell formation (Cebon J et al. 2007; Kantoff et al. 2010). Comparatively this type of therapy causes fewer side effects than chemotherapy.

1.2.3.1 Monoclonal Antibodies and Cytokines

Monoclonal antibodies (MAbs) are engineered proteins (antibodies) which may be of murine origin or chimeric or humanized (Presta 2006; Ma and Adjei 2009). Several such immunotherapeutic proteins have been developed which are in the form of monoclonal antibodies, cytokines, and hormones (Dranoff 2004; Tayal and Kalra 2008). Among this, MAbs play a remarkable role in the treatment of various cancers. These are widely introduced through parenteral route and few in oral route (Chung 2008). MAbs are generally categorized as naked, radioactive, conjugated, and antibody-drug conjugates (Dubel 2007). Earlier, these were extensively used to treat only hematological malignancies (Mahapatra and Kapoor 2011). However, advanced research studies on these types of proteins showed that they are capable of regulating the signaling through epidermal growth factor receptors (EGFRs) and vascular endothelial growth factor receptors (VEGFRs) and thereby trigger the apoptosis or may produce cellular cytotoxicity (Tonini et al. 2005) in cancer cells. Similarly, the cytokines in the form of interleukins such as interleukins-2 (Liu 2003) and interferons like interferons-α, β, γ (Shankaran et al. 2001) have a major role in various cancer treatment (Besedovsky and Del Rey 2001; Kelley et al. 2003). These nonspecific immunotherapies are mostly administered by parenteral route through the muscle, skin, and vein (Liu 2003).

Development of immunotherapeutic agents is still remaining a challenge for the researchers due to the difficulty in recognizing the relationship between protein complex structure and its therapeutic activity. The lack of knowledge about MAbs and cytokines may affect the immunogenicity of the product and may become toxic. Different kinds of serious side effects were found in the MAbs immunotherapy based on the antigen it binds (Kelley et al. 2003). Several such used in immunotherapy are known to cause serious problems, for instance, bevacizumab used to treat metastatic conditions of colon and rectal cancer will likely to cause kidney impairment (Miller et al. 2007) and cetuximab used in colorectal cancer treatment is known to induce serious rashes and ocular toxicity (Tonini et al. 2005; Chan et al. 2009; Giro et al. 2009). Moreover, rituximab used in follicular non-Hodgkin lymphoma treatment may cause some serious side effects such as cytokine-related syndrome (CRS), autoimmune hematological disorders, prominent acute infusion reactions, tumor lysis syndrome, progressive multifocal leukoencephalopathy (PML), and renal toxicity (Gianni et al. 2003); interleukins-2 may cause chills and fever (Dutcher et al. 2001; Schwartzentruber 2001). Interferon may cause severe side effects which are intolerable by the patients which include reduction in the blood cell counts, thereby increasing the risk of microbial infections; moreover, the management of side effects caused by interferon is very hard (Tayal and Karla 2008).

1.2.3.2 Vaccines

Vaccines are one of the immunotherapeutic agents used in cancer therapy (Martin A. 'Mac' Cheever 2008; Intlekofer et al. 2012). The word "cancer vaccine" denotes substances that prevent cancer-causing infections, for example, HPV vaccine in cervical cancer and hepatitis B vaccine in liver cancer development, and are also used to cure existing cancers such as breast and prostate cancer, etc. However, the endurable immunogenicity and the duration of activity depend on the type of antigen and adjuvants used in the formulation. Accordingly, the vaccines are categorized into different types (Table 1.1). Furthermore, tumor physiology, patient's age, and health conditions may also modify the effectiveness of vaccines (Dunn et al. 2004; Lollini et al. 2006; Pejawar-Gaddy and Finn 2008). Vaccines need to be produced with long-term memory without causing any autoimmune disorders for effective use in therapy (Lotze and Rees 2004). Last decades of hypothetical reports on vaccines in various cancers reveal that vaccines help in maintaining a healthy immune system in individuals and destroy cancer cells consistently whenever they arise (Dunn et al. 2004). Several clinical trials are in process to investigate the vaccines, as prospective treatment in non-small cell lung cancer (NSCLCs) and also against other types of cancer (Jaffee et al. 2001; Ringden et al. 2009). Examples of vaccine in clinical trials include tumor-cell vaccine (belagenpumatucel-L), melanoma antigen, and mucin-1vaccine/liposomes (Goodyear and Hossein 2011). Several other advanced vaccines such as DNA vaccines, antigen vaccines, viral-vector-based vaccines, tumor-cell vaccines, and dendritic cell vaccines are also in the clinical trials aimed to improve the effectiveness and minimize the toxicity. For instance, sipuleucel-T is a type of therapeutic cancer vaccine used in cellular immunotherapy consisting of

PBMCs (autologous peripheral-blood mononuclear cells) like APCs (antigen-presenting cells) activated by recombinant fusion proteins (PA2024) ex vivo. PA2024 protein contains prostate-specific antigens (prostatic acid phosphatase) which are attached to an immune-cell activator (granulocyte-macrophage colony-stimulating factor) (Goldman and De Francesco 2009; Kantoff et al. 2010). The common side effects of immunotherapy include fatigue, chills, dizziness, fever, and many more. However, additional serious side effects like systemic lupus erythematosus, pelvic inflammatory disease, and autoimmune diseases, including arthritis, asthma, and appendicitis, have been reported in patients undergoing immunotherapy. Moreover, vaccines may also interact with similar kind of drugs and cause serious life-threatening effects which are quite difficult to manage (Becker et al. 2010; Allen and Giridhar 2014).

1.2.3.3 Stem Cells and Stem Cell Therapy

Stem cells are immature blood cells, formed by the process of hematopoiesis. Generally, these are abundant in the bone marrow and few are in the bloodstream. After maturation into a complete cell in the bone marrow, they move into the systemic circulation. These stem cells have been identified to play significant role in the treatment of certain cancer and are commonly called as stem cell therapy and bone marrow transplant therapy (Bashey 2000). Generally, stem cells comprise three diverse components such as red blood cells (RBCs), white blood cells (WBCs), and platelets, which have a definite role. RBCs help in the respiration process, while WBCs is comprised of various cell components such as neutrophils and lymphocytes (natural killer cells, T and B lymphocytes) which have a wide role against various infections (Bacigalupo 2003). Platelets, one of the important blood components, help to stop bleeding if it occurs either from known or unknown injuries (George 2000; George and Colman 2006). In most chemotherapy or radiation therapy cases, bone marrow destruction is vivid. For instance, in multiple myeloma or in leukemia after the combination of chemotherapy and radiation therapy, huge suppression in bone marrow cells takes place (Barkholt et al. 2008).

In these circumstances, the right choice of therapy is stem cell transplant. During stem cell transplant, individual's system stem cells can be replaced with new body stem cells or also it may work as graft-versus-cancer effect, i.e., it can find and kill the cancerous cell than the individual's immune cells. However, patients with stem cell transplant are under high risk due to related complications. Generally, the physician will decide the necessity of stem cell transplant depending upon the age factor of an individual, condition of disease, type of donor, etc. Collectively some of the common side effects include mouth and throat pain (short period), nausea and vomiting, and infection caused by certain bacteria, virus, and fungi (Bashey 2000). Bacterial and fungal infections can be generally managed with some antibiotics and antifungals (Perrotta and Synder 2001). Cytomegalovirus can often cause infection such as pneumocystis pneumonia in stem cell transplanted patients, and this is a little risky to manage. Hence, patients are advised with precautionary/prevention methods for the viral infections. All these serious side effects may also occur due to graft rejection (tissue mismatching of donor and recipient). If the patients are not treated on time during infection, they may further have the risk of graft-versus-host disease (Brecher et al. 2000; Firestone and Pitocco 2005; Villa et al. 2016). Platelet counts are reduced generally after the first 3 weeks of stem cell transplant; nonetheless, if the number of platelet in the blood becomes very low in the patient, it may be difficult to treat internal organ bleeding or bruising, which needs platelet transplant (Schiffer et al. 2001). Various other infections such as infections in the skin, lungs, and gastrointestinal tract are also observed due to graft-versus-host disease (Chemaitilly and Sklar 2007). Some of the other serious side effects of stem cell transplants include secondary cancers (Holman 2000), imbalanced hormone secretion in the thyroid and pituitary gland, vision loss (Firestone and Pitocco 2005), and many more.

1.2.4 Photodynamic Therapy

Photodynamic therapy also termed as photo-chemotherapy involves injection of photosensitizing agents through the vein or administered under the skin. After certain duration of intervals from hours to days, these agents are activated in the system by means of passing different wavelengths of light depending upon the type of photosensitizing agent leading to sensitization of cancerous as well as normal cells. The normal cells get rid of these agents within a couple of days; however cancer cells may hold these agents for many days to weeks. This therapy plays a significant supportive role in treating mouth and skin cancers as well as precancerous cells (van Straten et al. 2017). Specificity photodynamic therapy allows it to be used in many localized cancers such as skin and mouth cancers without pain (Querfeld et al. 2003). At present, there are only few photodynamic therapeutic agents such as porfimer sodium (aminolevulinic acid) that have been approved for the treatment of Barret's esophagus (Posner et al. 2005), NSCLCs (Ross 2003), and acute keratosis (De Berker et al. 2007). As mentioned earlier, various wavelengths of lights have been used to activate the photosensitizing agent, for instance, blue light for the activation of aminolevulinic acid (Piacquadio et al. 2004) and red laser light to activate the porfimer sodium, through a thin fiber-optic glass filament. In Barrett's esophagus the glass filament has been introduced through a flexible tube called endoscope (Phan et al. 2005). Similarly, bronchoscope is used in lung cancer (Vachani et al. 2008). Photodynamic therapy has shown more patient compliance than radiation therapy, chemotherapy, and surgery. The advantages of photodynamic therapy include the following: it is less intrusive, it has no extended period of side effects, and only minute scars occur after the treatment (Phan et al. 2005). Some of the current research studies on photodynamic therapy showed that this therapy is useful even for deeper tissues and the organs. Phototherapeutic agent such as Photochlor is found to be very effective against throat, lung, skin, oral, and esophageal cancers.

Advanced level studies on photodynamic therapy may be focused on combinatorial therapy, as well as imaging and treatment of solid tumors (Bellnier et al. 2006).

Though the therapy has wide range of advantages, it carries certain limitations such that phototherapeutic agents cannot pass into larger and deeper tissue effectively and can act only at the peripheral level. In certain conditions photodynamic therapy is not suitable to patients who have past history of illness like fistula (between the trachea and bronchus), porphyrias (the presence of tumors at the foremost blood vessels in the esophagus and stomach increases vein size above the normal level), etc. Depending upon the part of cells/tissue treated, several common specific side effects were observed post porfimer sodium treatment, for example, severe cough, difficulties in swallowing in the case of esophageal cancer treatment with porfimer sodium (Phan et al. 2005), pneumonia, coughing with blood, and fluid collection in the case of lung cancer treatment (Karp and Thurer 2003). Photodynamic agents such as aminolevulinic acid cause side effects like photosensitivity reactions on the skin (swelling and reddening of the skin) induced by exposure of sunlight, immunosuppression, porphyria, flaking, and itching (usually gets rid third week after treatment). Some of the other side effects caused by aminolevulinic acid are eczema and hives which sometimes turn into serious allergic reactions.

1.2.5 Hormone Therapy

Modulations in hormone secretion are responsible for some cancer formations such as breast, prostate (Group PC 2000), and uterine cancer. Therapeutic agents belonging to this category will either stop production of such sex hormones or block the cells to utilize this hormone for their further growth. For example, tamoxifen is one of the most common hormone therapies for breast cancer which controls cell proliferation by blocking the estrogen receptors (Ellis and Perou 2013). In prostate cancer treatment, GnRH agonists such as goserelin, leuprorelin, and triptorelin stop the testicles from producing testosterone (Jaiyesimi et al. 1995; Fisher et al. 2005). Moreover, earlier reports show that hormonal treatments also work well with uterine cancers, such as endometrial stromal sarcoma, and these include tamoxifen (Jaiyesimi et al. 1995), letrozole (Elisaf et al. 2001), megestrol acetate, medroxyprogesterone (Loprinzi et al. 1990); these drugs either stop the production of respective hormones or block the cells to utilize the hormone for further growth. The side effects of hormone therapy are specific to individuals and are distinctive to the sex of the patient. Few side effects seen in women are menopausal symptoms such as genital irritation and vaginal discharge, fatigue, gastroenterological disorders, alopecia (hair thinning), muscle and bone change, increase in body weight, depression, and blood coagulation (Elisaf et al. 2001). In men the side effects observed were tiredness and impotence (Hellestedt and Pienta 2002). In women, some worst side effects seen in high dose of bicalutamide (casodex) are tumor flare and breast tenderness. In men, certain drugs like leuprorelin and goserelin may lead to depression and mood swings, osteoporosis (Elisaf et al. 2001), etc. Physiological side effects are managed with counseling, while bone-related effects may be treated with vitamin D and calcium (Estilo et al. 2008; Mora et al. 2008).

1.2.6 Radiation Therapy

Radiation is a choice of treatment to shrink the solid tumors and kill the cancer cells using high radiation with various radioactive substances (Halperin et al. 2004). Radioactive substances are physical agents which radiate the electrically charged ions. This high energy radiation such as gamma rays and X-rays causes huge damage in the genetic content of the cell/DNA thereby killing the cancer cell. Radiation can be activated through radioactive substances in the human system, for example, brachytherapy (Akbal et al. 2008) or by external source, i.e., by external-beam radiation therapy (Hall and Cox 2003). Some of the radioactive substances that are used in the treatment of different cancer are calcium (^{44}Ca and ^{45}Ca) in bone cancer, ^{131}I in thyroid cancer (Davies and Gilbert Welch 2014), ^{59}Fe in blood cancer, and many more.

The side effects of radiotherapy may be acute/chronic and may vary in every radiation process. Radiation therapy used in the stomach or abdomen region may cause diarrhea, nausea, and vomiting, whereas patients undergoing radiation in the head and neck area may develop mouth blisters (Bensinger et al. 2008). Acute side effects may be shorter or last for weeks and may become extinct. For instance, during head and neck radiation therapy, patients may experience hair loss. Some of chronic effects caused after radiation therapy will last for weeks to years or may persist for a lifetime (Carper 2007). The side effects observed after the radiation therapy in the abdomen or pelvis are bowel dysfunction, infertility, or sexual-related complications (Miller et al. 2007). In certain cases lymphedema (fluid buildup and swelling in parts of the body) was also observed after the radiation therapy (Cheville 2007). Radiation therapy may also lead to the development of secondary cancers, for instance, the radiation treatment for Hodgkin lymphoma in young women was later found to develop secondary breast cancers (Horning 2008).

1.2.7 Targeted/Nanoscale Therapy

Targeted delivery involves delivery of drugs specifically to cancer cells rather than the normal cells. The targeted drugs work on cancer cells by means of blocking the signals, activating the immune system, altering the 3D structure of cancer cell protein molecules, blocking the angiogenesis (new blood vessel formation), and taking the toxins from the cancer cells (Croce 2008). The specific delivery has allowed this therapy to be used as main a course of treatment or as co-therapy. Moreover, pathophysiological conditions of cancer tumors such as damaged lymphatic drainage, leaky vascular system, and enhanced permeability and retention effect affect the traditional chemotherapeutic drug efficacy (Jong and Borm 2008). Drug distribution and drug specificity can be achieved by means of reducing the particle size ranging from micro- to nanoscale known as nano-medicines (Couvreur 2013). Nanotechnology includes designing, synthesis, and characterization of bulk material into nanoscale range which involves engineering, and technology (Akhtar et al. 2015; Swamy et al. 2015a,b; Rudramurthy et al. 2016). These engineered nanomaterials must be safe and biocompatible to use in biology and medicine. Several different techniques have been developed for the synthesis and characterization of different nanoparticles (Akhtar et al. 2013; Akhtar and Ahmad 2014; Rudramurthy et al. 2016). Nanomaterials or nanoparticles used in diagnostic images, or in cancer treatment act by regulating the cell proliferation or kill the cell and/or block the new blood vessel formation in cancer cells (Lila et al. 2010). Nano-therapeutic agents are widely used as biosensors, diagnostic imaging and to deliver the drug specifically to cancer cells. Nano-engineered drug substances can withstand for longer period in the cell and enhance the biological half-lives of therapeutic effects which help to reduce the dose and dosage regimen (multiple administration). Moreover, this also assists the nano-sized material to avoid quick clearance and metabolism from the individual's system (Ruoslahti et al. 2010; Das et al. 2015).

Cancer therapy using chemotherapeutic agents may damage the healthy cells also; hence, dose and dose frequency during cancer chemotherapy should be optimum and otherwise may lead to lower the therapeutic efficacy or increase the toxicity (Mukherjee et al. 2008). To balance these difficulties, nanoscale delivery is a favorable choice of therapy to target the cancer cells. Nanoscale delivery systems have been well progressed and developed as a promising tool in the improvement of the therapeutic efficacy, minimization of side effects, and lowering of the doses of drugs (Saad et al. 2008; Subbiah et al. 2010). Various types of nanoscale drug delivery systems have been developed to use in cancer therapy and are in the form of quantum dots (Fang et al. 2012), nanoliposomes (Rudra et al. 2010), nanoparticles (Desai et al. 2003; Kanagesan et al. 2016; Aziz et al. 2016), nanocarriers (Kumari et al. 2016), nanorobotics (da Silva Luz et al. 2016), nanospheres (Dhanaraj et al. 2014), nanoshells (Zhao et al. 2014), dendrimers (Vieweg and Dannull 2003), and fullerenes (Prylutska et al. 2014). Nanomaterials differ from one another on the basis of method of fabrication, such as site specificity, dose, and dose frequency.

Similar to regular chemotherapeutic agents, targeted/nanoscale delivery do not harm the healthy cells. However, some side effects have been observed depending upon the drug and its target site and, may vary from individual to individual. Since the past two decades of research and clinical trials on targeted cancer therapies, very few drugs have been approved by US FDA for safe therapy. Consequently, the serious side effects caused by each nanoscale system are not exactly observed as same as in traditional chemotherapeutic drugs. Nanoscale/targeted therapy is also known to cause some side effects, for instance, the drugs act as epidermal growth factor receptor inhibitors that block the protein production both in normal and cancer cells (Kuan et al. 2001). Hence, this drug may affect the normal skin protein formation and may change the skin color of patients. Likewise, angiogenesis inhibitors may block the blood vessel formation in and around the cancer tissue and thereby affect the small blood vessels present in the foot and hands; thereby it may cause skin rashes like foot, hand, and mouth syndrome (Cooney et al. 2005). Some of the common side effects are changes in hair growth and swelling around the lips and eyes in certain targeted therapies (Basti 2007). Skin rashes have been managed with broad-spectrum sunscreen lotion with SPF of at least 30, zinc oxide, or titanium dioxide, and foot and hand syndrome has been treated with antibiotic gel (Timoney et al. 2003).

1.3 Management of Cancer Therapy Side Effects

Pain is one of the major challenges to treat as it can be due to the disease process or can also arise from the side effects of treatment (Shute 2013). Pain relief can be achieved by medications which are administered by various routes like oral, parenteral, or transdermal opiates. In addition sedatives, nerve blocks, and relaxation therapy may also be helpful (Lang and Patt 2004; Campbell 2011; Shute 2013). Nausea and vomiting is the second most common side effect and is managed by the use of antiemetics (Rao and Faso 2012). Various molecules like metoclopramide, megestrol acetate, and cannabinoids are used in this treatment. Adequate nutritional supplement intake can help in the patient's speedy recovery and in turn will reduce the hospital stays, reduce costs and severity of the side effects, and improve patient's well-being. Somatostatin and its analogues are used in the treatment, as it reduces intestinal secretions; pro-kinetic agents may also be of help in such patients. Opiate-induced bowel disturbance in the form of constipation is quite common as it is used regularly for pain relief. It is managed by the use of laxatives, increase of water intake/hydration, and regular bowel habit. During or post radiation therapy, local effect on the skin is associated in the form of dry skin, hyperpigmentation, and severe pruritus. Care of the skin is important to avoid ulcer formations; patients are educated on this front by using mild soap and oil-based lotions to reduce dryness. In pruritus, patients are advised to use calamine lotion, or cocoa butter, or to bathe in sodium bicarbonate (Bolderston et al. 2006; Chan et al. 2009; McQuestion 2011). Mostly chemotherapy is administered by the intravenous route; as a result of which, care of catheter/venous line is important to avoid inflow failure, catheter kinks, fibrin blood in catheter, and catheter migration. Using heparin flushing before and after drug administration can avoid blocks, and it is mandatory to check the catheter placement by taking an x-ray or dye studies to avoid associated complication (Goossens 2015). Moreover, post-chemotherapy patients become very weak and are very susceptible to systemic infection due to the development of neutropenia.

Diagnosing infections in cancerous patients is problematic because distinguishing between infections and fever of unknown origin is very difficult. This clinical challenge can be detected by identifying some biochemical markers, for example, increased concentration of blood procalcitonin (PCT) is linked with microbial infection. Thus, detecting PCT levels (normal = 0.5 ng/ml) will be a very useful parameter to detect severe infections, specifically a bloodstream infection in cancer patients (Durnas et al. 2016). Patients must be educated about the red flag signs and symptoms and also consult the doctor immediately. Patient education about adverse effects plays a major role in the management and also alleviates the serious or life-threatening treatment complications (Green et al. 2015; Tudor Car et al. 2016).

1.4 Conclusions and Future Prospects

The chemotherapeutic agents (either synthesized or from plants) may cause cell death both in healthy and cancerous cell. If normal cell proliferation is affected, it may produce unwanted effects which may be a little and/or very common after the therapy. However, most of the synthesized chemotherapeutic drugs alone or in combination with therapy lead to serious side effects. Advancement in the field of science and medicine led to the development of innovative treatment methodology known as nanoscale or targeted delivery. Many recent clinical trials have proved that targeted drug delivery system may target the cancerous cell with less harm and may help patient to live quality life. Several naturally derived anticancer molecules with minor side effects have been under consideration for clinical treatment of cancers. Even then, oncologists face the problems of new adverse effects linked to these new therapies, thus forcing to develop novel strategies against cancers. The knowledge of the pathophysiological effect of new therapies and their side effects are yet to be understood in detail. Moreover, the recommended use of new therapies still requires validation. Information about the clinical management of cancer is inadequate, hence, requiring more clinical investigations in this regard. Future direction of research should focus on identifying and overcoming adverse side effects in cancer patients by adopting different strategies. Thus, added preclinical and clinical investigations on toxicity and cancer management are immediately needed. Promoting public awareness, educating patients, effective diagnosis, and easy access to oncologists are the best ways to control and prevent cancers. Further, educating patients, caretakers, or medical staffs on the possible severe complications of cancer therapies and adopting suitable measures to overcome during the course of treatment process would be more useful.

Acknowledgment The authors are highly thankful to the editors for giving us this opportunity and to the innovators in the field of cancer treatment.

References

Akbal C, Tinay İ, Şimşek F, Turkeri LN (2008) Erectile dysfunction following radiotherapy and brachytherapy for prostate cancer: pathophysiology, prevention and treatment. Int Urol Nephrol 40(2):355–363

Akhtar MS, Ahmad U (2014) Biosynthesis of nanomaterials and their applications. Rev Adv Sci Eng 3:197–198

Akhtar MS, Panwar J, Yun YS (2013) Biogenic synthesis of metallic nanoparticles by plant extracts. ACS Sust Chem Eng 1:591–602

Akhtar M, Swamy MK, Umar A, Sahli A, Abdullah A (2015) Biosynthesis and characterization of silver nanoparticles from methanol leaf extract of *Cassia didymobotyra* and assessment of their antioxidant and antibacterial activities. J Nanosci Nanotechnol 15:9818–9823
Allen T, Giridhar MN (2014) A review on role of peptide vaccines in cancer therapy treatment of cancer disease by peptide vaccine. Adv Pharmacol Pharm 2:96–100
Amant F, Moerman P, Neven P, Timmerman D, Van Limbergen E, Vergote I (2005) Endometrial cancer. The Lancet 366(9484):491–505
Apgar B, Brotzman G (2002) High gradesquamous intraepithelial lesion, pp 249–263. isbn:9781437720730
Aziz SBA, Hashim M, Ismail I, Tamilselvan S, Alitheen NBBM, Swamy MK, Purna Chandra Rao B (2016) Synthesis, characterization and *in vitro* evaluation of manganese ferrite ($MnFe_2O_4$) nanoparticles for their biocompatibility with murine breast cancer cells (4T1). Molecules 21:312. https://doi.org/10.3390/molecules21030312
Bacigalupo A (2003) Haemopoietic stem cell transplants: the impact of haemorrhagic complications. Blood Rev 17:S6–10
Balducci L, Al-Halawani H, Charu V, Tam J, Shahin S, Dreiling L, Ershler WB (2007) Elderly cancer patients receiving chemotherapy benefit from first-cycle pegfilgrastim. Oncologist 12:1416–1424
Barkholt L, Remberger M, Hassan Z, Fransson K, Omazic B, Svahn B-M, Karlsson H, Brune M, Hassan M, Mattsson J, Ringde O (2008) A prospective randomized study using N-acetyl-L-cysteine for early liver toxicity after allogeneic hematopoietic stem cell transplantation. Bone Marrow Transplant 41:785–790
Bashey A (2000) Infection. In: Ball ED, Lister J, Law P (eds) Hematopoietic stem cell therapy. Churchill Livingstone, Philadelphia, pp 510–520
Basti S (2007) Ocular toxicities of epidermal growth factor receptor inhibitors and their management. Cancer Nurs 30:S10–S16
Becker MD, Dally J, Martini J (2010) Cancer vaccine therapies: failures and future opportunities. MD Becker Partners LLC, Newtown, pp 34–37
Bellnier DA, Greco WR, Loewen GM, Nava H, Oseroff AR, Dougherty TJ (2006) Clinical pharmacokinetics of the PDT photosensitizers porfimer sodium (Photofrin), 2-[1-hexyloxyethyl]-2-devinyl pyropheophorbide-a (Photochlor) and 5-ALA-induced protoporphyrin IX. Lasers Surg Med 38:439–444
Bensinger W, Schubert M, Ang KK, Brizel D, Brown E, Eilers JG, Elting L, Mittal BB, Schattner MA, Spielberger R, Treister NS, Trotti AM (2008) NCCN task force report: prevention and management of mucositis in cancer care. J Natl Compr Cancer Netw 1:S1–30
Besedovsky HO, Del Rey A (2001) Cytokines as mediators of central and peripheral immunoneuroendocrine interactions. In: Ader R, Fenton DL, Cohen N (eds) Psychoneuroimmunology, vol 1. Academic, San Diego, pp 1–20
Bolderston A, Lloyd NS, Wong RK, Holden L, Robb-Blenderman L (2006) The prevention and management of acute skin reactions related to radiation therapy: a systematic review and practice guideline. Support Care Cancer 14:802–817
Brecher ME, Holland PV, Pineda AA, Tegtmeier GE, Yomtovian R (2000) Growth of bacteria in inoculated platelets: implications for bacteria detection and the extension of platelet storage. Transfusion 40:1308–1312
Caldas C (2012) Cancer sequencing unravels clonal evolution. Nat Biotechnol 30:408–410
Campbell V (2011) The challenges of cancer pain assessment and management. Ulster Med J 80:104–106
Carper E (2007) Head and neck cancers. In: Haas M, Hogle W, Moore-Higgs G, Gosselin-Acomb T (eds) Radiation therapy: a guide to patient care. Mosby Elsevier, St. Louis, pp 84–117
Cebon J, Gedye C, John T, Davis ID (2007) Immunotherapy of advanced or metastatic melanoma. Clin Adv Hematol Oncol 5:994–1006

Chan A, Teoh D, Sanghera P, Hartley A (2009) Radiotherapy compliance is maintained with hypofractionation and concurrent cetuximab in locally advanced head and neck cancer. Radiother Oncol 93:654–661
Chemaitilly W, Sklar C (2007) Endocrine complications of hematopoietic stem cell transplantation. Endocrinol Metab Clin N Am 36:983–998
Cheville A (2007) Current and future trends in lymphedema management: implications for women's health. Phys Med Rehabil Clin North Am 18:540–543
Chorawala MR, Oza PM, Shah GB (2012) Mechanisms of anticancer drugs resistance: an overview. Int J Pharmaceut Sci Drug Res 4:1–9
Chung CH (2008) Managing premedications and the risk for reactions to infusional monoclonal antibody therapy. Oncologist 13:725–732
Cooney MM, Remick SC, Vogelzang NJ (2005) Promising systemic therapy for renal cell carcinoma. Curr Treat Options in Oncol 6:357–365
Couvreur P (2013) Nanoparticles in drug delivery: past, present and future. Adv Drug Deliv Rev 65:21–23
Croce CM (2008) Oncogenes and cancer. New Eng J Med 358:502–511
da Silva Luz GV, Gomes Barros KV, de Araujo FBC, da Silva GB, da Silva PAF, Condori RCI, Brasil LM (2016) Nanorobotics in drug delivery systems for treatment of cancer: a review. J Mat Sci Eng A 6:167–180
Das PJ, Paul P, Mukherjee B, Mazumder B, Mondal L, Baishya R, Debnath MC, Kumar SD (2015) Pulmonary delivery of voriconazole loaded nanoparticles providing a prolonged drug level in lungs: a promise for treating fungal infection. Mol Pharm 12:2651–2664
Davies L, Gilbert Welch H (2014) Current thyroid cancer trends in the United States. JAMA Otolaryngol Head Neck Surg 140:317–322
de Berker D, McGregor JM, Hughes BR (2007) Guidelines for the management of actinic keratosis. Brit J Dermatol 156:222–230
Desai N, Yao Z, Trieu V, Soon-Shiong P, Dykes D, Noker P (2003) Evidence of greater tumor and red cell partitioning and superior antitumor activity of Cremophor-free nanoparticle paclitaxel (ABI-007) compared with Taxol. Breast Cancer Res Treat 82:S82–S83
Dewey WC, Ling CC, Meyn RE (1995) Radiation-induced apoptosis: relevance to radiotherapy. Int J Radiat Oncol Biol Phys 33:781–796
Dhanaraj SA, Muralidharan S, Santhi K, SzeHui AL, Wen CJ, Teng HC (2014) Targeted drug delivery system formulation and evaluation of chitosan nanospheres containing doxorubicin hydrochloride. Int J Drug 6:186–193
Disaia PJ, Creasman WT (2002) Germ cell, stromal, and other ovarian tumors. In: Clinical gynecologic oncology, 6th edn. Mosby, St. Louis, pp 362–363
Dranoff G (2004) Cytokines in cancer pathogenesis and cancer therapy. Nat Rev Cancer 4:11–22
Dubel S (2007) Recombinant therapeutic antibodies. Appl Microbiol Biotechnol 74:723–729
Dunn GP, Old LJ, Schreiber RD (2004) The three Es of cancer immunoediting. Annu Rev Immunol 22:329–360
Durnas B, Wątek M, Wollny T, Niemirowicz K, Marzec M, Bucki R, Gozdz S (2016) Utility of blood procalcitonin concentration in the management of cancer patients with infections. Onco Target Ther 9:469–475
Dutcher J, Atkins MB, Weiss G, Margolin K, Clark J, Sosman J, Logan T, Aronson F, Mier J (2001) Kidney cancer: the cytokine working group experience (1986-2001): Part I. IL-2-based clinical trials. Med Oncol 18:197–207
Elisaf MS, Bairaktari ET, Nicolaides C, Kakaidi B, Tzallas CS, Katsaraki A, Pavlidis NA (2001) Effect of letrozole on the lipid profile in postmenopausal women with breast cancer. Eur J Cancer 37:1510–1513
Ellis MJ, Perou CM (2013) The genomic landscape of breast cancer as a therapeutic roadmap. Cancer Discov 3:27–34
Elmore S (2016) Apoptosis: A Review of Programmed Cell Death. Toxicol Pathol 35(4):495–516
Estilo CL, Van Poznak CH, Wiliams T, Bohle GC, Lwin PT, Zhou Q, Riedel ER, Carlson DL, Schoder H, Farooki A, Fornier M, Halpern JL, Tunick SJ, Huryn JM (2008) Osteonecrosis

of the maxilla and mandible in patients with advanced cancer treated with bisphosphonate therapy. Oncologia 13:911–920
Fang M, Peng CW, Pang DW, Li Y (2012) Quantum dots for cancer research: current status, remaining issues, and future perspectives. Cancer Biol Med 9:151–163
Faries MB, Morton DL (2007) Surgery and sentinel lymph node biopsy. Semin Oncol 34:498–508
Firestone DT, Pitocco C (2005) Adverse effects of blood transfusion. In: Rudmann (ed) Textbook of Bloodbanking and transfusion medicine, 2nd edn. Elsevier Saunders, Philadelphia, pp 396–418
Fisher B, Costantino JP, Wickerham DL, Cecchini RS, Cronin WM, Robidoux A, Bevers TB, Kavanah MT, Atkins JN, Margolese RG, Runowicz CD, James JM, Ford LG, Wolmark N (2005) Tamoxifen for prevention of breast cancer: report of the National Surgical Adjuvant Breast and Bowel Project P-1 Study. J Natl Cancer Inst 97:1652–1662
Garraway LA, Lander ES (2013) Lessons from the cancer genome. Cell 153:17–37
George JN (2000) Platelets. Lancet 355:1531–1539
George JN, Colman RW (2006) Overview of platelet structure and function. In: Colman RW, Clowes AW, Goldhaber SZ (eds) Hemostasis and thrombosis: basic principles and clinical practice, 5th edn. Lippincott Williams & Wilkins, Philadelphia, pp 437–442
Gianni AM, Magni M, Martelli M, Barbui T (2003) Long-term remission in mantle cell lymphoma following high-dose sequential chemotherapy and *in vivo* rituximab-purged stem cell autografting (R-HDS regimen). Blood 102:749–755
Gilbert DN, Mollerig RC, Elliopoulos GM, Sande MA, Chambers HF, Saag MS (2008) Recommended antimicrobial agents against selected bacteria. In: Gilbert DN, Robert C, Moellering RC, Elliopoulos GM, Sande MA (eds) The Sanford guide to antimicrobial therapy, 38th edn. Antimicrobial Therapy Inc, Sperryville, pp 4–130
Giro C, Berger B, Bolke E, Giernik F, Duprez F, Locati L, Maillard S, Ozshin M, Pfeffer R, Robertson AG, Langendijk JA, Budach W (2009) High rates of severe radiation dermatitis during radiation therapy with concurrent cetuximab in head and neck cancer: results of survey in EORTC institutes. Radiother Oncol 90:166–171
Goldman B, De Francesco L (2009) The cancer vaccine roller coaster. Nat Biotechnol 27:129–139
Goodyear SR, Hossein B (2011) Lung cancer vaccines: a review of three ongoing trials. Clin Invest 1:849–858
Goossens GA (2015) Flushing and locking of venous catheters: available evidence and evidence deficit. Nurs Res Pract 2015:985686. https://doi.org/10.1155/2015/985686
Grant RW, Mariani RA, Vieira VJ, Fleshner M, Smith TP, Keylock KT, Lowder TW, McAuley E, Hu L, Chapman-Novakofski K, Woods JA (2008) Cardiovascular exercise intervention improves the primary antibody response to keyhole limpet hemocyanin (KLH) in previously sedentary older adults. Brain Behav Immunol 22:923–932
Green T, Atkin K, Macleod U (2015) Cancer detection in primary care: insights from general practitioners. Br J Cancer 112:S41–S49
Grossman D, Leffell DJ (2009) Squamous cell carcinoma. In: Wolff K, Goldsmith LA, Katz SI, Gilchrest BA, Paller AS, Leffell DJ, Fitzpatrick's (eds) Dermatology in general medicine, 7th edn. McGraw-Hill, New York, pp 1028–1036
Group PC (2000) Maximal androgen blockade in advanced prostate cancer: an overview of randomised trials. Lancet 355:1491–1498
Hall EJ, Cox JD (2003) Physical and biological basis of radiation therapy. In: Cox JD, Ang KK (eds) Radiation oncology. Mosby, St. Louis, pp 3–62
Halliwell B (2009) The wanderings of a free radical. Free Radic Biol Med 46(5):531–542
Halperin EC, Schmidt-Ullrich RK, Perez CA, Brady LW (2004) Overview and basic science of radiation oncology. In: Perez CA, Brady LW, Halperin EC, Schmidt-Ullrich RK (eds) Principles and practice of radiation oncology, 4th edn. Lippincott Williams and Wilkins, Philadelphia, pp 1–95
Hart SL, Latini DM, Cowan JE, Carroll PR (2008) Fear of recurrence, treatment satisfaction, and quality of life after radical prostatectomy for prostate cancer. Support Care Cancer 16:61–169

Hausheer FH, Schilsky RL, Bain S, Berghorn EJ, Lieberman F (2006) Diagnosis, management and evaluation of chemotherapy induced peripheral neuropathy. Semin Oncol 33:15–49
Hellestedt BA, Pienta KJ (2002) The current state of hormonal therapy for prostate cancer. CA Cancer J Clin 52:154–179
Holman PR (2000) Myelodysplasia and second malignancies. In: Ball ED, Lister J, Law P (eds) Hematopoietic stem cell therapy. Churchill Livingstone, Philadelphia, pp 647–657
Honea N, Brant J, Beck SL (2007) Treatment related symptom clusters. Semin Oncol Nurs 23:142–151
Horning S (2008) Hodgkin's lymphoma. In: Abeloff MD, Armitage JO, Niederhuber JE, Kastan MB, MCKenna WG (eds) Abeloff's clinical oncology, 4th edn. Churchill Livingstone, Philadelphia, pp 2253–2268
Housman G, Byler S, Heerboth S, Lapinska K, Longacre M, Snyder N, Sarkar S (2014) Drug resistance in cancer: an overview. Cancers 6:1769–1792
Intlekofer KA, Cunningham MJ, Caplan AL (2012) The HPV vaccine controversy. Virtual Mentor 14:39–49
Jaffee EM, Hruban RH, Biedrzycki B, Laheru D, Schepers K, Sauter PR, Goemann M, Coleman J, Grochow L, Donehower RC, Lillemoe KD, O'Reilly S, Abrams RA, Pardoll DM, Cameron JL, Yeo CJ (2001) Novel allogeneic granulocyte-macrophage colony-stimulating factor-secreting tumor vaccine for pancreatic cancer: a phase I trial of safety and immune activation. J Clin Oncol 19:145–156
Jaiyesimi IA, Buzdar AU, Decker DA, Hortobagyi GN (1995) Use of tamoxifen for breast cancer: twenty-eight years later. J Clin Oncol 13:513–529
Jong WHD, Borm PJA (2008) Drug delivery and nanoparticles applications and hazards. Int J Nanomedicine 3:133–149
Joshi TG, Ehrenberger HE (2001) Cancer clinical trials in the new millennium: novel challenges and opportunities for oncology nursing. Clin J Oncol Nurs 5:147–152
Kampan NC, Madondo MT, McNally OM, Quinn M, Plebanski M (2015) Paclitaxel and its evolving role in the management of ovarian cancer. Biomed Res Int 2015:413076. https://doi.org/10.1155/2015/413076
Kanagesan S, Hashim M, AB Aziz S, Ismail I, Tamilselvan S, Alitheen NB, Swamy MK, Purna Chandra Rao B (2016) Evaluation of antioxidant and cytotoxicity activities of copper ferrite ($CuFe_2O_4$) and zinc ferrite ($ZnFe_2O_4$) nanoparticles synthesized by sol-gel self-combustion method. Appl Sci 6:184. https://doi.org/10.3390/app6090184
Kane MA (2012) Preventing cancer with vaccines: progress in the global control of cancer. Cancer Prev Res 5:24–29
Kantoff PW, Higano CS, Shore ND, Roy Berger E, Small EJ, Penson DF, Redfern CH, Ferrari AC, Dreicer R, Sims RB, XuY FMW, Schellhammer PF (2010) Sipuleucel-T immunotherapy for castration-resistant prostate cancer. New Eng J Med 363:411–422
Karp DD, Thurer RL (2003) Non-small cell lung cancer. In: Furie B, Cassileth PA, Atkins MB, Mayer RL (eds) Clinical hematology and oncology: presentation, diagnosis, and treatment. Churchill Livingstone, Philadelphia, pp 958–982
Kelley KW, Bluthe RM, Dantzer R, Zhou JH, Shen WH, Johnson RW, Broussard SR (2003) Cytokine-induced sickness behavior. Brain Behav Immun 17:S112–S118
Kimberlin C, Brushwood D, Allen W, Radson E, Wilson D (2004) Cancer patient and caregiver experiences: communication and pain management issues. J Pain Sympt Manag 28:566–578
Kinzler K, Vogelstein B (1996) Lessons from hereditary colorectal cancer. Cell 87:159–170
Kroschinsky F, Stölzel F, von Bonin S, Beutel G, Kochanek M, Kiehl M, Schellongowski P (2017) New drugs, new toxicities: severe side effects of modern targeted and immunotherapy of cancer and their management. Crit Care 21:89. https://doi.org/10.1186/s13054 -017-1678-1
Kuan CT, Wikstrand CJ, Bigner DD (2001) EGF mutant receptor VIII as a molecular target in cancer therapy. Endocr Relat Cancer 8:83–96
Kumari P, Ghosh B, Biswas S (2016) Nanocarriers for cancer-targeted drug delivery. J Drug Target 4:179–191

Schiffer CA, Anderson KC, Bennett CL, Bernstein S, Elting LS, Goldsmith M, Goldstein M, Hume H, McCullough JJ, McIntyre RE, Powell BL, Rainey JM, Rowley SD, Rebulla P, Troner MB, Wagnon AH (2001) Platelet transfusion for patients with cancer: clinical practice guidelines of the American Society of Clinical Oncology. J Clin Oncol 19:1519–1538

Schwartzentruber DJ (2001) Guidelines for the safe administration of high dose Interleukin-2. J Immunother 24:287–293

Shankaran V, Ikeda H, Bruce AT, White JM, Swanson PE, Old LJ, Schreiber RD (2001) IFN gamma and lymphocytes prevent primary tumour development and shape tumor immunogenicity. Nature 4:1107–1111

Sharma S, O'Keefe SJ (2007) Environmental influences on the high mortality from colorectal cancer in African Americans. Postgrad Med J 83:583–589

Shayne M, Culakova E, Poniewierski MS, Dale DC, Crawford J, Lyman GH (2009) Dose intensity and hematologic toxicity in older cancer patients receiving systemic chemotherapy. Cancer 110:5319–5328

Sheridan AT, Dawber RP (2000) Curettage, electrosurgery and skin cancer. Aust J Dermatol 41:19–30

Shewach DS, Kuchta RD (2009) Introduction to cancer chemotherapeutics. Chem Rev 109:2859–2861

Shute C (2013) The challenges of cancer pain assessment and management. Ulster Med J 82:40–42

Subbiah R, Veerapandian M, Yun KS (2010) Nanoparticles: functionalization and multifunctional applications in biomedical sciences. Curr Med Chem 17:4559–4577

Swamy MK, Akhtar MS, Mohanty SK, Sinniah UR (2015a) Synthesis and characterization of silver nanoparticles using fruit extract of *Momordica cymbalaria* and assessment of their *in vitro* antimicrobial, antioxidant and cytotoxicity activities. Spectrochim Acta A 151:939–944

Swamy MK, Sudipta K, Jayanta K, Balasubramanya S (2015b) The green synthesis, characterization, and evaluation of the biological activities of silver nanoparticles synthesized from *Leptadenia reticulata* leaf extract. Appl Nanosci 5:73–81

Tayal V, Kalra BS (2008) Cytokines and anti-cytokines as therapeutics-an update. Eur J Pharmacol 579:1–12

Timoney JP, Eagan MM, Sklarin NT (2003) Establishing clinical guidelines for the management of acute hypersensitivity reactions secondary to the administration of chemotherapy/biologic therapy. J Nurs Care Qual 18:80–86

Tonini G, Vincenzi B, Santini D, Olzi D, Lambiase A, Bonini S (2005) Ocular toxicity related to cetuximab monotherapy in an advanced colorectal cancer patient. J Natl Cancer Inst 97:606–607

Travis LB, Curtis RE, Glimelius B, Holowaty EJ, Van Leeuwen FE, Lynch CF, Hagenbeek A, Stovall M, Banks PM, Adami J, Gospodarowicz MK (1995) Bladder and kidney cancer following cyclophosphamide therapy for non-Hodgkin's lymphoma. J Natl Cancer Inst 87:524–530

Tu Y, Cheng S, Zhang S, Sun H, Xu Z (2013) Vincristine induces cell cycle arrest and apoptosis in SH-SY5Y human neuroblastoma cells. Int J Mol Med 31:113–119

Tudor Car L, Papachristou N, Urch C, Majeed A, El-Khatib M, Aylin P, Atun R, Car J, Vincent C (2016) Preventing delayed diagnosis of cancer: clinicians views on main problems and solutions. J Glob Health 6:020901. https://doi.org/10.7189/jogh.06.020901

Vachani A, Seijo L, Unger M, Sterman D (2008) Bronchoscopy, transthoracic needle aspiration, and related procedures. In: Fishman AP, Elias JA, Fishman JA, Grippi MA, Senior RM, Pack AI (eds) Fishman's pulmonary diseases and disorders, 4th edn. The McGraw-Hill Companies, China, pp 629–648

van Straten D, Mashayekhi V, de Bruijn HS, Oliveira S, Robinson DJ (2017) Oncologic photodynamic therapy: basic principles, current clinical status and future directions. Cancers 9:19. https://doi.org/10.3390/cancers9020019

Vieweg J, Dannull J (2003) Tumor vaccines: from gene therapy to dendritic cells-the emerging frontier. Urol Clin North Am 30:633–643

Villa NY, Rahman MM, McFadden G, Cogle CR (2016) Therapeutics for graft-versus-host disease: from conventional therapies to novel virotherapeutic strategies. Virus 8:85

Visvader JE, Lindeman GJ (2008) Cancer stem cells in solid tumors: accumulating evidence and unresolved questions. Nat Rev Cancer 8:755–768

Volker DL (2001) Carcinogenesis: application to clinical practice. Clin J Oncol Nurs 5:225–226

Weinberg RA (2007) The biology of cancer, 1st edn. Garland Science, Taylor and Francis Group, LLC, New York

Zhao J, Wallace M, Marites PM (2014) Cancer theranostics with gold nanoshells. Nanomedicine 9:2041–2057

Chapter 2
Nano-natural Products as Anticancer Agents

Atish Tulshiram Paul and Anil Jindal

Abstract Cancer is one of the noxious diseases and is a major public health problem worldwide. The clinical management of cancer involves various approaches, but the most common one is chemotherapy. Natural products such as paclitaxel, camptothecin, podophyllotoxin, etc. have been used as major sources of anticancer drugs in many clinical trials. In spite of availability of these drugs for treatment of cancer, failure in chemotherapy is very common due to dose-limited toxicities and occurrence of drug resistance. In this regard, the nano-delivery systems directly target and deliver the selective drug to the cancerous sites and increase the permeability and intracellular accretion of anticancer drugs. Thus, the aim of this chapter is to focus on the application of nanotechnology to develop nano-natural products for effective treatment of cancer.

Keywords Cancer · Cell permeability · Nanoparticle · Nanotechnology · Natural products

2.1 Introduction

Cancer is one of the noxious diseases and a major public health problem worldwide. Despite genetic causes, there are several reasons that cause cancer in an individual which include lifestyle (such as consumption of tobacco and alcohol), exposure to infectious agents (e.g., *H. pylori*), exposure to carcinogens (e.g., chemicals and radiation), ethnicity, etc. As per World Health Organization (WHO) statistics, cancer has been responsible for 8.8 million deaths in 2015 (WHO 2017). As predicted, by 2030, the figure of deaths due to cancer around the globe will be 13.2 million with global burden of new cancer cases being estimated to rise to 20.3 million by 2030 (Ferlay et al. 2010; Bray et al. 2012). Even though many advances have been

A. T. Paul (✉) · A. Jindal
Department of Pharmacy, Birla Institute of Technology and Science Pilani, (Pilani campus), Pilani 333031, Rajasthan, India
e-mail: atish.paul@pilani.bits-pilani.ac.in

M. S. Akhtar, M. K. Swamy (eds.), *Anticancer Plants: Clinical Trials and Nanotechnology*, https://doi.org/10.1007/978-981-10-8216-0_2

made in tackling cancer, but still it remains as a major challenge in developed countries. In the case of low- and middle-income countries, the cases of cancer increased due to the change in lifestyle. Due to its devastating effect in terms of premature deaths and morbidity, it has caused a 20% global economic burden when compared to other diseases. Hence, in 2011, United Nations has included cancer in the list of noncommunicable diseases and has pledged to reduce death by 25% till the year 2025. There are many hurdles and challenges in cancer treatment with respect to clinical as well as from drug discovery point of view. After many decades of basic research work, we have been able to understand cancer. The modern-day technology has provided insights into the numerous factors that are relevant to cancer development and progression. However, still today in most of the cases, diagnosis of cancer is found only in its late phase. This issue of late diagnosis has defeated modern-day research, wherein newly developed drugs are not able to save lives of people even after their availability. The recent data, research, and facts need to be translated into new clinical approaches, as well as public education and policy initiatives. People need to be made aware about various factors that may lead to cancer development. Also there is an urgent need to strengthen the global cancer surveillance and cancer control programs (John and Ross 2010). At drug discovery and development levels, newer classes of compounds, biological methods, etc. need to be developed for achieving higher success rate of clinical useful drugs. Natural products, in spite of being the major source of drugs for millennia, face higher failure rate due to poor pharmacokinetics, bioavailability, poor efficacy, chemical instability, first-pass metabolism, and unwanted biodistribution in the organs of the body.

In the recent era, the nanotechnology has become a mainstay for the field of drug discovery and development. Nanoparticles (NPs) are defined as particulate matter ranging in the size of 1–100 nm, with an ability to carry or deliver medicinally active drugs, nutritional materials, diagnostic aids, etc. (Akhtar et al. 2013; Akhtar and Ahmad 2014; Swamy et al. 2015a, b). Major reasons for the development of nanotechnology platforms for pharmaceutical purpose have been attributed to the following reasons: reduction of toxicity of existing drugs, improvement of pharmacokinetic properties, formulation stability (shelf life), biocompatibility, etc. Various structural NPs such as liposomes (natural and synthetic lipids), micelles (amphiphilic biomaterials), dendrimers (branched biomaterials), exosomes (biomembrane), polymers (polylactic acid, polycaprolactam, polycyanoacrylates), carbon derivatives (various carbon allotropes), and inorganic derivatives (gold, silver, silica, Cd/Zn-selenides) have been used for development of nanoparticulate delivery (Fig. 2.1). Over a decade, nanotechnology has significantly contributed in the development of new anticancer products. The first nanoanticancer product to be approved by FDA was PEGylated liposomal doxorubicin in 1995 for treatment of HIV-related Kaposi sarcoma, ovarian cancer, and multiple myeloma. Since then many nanoparticulate delivery systems have been discovered and developed for tackling the menace of cancer (Table 2.1). Thus, the aim of the present chapter is to highlight the application of nanotechnology as an anticancer agent and also to develop nano-natural products for effective treatment of cancer.

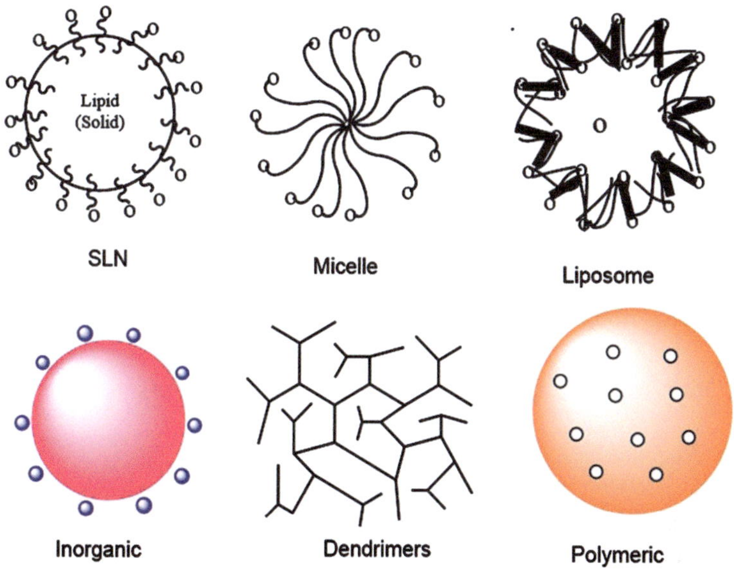

Fig. 2.1 Nanoparticulate systems available for delivery of chemicals

Table 2.1 Summary of some approved nanoparticulate systems used in cancer therapy

Active ingredient	Brand name	Platform technology	Cancer types
Doxorubicin	Doxil	PEGylated liposome	Kaposi sarcoma, ovarian cancer, and multiple myeloma
	DaunoXome	Liposome	Kaposi sarcoma
	Myocet	Liposome	Metastatic breast cancer
Irinotecan	Onivyde	PEGylated liposome	Metastatic pancreatic cancer
Paclitaxel	Abraxane	Albumin nanoparticles	Breast, lung, and pancreatic cancer
Vincristine sulfate	Marqibo	Liposome	Acute lymphoblastic leukemia

2.2 Natural Products and Anticancer Drug Discovery

Since primeval times, nature is recognized for being a source of diverse and unique chemical compounds used by mankind. The natural products or secondary metabolites are obtained from a wide range of naturally occurring sources like plants, minerals, animals, microbes, and marine fauna. These may serve as the basis of drug discovery programs for the past several decades due to their structural uniqueness and functional diversity (Newman 2008). The total number of natural products that have been identified from the various sources is approximately ~200,000 or a bit more. However, these may not be the actual figure, and a very small percentage (about 15%) of higher plants have been explored, and a vast majority still remains to be explored. The major contribution of the above natural sources in drug development has been mainly in the area of cancer (60%) and infectious diseases (75%) (Cragg et al. 1997; Newman et al. 2003; Newman and Cragg 2007, 2012). Recently, Newman and Cragg (2016) have highlighted the contribution of natural products in the area of anticancer drugs. Among the various anticancer drugs discovered till date, the most successful are Taxol®, vincristine, camptothecin, etc. Some of the other molecules include temsirolimus (Torisel™), ixabepilone (Ixempra™), romidepsin (Istodax®), and amrubicin hydrochloride (Calsed®).

2.3 Nano-natural Products as Anticancer Agents

2.3.1 Phenolics

Phenolic compounds are one of the most abundant classes of secondary metabolite (Fig. 2.2). These compounds bear one or more hydroxyl groups on aromatic ring(s) and are referred to as polyphenolics at times. They are abundantly present in fruits and vegetables and exhibit a wide range of different biological activities such antioxidant, anti-inflammatory, antimicrobial, etc. They are known to possess the ability to prevent different types of cancers. Some of them are described below.

2.3.1.1 Quercetin and Citropten

Gismondi et al. (2015) reported the nanodiamond conjugation of citropten and quercetin to improve their antiproliferative activity on human (HeLa) and murine (B16F10) tumor cells. Nanodiamonds were synthesized by three different protocols (oxidation, chemical, and plasma reduction). The citropten and quercetin content in each mg of nanodiamond adduct was 0.5 mmol, respectively. However, citropten-loaded samples that were prepared by plasma reduction method contained exceptionally higher content of the drug. Treatment of HeLA cell line for 48 and 72 h resulted in great antiproliferative effects especially for citropten adduct prepared by oxidation and plasma reduction, which reduced cell growth better than pure citropten. On the other hand, nanodiamonds functionalized with quercetin similarly inhibited proliferation rate with respect to pure quercetin. In both the cell lines, citropten bioactivity was enhanced after conjugation with nanodiamonds prepared by oxidation and plasma reduction, while antiproliferative effects of quercetin were increased after conjugation with nanodiamonds prepared by chemical and plasma reduction. The observed effect may be due to the ability of nanodiamond adducts to stimulate cell processes, such as cell cycle differentiation, cell cycle arrest, etc. Various nanoparticulate delivery systems have been developed for effective delivery of quercetin for anticancer effect (Table 2.2).

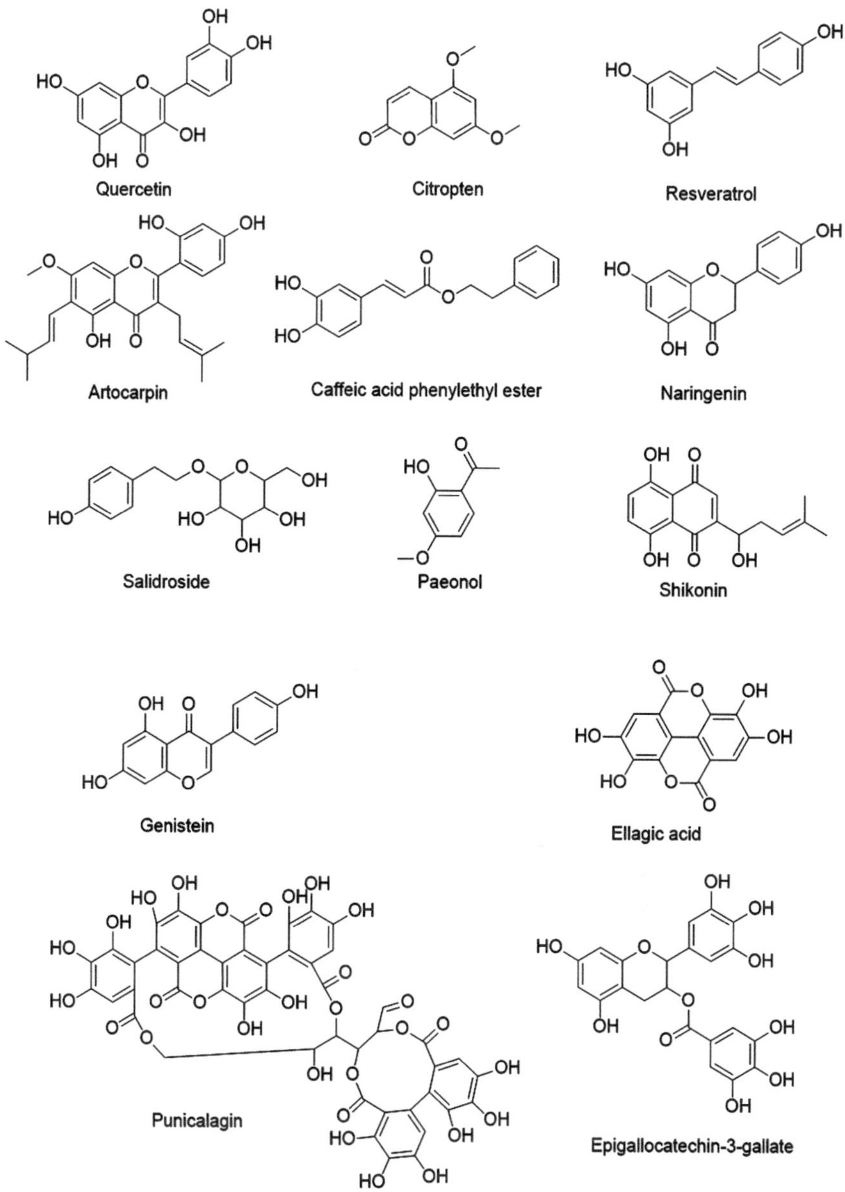

Fig. 2.2 Structures of polyphenolic class of compounds

Table 2.2 Quercetin-based nanoparticulate delivery systems and their anticancer effect in various cancer models

Carrier type	Modification	Cancer models	References
Nanoliposomes	Nil	C6 glioma cell	Wang et al (2012)
Nanomicelles	Nil	Lung tumor mice	Tan et al. (2012)
Folic acid-PEG-PLGA	Folic acid conjugation	HeLa cells; HeLa or IGROV-1tumor-bearing mice	El-Gogary et al. (2014)
NLCs	Nil	MCF-7 and MDA-MB-231 breast cancer cells	Sun et al. (2014)
Liposomes	Nil	MCF-7 breast cancer cells	Rezaei-Sadabady et al. (2016)

Table 2.3 Resveratrol-based nanoparticulate delivery systems and their anticancer effect in various cancer models

Carrier type	Modification	Cancer models	References
Bovine serum albumin	Nil	Human primary ovarian carcinoma cells in nude mice	Guo et al. (2010)
Dequalinium polyethylene glycol-distearoylphosphatidylethanolamine conjugate	Nil	Human lung adenocarcinoma A549 cells, resistant and tumor spheroid A549/cDDP cells, and xenografted resistant A549/cDDP cancers in nude mice	Wang et al. (2011a)
Chitosan	Nil	HepG2 cells and Kunming mice	Bu et al. (2013)
Gelatin	Nil	Lung cancer (NCI-H460) cells	Karthikeyan et al. (2013)
PEG-PLA	Nil	CT26 colon cancer cells and CT26 tumor-bearing mice	Sanna et al. (2013)
SLN	Nil	C6 glioma cells	Jose et al. (2014)

2.3.1.2 Resveratrol

Teskac and Kristl (2010) have investigated solid lipid NPs (SLNs) of trans-resveratrol. The prepared SLNs were compared for their effects on the internalization, morphology, growth, metabolic activity, and genetic material of keratinocytes using NCTC2544 cell line. It was observed that SLNs with a size below 180 nm move promptly through the cell membrane, distributing throughout the cytosol and finally localizing in the perinuclear region without inducing cytotoxicity. An increase in solubility, stability, and intracellular delivery was observed for SLNs of resveratrol, wherein resveratrol was released in a biphasic pattern. Resveratrol SLNs were found to concentrate around nuclei and released resveratrol in a constant way to express its cytostatic effect with prominent S-arrest of cell cycle and a large drop of G2/M phase. Resveratrol has been delivered effectively through different NPs as summarized in Table 2.3.

2.3.1.3 Artocarpin

Artocarpin, a prenylated flavonoid isolated from *Artocarpus communis* extracts, has been found to induce death in human hepatocellular carcinoma cells. Tzeng et al. (2016) developed the PVP nanoparticulate system to overcome the poor water solubility and investigated the improvement in therapeutic efficacy of artocarpin against HepG2 and PLC/PRF/5 cell line. Artocarpin-PVP NPs enhanced the solubility of artocarpin by reducing particle size and forming hydrogen bonds. The NPs also showed superior autophagic cytotoxicity as indicated by increased autophagy-related protein LC3-II expression and the ratio of LC3-II/LC3-I after 24 h treatment. However, artocarpin was unable to induce autophagy in hepatoma cells and did not alter the LC3 expression.

2.3.1.4 Caffeic Acid Phenethyl Ester

Caffeic acid phenethyl ester (CAPE) NPs were prepared using methoxy poly(ethylene glycol)-*b*-poly(ε-caprolactone) copolymer, and their antitumor activity against pulmonary metastasis model of CT26 colon carcinoma cells was studied. CAPE NPs had particle size of <300 nm, and they inhibited proliferation, induced apoptosis, and showed anti-invasive and antimigrative effect against CT26 cells as compared to CAPE. CAPE NPs also showed superior antimetastatic efficacy in pulmonary metastasis model of CT26 cells in nude mouse (Lee et al. 2015).

2.3.1.5 Naringenin

Chemopreventive efficacy of naringenin-loaded NPs was described in 7,12-dimethylbenz[a]anthracene (DMBA)-induced experimental oral carcinogenesis in Syrian hamsters. Oral administration of naringenin-loaded NPs completely prevented tumor formation as compared to free naringenin and significantly reduced the degree of histological lesions. Additionally, in the buccal mucosa of DMBA-exposed animals, NPs downregulated the expression of proliferating cell nuclear antigen (PCNA) and p53 and exerted higher anti-lipid peroxidative and antioxidant activities, when compared with free naringenin (Sulfikkarali et al. 2013).

2.3.1.6 Salidroside and Paeonol

Peng et al. (2014) constructed a novel nanosphere gel for sequential delivery of salidroside and paeonol for anti-melanogenic activity. Nanospheres containing both paeonol and salidroside were incorporated in carbomer hydrogel to form a dual drug releasing nanosphere gel. Dynasan 116 and Miglyol 812 in a ratio of 6:4 gave the best nanoparticle (particle size was 275 nm) with a low PDI index (0.208). With this nanosphere gel, rapid release of salidroside from the hydrogel followed by

Table 2.6 (continued)

Carrier type	Modification	Cancer models	References
Nanohybrids	Nano-Ag and chitosan	Mouse fibroblast (L929) and human colorectal adenocarcinoma (HT-29) cells	Barbinta et al. (2016)
Dendritic mesoporous silica NPs	Calcium doping of dendritic mesoporous silica NPs modified with folic acid	Breast cancer (A549 and MCF-7) cells; BALB/c nude mice	Wang et al. (2016)
PLGA	Nil	HPV-infected cervical cancer (Caski and SiHa) cell lines and NOD-SCID gamma (NSG) orthotopic xenograft mouse model	Zaman et al. (2016)
PEG	Transferrin conjugation	Breast cancer (MCF-7) cells and BALB/c nude mice	Cui et al. (2017)
Polyamidoamine dendrimer	Nil	Breast cancer cells (MCF-7) cells	Falconieri et al. (2017)
Curcumin nanosuspension	mPEG2000-DSPE and soybean lecithin as stabilizer	Hela, 4T1, HCT-8, and HepG2 cell lines and H22 tumor-bearing mice	Hong et al. (2017)

2.3.4.2 3α,24-Dihydroxyurs-12-Ene and 3α, 24-Dihydroxyolean-12-Ene (TPD)

It is obtained from boswellic acids loaded in PLGA (TPDPLGA-NPs) to develop NPs using emulsion-solvent evaporation method. The developed TPDPLGA-NPs had a particle size distribution of 152.56 nm and zeta potential of −17.36 mV. 51.03 μg/mg of TPD could be loaded in the optimized formulation, which exhibited a sustained release profile. TPDPLGA-NPs inhibited OVCAR-5 cells with IC_{50} values 32.8 μM after 48 h. The developed TPDPLGA-NPs also exhibited synergistic interaction effect in combination with cisplatin (CI <1), where it caused a 3.8-fold reduction of cisplatin (Alam et al. 2016). In another study, the same TPD from *Boswellia serrata* was formulated into solid lipid NPs (SLNs) by the microemulsion method. The TPD-SLNs obtained were solid, and spherical particles in the range of 100–200 nm, which inhibited the HL-60 cells, cells with an IC_{50} value of 9 μg/ml. TPD-SLNs induced apoptosis by intrinsic and extrinsic apoptotic pathways via activation of both TNF-R1 and cytochrome-c in HL-60 cells. TPD-SLNs were 12 times more active than TPD alone in sarcoma-180 solid tumor-bearing mice exhibiting and proving their *in vivo* potential (Bhushan et al. 2013).

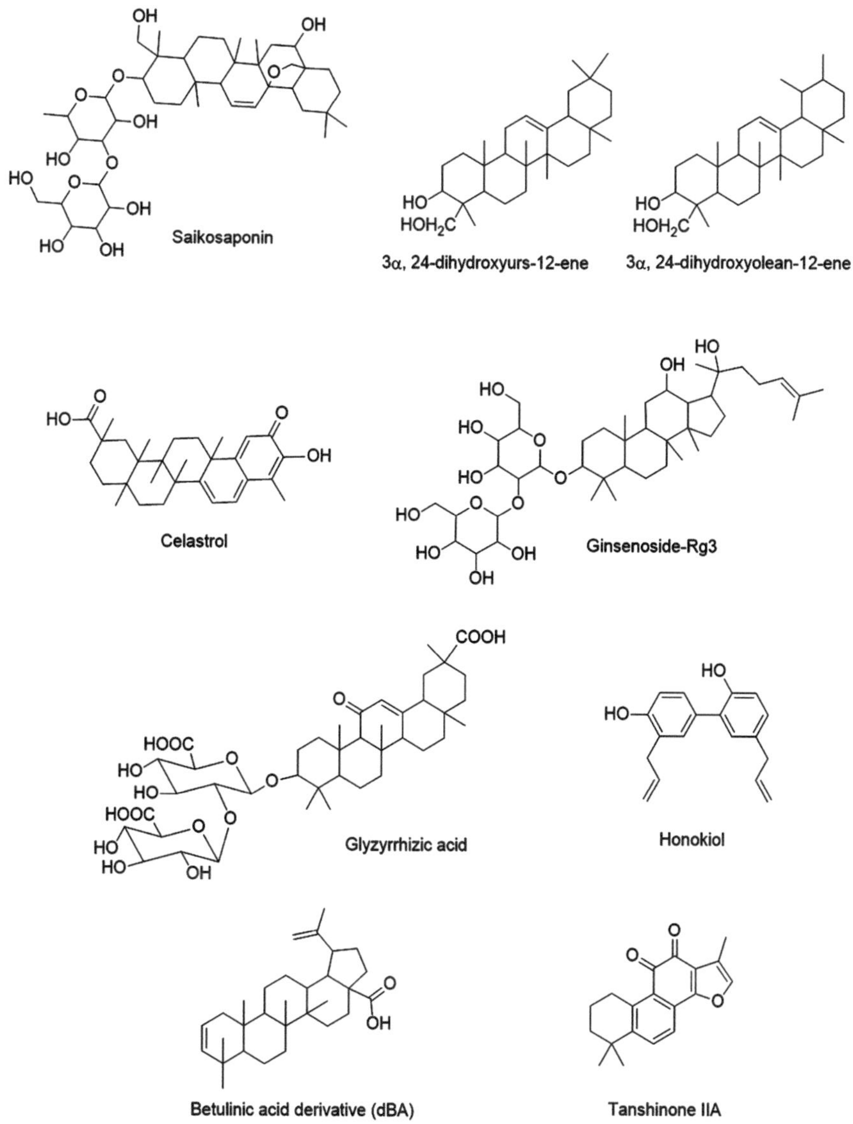

Fig. 2.5 Structures of saponin class of compounds

2.3.4.3 Celastrol

Celastrol is a triterpene, which is extracted from *Tripterygium wilfordii* Hook F. Poly(ethylene glycol)-block-poly(ε-caprolactone) nanopolymeric micelles were developed incorporating celastrol to study the inhibitory effect of these NPs on SO-Rb 50 cell proliferation. It was found that CNPs inhibited the proliferation of

SO-Rb 50 cells in a dose- and time-dependent manner with an IC_{50} of 17.73 μg/mL via decreasing the expression of Bcl-2, NF-κB p65, and phospo-NF-κB p65 at a concentration of 54.4 μg/ml for 48 h. However, the Bax/Bcl-2 ratio was found to increase, while the expression of Bax was not significantly changed. The *in vivo* potential of CNPs was evaluated using a human retinoblastoma xenograft model in NOD-SCID mice. CNP at a dose of 27.2 mg/kg/2 days successfully inhibited the growth of retinoblastoma and induced apoptosis in retinoblastoma cells in mice (Li et al. 2012).

2.3.4.4 Ginsenoside Rg3

Ginsenoside Rg3 is a saponin derived from the Chinese herb *Panax ginseng*. A VEGFR-3 antibody-conjugated ginsenoside Rg3 nanoemulsion (VRIN) was prepared and evaluated in an orthotopic mouse model of human gastric cancer. VRIN (1 mg/kg every other day, i.v.) was administered to the tumor-bearing mice. Both VRIN and the standard drug 5-FU significantly inhibited primary tumor growth as compared to vehicle control, while VRIN-treated group showed significant inhibition of lymph-node metastasis. VRIN-treated group also showed inhibition of VEGF-C, VEGF-D, and VEGFR-3 expression in the tumor (Dai et al. 2017).

2.3.4.5 Glycyrrhizic Acid

Chandrasekharan et al. (2011) developed a silver NP (SN)-complexed glycyrrhizic acid (GLY)-based delivery system for radioprotective application. On oral administration, the SN-GLY complex (50 mg GLY/kg body weight) was found to reduce the radiation-induced damage in peripheral blood leucocytes, bone marrow cells, and spleen cells of Swiss albino mice after 1 h of radiation exposure. Also SN-GLY administration resulted in reduction in micronucleus formation and chromosomal aberrations induced by exposure of mice to whole-body gamma irradiation.

2.3.4.6 Honokiol

Honokiol is a constituent of Chinese medicinal herb *Magnolia officinalis* or *M. grandiflora*. PCL-PEG-PCL (PCEC)-based monodispersed NPs (<200 nm) were prepared by solvent diffusion method without using any surfactants. The prepared PCL-PEG-PCL NPs showed poor cytotoxicity (IC_{50} > 5 mg/mL) toward HEK293 cells. They also inhibited the growth of cisplatin-sensitive (A2780s) and cisplatin-resistant (A2780cp) human ovarian cancer cells at various concentrations used (Gou et al. 2009).

2.3.4.7 *Phytolacca decandra*

PLGA-based NPs of a betulinic-acid derivative (dBA) that was isolated from an ethanolic extract of *Phytolacca decandra* were developed. The developed NPs of dBA (NdBA) were tested for anticancer activity using lung adenocarcinoma A549 cancer cells. NdBA induced apoptosis in A549 cancer cells via ER stress and Ca^{+2}-associated mitochondria-dependent pathway at IC_{50} values of 50 mg/ml and 100 mg/ml, respectively. NdBA also caused the breakdown of the oxidative phosphorylation system by depletion of ATP/ADP ratio and elevation of ROS (Das et al. 2014).

2.3.4.8 Tanshinone IIA

Tanshinone IIA (TA) is a major active component of Danshen (*Salvia miltiorrhiza* bunge), a well-known traditional Chinese medicine. Various NEs (2^3 factorial design) of TA were prepared using emulsification/high-pressure homogenization method and ratios of Tween 80, lecithin, and water. TA-NE-F4 (particle size 95 nm, spherical and intact; entrapment efficiency 99.3%) exhibited time- and dose-dependent manner cytotoxicity against T24 human bladder cancer cells that was 103-fold greater than pure TA alone (Chang et al. 2011).

2.4 Conclusions and Future Prospects

Various factors such as environmental pollution and changed lifestyle are always associated with the increased risk of cancer. In this regard, the natural products have proven their anticancer potential for the drug discovery. Nowadays, these drugs are successful at clinical level. In addition to these drugs, many candidates are still in clinical trial phases. In spite of such contribution, there has not been a decline in the incidences and demands to detailed study in both the disease understanding and drugs acting on them. Some major hurdles in the successful development of anticancer agents are poor solubility, development of resistance, unwanted side effects, etc. Nanotechnology platform has been successful in recent times to answer and solve the above hurdles via "nano-natural products" and has provided a new dimension and face to natural products in treatment of cancer.

References

Aditya N, Shim M, Lee I, Lee Y, Im MH, Ko S (2013) Curcumin and genistein coloaded nano-structured lipid carriers: *in vitro* digestion and antiprostate cancer activity. J Agric Food Chem 61:1878–1883

Akhtar MS, Ahmad U (2014) Biosynthesis of nanomaterials and their applications. Rev Adv Sci Eng 3:197–198

Akhtar MS, Panwar J, Yun YS (2013) Biogenic synthesis of metallic nanoparticles by plant extracts. ACS Sustain Chem Eng 1:591–602

Alam N, Qayum A, Kumar A, Khare V, Sharma PR, Andotra SS, Singh SK, Koul S, Gupta PN (2016) Improved efficacy of cisplatin in combination with a nano-formulation of pentacyclic triterpenediol. Mater Sci Eng C 68:109–116

Anitha A, Deepa N, Chennazhi K, Lakshmanan VK, Jayakumar R (2014) Combinatorial anticancer effects of curcumin and 5-fluorouracil loaded thiolated chitosan nanoparticles towards colon cancer treatment. Biochim Biophys Acta 1840:2730–2743

Babaei E, Sadeghizadeh M, Hassan ZM, Feizi MA, Najafi F, Hashemi SM (2012) Dendrosomal curcumin significantly suppresses cancer cell proliferation *in vitro* and *in vivo*. Int Immunopharmacol 12:226–234

Banerjee S, Sahoo AK, Chattopadhyay A, Ghosh SS (2014) Recombinant IκBα-loaded curcumin nanoparticles for improved cancer therapeutics. Nanotechnology 25:345102. https://doi.org/10.1088/0957-4484/25/34/345102

Barbinta-Patrascu ME, Badea N, Pirvu C, Bacalum M, Ungureanu C, Nadejde PL, Ion C, Rau I (2016) Multifunctional soft hybrid bio-platforms based on nano-silver and natural compounds. Mater Sci Eng C 69:922–932

Bhushan S, Kakkar V, Pal HC, Guru SK, Kumar A, Mondhe DM, Sharma PR, Taneja SC, Kaur IP, Singh J, Saxena AK (2013) Enhanced anticancer potential of encapsulated solid lipid nanoparticles of TPD: a novel triterpenediol from *Boswellia serrata*. Mol Pharm 10:225–235

Bray F, Jemal A, Grey N, Ferlay J, Forman D (2012) Global cancer transitions according to the human development index (2008–2030): a population-based study. Lancet Oncol 13:790–801

Bu L, Gan LC, Guo XQ, Chen FZ, Song Q, Qi-Zhao GXJ, Hou SX, Yao Q (2013) Trans-resveratrol loaded chitosan NPs modified with biotin and avidin to target hepatic carcinoma. Int J Pharm 452:355–362

Castillo PM, de la Mata M, Casula MF, Sánchez-Alcázar JA, Zaderenko AP (2014) PEGylated versus non-PEGylated magnetic nanoparticles as camptothecin delivery system. Beilst J Nanotechnol 5:1312–1319

Chandrasekharan DK, Khanna PK, Nair CK (2011) Cellular radioprotecting potential of glyzyrrhizic acid, silver nanoparticle and their complex. Mutat Res 723:51–57

Chang LC, Wu CL, Liu CW, Chuo WH, Li PC, Tsai TR (2011) Preparation, characterization and cytotoxicity evaluation of tanshinone IIA nanoemulsions. J Biomed Nanotechnol 7:558–567

Chung JE, Tan S, Gao SJ, Yongvongsoontorn N, Kim SH, Lee JH, Choi HS, Yano H, Zhuo L, Kurisawa M, Ying JY (2014) Self-assembled micellar nanocomplexes comprising green tea catechin derivatives and protein drugs for cancer therapy. Nat Nanotechnol 9:907–912

Cragg GM, Newman DJ, Snader KM (1997) Natural products in drug discovery and development. J Nat Prod 60:52–60

Cui T, Zhang S, Sun H (2017) Co-delivery of doxorubicin and pH-sensitive curcumin prodrug by transferrin targeted nanoparticles for breast cancer treatment. Oncol Rep 37:1253–1260

Dai X, Liu D, Liu M, Zhang X, Wang W, Jin F, Qian Y, Wang X, Zhao J, Wu Y, Xiong F, Chang NA, Sun YU, Yang Z, Hoffman RM, Liu Y (2017) Anti-metastatic efficacy of traditional Chinese medicine (TCM) ginsenoside conjugated to a VEFGR-3 antibody on human gastric cancer in an orthotopic mouse model. Anticancer Res 37:979–986

Das J, Das S, Paul A, Samadder A, Khuda-Bukhsh AR (2014) Strong anticancer potential of nano-triterpenoid from *Phytolacca decandra* against A549 adenocarcinoma via a Ca^{2+}-dependent mitochondrial apoptotic pathway. J Acupunct Merid Stud 7:140–150

El-Gogary RI, Rubio N, Wang JTW, Al-Jamal WT, Bourgognon M, Kafa H, Naeem M, Klippstein R, Abbate V, Leroux F, Bals S, Van Tendeloo G, Kamel AO, Awad GA, Mortada ND, Al-Jamal KT (2014) Polyethylene glycol conjugated polymeric nanocapsules for targeted delivery of quercetin to folate-expressing cancer cells *in vitro* and *in vivo*. ACS Nano 8:1384–1401

Falconieri MC, Adamo M, Monasterolo C, Bergonzi MC, Coronnello M, Bilia AR (2017) New dendrimer-based nanoparticles enhance curcumin solubility. Planta Med 83:420–425

Ferlay J, Shin HR, Bray F, Forman D, Mathers C, Parkin DM (2010) Estimates of worldwide burden of cancer in 2008: GLOBOCAN 2008. Int J Cancer 127:2893–2917

Ghoneum M, Gollapudi S (2011) Synergistic apoptotic effect of Arabinoxylan rice bran (MGN-3/biobran) and curcumin (turmeric) on human multiple myeloma cell line U266 *in vitro*. Neoplasma 58:118–123

Gismondi A, Reina G, Orlanducci S, Mizzoni F, Gay S, Terranova ML, Canini A (2015) Nanodiamonds coupled with plant bioactive metabolites: a nanotech approach for cancer therapy. Biomaterials 38:22–35

Gou M, Zheng L, Peng X, Men K, Zheng X, Zeng S, Guo G, Luo F, Zhao X, Chen L, Wei Y, Qian Z (2009) Poly(epsilon-caprolactone)-poly(ethylene glycol)-poly(epsilon-caprolactone) (PCL-PEG-PCL) nanoparticles for honokiol delivery *in vitro*. Int J Pharm 375:170–176

Gou M, Men K, Shi H, Xiang M, Zhang J, Song J, Long J, Wan Y, Luo F, Zhao X, Qian Z (2011) Curcumin-loaded biodegradable polymeric micelles for colon cancer therapy *in vitro* and *in vivo*. Nanoscale 3:1558–1567

Guo L, Peng Y, Yao J, Sui L, Gu A, Wang J (2010) Anticancer activity and molecular mechanism of resveratrol-bovine serum albumin NPs on subcutaneously implanted human primary ovarian carcinoma cells in nude mice. Cancer Biother Radiopharm 25:471–477

Half E, Arber N (2009) Colon cancer: preventive agents and the present status of chemoprevention. Expert Opin Pharmacother 10:211–219

Hatcher H, Planalp R, Cho J, Torti FM, Torti SV (2008) Curcumin: from ancient medicine to current clinical trials. Cell Mol Life Sci 65:1631–1652

Hong J, Liu Y, Xiao Y, Yang X, Su W, Zhang M, Liao Y, Kuang H, Wang X (2017) High drug payload curcumin nanosuspensions stabilized by mPEG-DSPE and SPC: *in vitro* and *in vivo* evaluation. Drug Deliv 24:109–120

Hsieh DS, Wang H, Tan SW, Huang YH, Tsai CY, Yeh MK, Wu CJ (2011) The treatment of bladder cancer in a mouse model by epigallocatechin-3-gallate-gold nanoparticles. Biomaterials 32:7633–7640

Hu SC, Lee IT, Yen MH, Lin CC, Lee CW, Yen FL (2016) Anti-melanoma activity of Bupleurum chinense, Bupleurum kaoi and nanoparticle formulation of their major bioactive compound saikosaponin-d. J Ethnopharmacol 179:432–442

Jin G, Jin M, Jin Z, Gao Z, Yin X (2016) Docetaxel-loaded PEG-albumin nanoparticles with improved antitumor efficiency against non-small cell lung cancer. Oncol Rep 36:872–876

Jithan A, Madhavi K, Madhavi M, Prabhakar K (2011) Preparation and characterization of albumin nanoparticles encapsulating curcumin intended for the treatment of breast cancer. Int J Pharm Investig 1:119–125

John RM, Ross H (2010) The global economic cost of cancer. American Cancer Society, Livestrong, Austin

Johnson JJ, Mukhtar H (2007) Curcumin for chemoprevention of colon cancer. Cancer Lett 255:170–181

Jose S, Anju S, Cinu T, Aleykutty NA, Thomas S, Souto EB (2014) *In vivo* pharmacokinetics and biodistribution of resveratrol-loaded solid lipid nanoparticles for brain delivery. Int J Pharm 474:6–13

Karthikeyan S, Prasad NR, Ganamani A, Balamurugan E (2013) Anticancer activity of resveratrol-loaded gelatin NPs on NCI-H460 non-small cell lung cancer cells. Biomed Prev Nutr 3:64–73

Khan N, Bharali DJ, Adhami VM, Siddiqui IA, Cui H, Shabana SM, Mousa SA, Mukhtar H (2014) Oral administration of naturally occurring chitosan-based nanoformulated green tea polyphenol EGCG effectively inhibits prostate cancer cell growth in a xenograft model. Carcinogenesis 35:415–423

Kim JE, Park YJ (2017) High paclitaxel-loaded and tumor cell-targeting hyaluronan-coated nanoemulsions. Colloids Surf B Biointerfaces 150:362–372

Lee HY, Jeong YI, Kim EJ, Lee KD, Choi SH, Kim YJ, Kim DH, Choi KC (2015) Preparation of caffeic acid phenethyl ester-incorporated nanoparticles and their biological activity. J Pharm Sci 104:144–154

Li Z, Wu X, Li J, Yao L, Sun L, Shi Y, Zhang W, Lin J, Liang D, Li Y (2012) Antitumor activity of celastrol nanoparticles in a xenograft retinoblastoma tumor model. Int J Nanomedicine 7:2389–2398
Li L, Xiang D, Shigdar S, Yang W, Li Q, Lin J, Liu K, Duan W (2014) Epithelial cell adhesion molecule aptamer functionalized PLGA-lecithin-curcumin-PEG nanoparticles for targeted drug delivery to human colorectal adenocarcinoma cells. Int J Nanomedicine 9:1083–1096
Li H, Shi L, Wei J, Zhang C, Zhou Z, Wu L, Liu W (2016) Cellular uptake and anticancer activity of salvianolic acid B phospholipid complex loaded nanoparticles in head and neck cancer and precancer cells. Colloids Surf B Biointerfaces 147:65–72
Liu Z, Zhang F, Koh GY, Dong X, Hollingsworth J, Zhang J, Russo PS, Yang P, Stout RW (2015) Cytotoxic and antiangiogenic paclitaxel solubilized and permeation-enhanced by natural product nanoparticles. Anti-Cancer Drugs 26:167–179
Lu J, Huang Y, Zhao W, Marquez RT, Meng X, Li J, Gao X, Venkataramanan R, Wang Z, Li S (2013) PEG-derivatized embelin as a nanomicellar carrier for delivery of paclitaxel to breast and prostate cancers. Biomaterials 34:1591–1600
Matthaiou EI, Barar J, Sandaltzopoulos R, Li C, Coukos G, Omidi Y (2014) Shikonin-loaded antibody-armed nanoparticles for targeted therapy of ovarian cancer. Int J Nanomedicine 9:1855–1870
Mendes LP, Gaeti MP, De Vila PH, de Sousa Vieira M, Dos Santos Rodrigues B, de Ávila Marcelino RI, Dos Santos LC, Valadares MC, Lima EM (2014) Multi-compartimental NPs for co-encapsulation and multimodal drug delivery to tumor cells and neovasculature. Pharm Res 31:1106–1119
Nair KL, Thulasidasan AKT, Deepa G, Anto RJ, Kumar GS (2012) Purely aqueous PLGA nanoparticulate formulations of curcumin exhibit enhanced anticancer activity with dependence on the combination of the carrier. Int J Pharm 425:44–52
Najlah M, Kadam A, Wan KW, Ahmed W, Taylor KM, Elhissi AM (2016) Novel paclitaxel formulations solubilized by parenteral nutrition nanoemulsions for application against glioma cell lines. Int J Pharm 506:102–109
Nekkanti V, Venkateswarlu V, Ansari KA, Pillai R (2011) Development and pharmacological evaluation of a PEG based nanoparticulate camptothecin analog for oral administration. Curr Drug Deliv 8:661–666
Newman DJ (2008) Natural products as leads to potential drugs: an old process or the new hope for drug discovery. J Med Chem 51:2589–2599
Newman DJ, Cragg GM (2007) Natural products as sources of new drugs over the last 25 years. J Nat Prod 70:461–477
Newman DJ, Cragg GM (2012) Natural products as sources of new drugs over the 30 years from 1981 to 2010. J Nat Prod 75:311–335
Newman DJ, Cragg GM (2016) Natural products as sources of new drugs from 1981 to 2014. J Nat Prod 79:629–661
Newman DJ, Cragg GM, Snader KM (2003) Natural products as sources of new drugs over the period 1981–2002. J Nat Prod 66:1022–1037
de Pace RC, Liu X, Sun M, Nie S, Zhang J, Cai Q, Gao W, Pan X, Fan Z, Wang S (2013) Anticancer activities of (−)-epigallocatechin-3-gallate encapsulated nanoliposomes in MCF7 breast cancer cells. J Liposome Res 23:187–196
Palange AL, Di Mascolo D, Carallo C, Gnasso A, Decuzzi P (2014) Lipid-polymer nanoparticles encapsulating curcumin for modulating the vascular deposition of breast cancer cells. Nanomedicine 10:991–1002
Park J, Ayyappan V, Bae EK, Lee C, Kim BS, Kim BK, Lee YY, Ahn KS, Yoon SS (2008) Curcumin in combination with bortezomib synergistically induced apoptosis in human multiple myeloma U266 cells. Mol Oncol 2:317–326
Peng LH, Xu SY, Shan YH, Wei W, Liu S, Zhang CZ, Wu JH, Liang WQ, Gao JQ (2014) Sequential release of salidroside and paeonol from a nanosphere-hydrogel system inhibits ultraviolet B-induced melanogenesis in guinea pig skin. Int J Nanomedicine 9:1897–1908

Pham J, Brownlow B, Elbayoumi T (2013) Mitochondria-specific pro-apoptotic activity of genistein lipidic nanocarriers. Mol Pharm 10:3789–3800
Phan V, Walters J, Brownlow B, Elbayoumi T (2013) Enhanced cytotoxicity of optimized liposomal genistein via specific induction of apoptosis in breast, ovarian and prostate carcinomas. J Drug Target 21:1001–1011
Pund S, Borade G, Rasve G (2014) Improvement of anti-inflammatory and anti-angiogenic activity of berberine by novel rapid dissolving nanoemulsifying technique. Phytomedicine 21:307–314
Punfa W, Yodkeeree S, Pitchakarn P, Ampasavate C, Limtrakul P (2012) Enhancement of cellular uptake and cytotoxicity of curcumin-loaded PLGA nanoparticles by conjugation with anti-P-glycoprotein in drug resistance cancer cells. Acta Pharmacol Sin 33:823–831
Ranjan AP, Mukerjee A, Helson L, Gupta R, Vishwanatha JK (2013) Efficacy of liposomal curcumin in a human pancreatic tumor xenograft model: inhibition of tumor growth and angiogenesis. Anticancer Res 33:3603–3609
Rezaei-Sadabady R, Eidi A, Zarghami N, Barzegar A (2016) Intracellular ROS protection efficiency and free radical-scavenging activity of quercetin and quercetin-encapsulated liposomes. Artif Cells Nanomed Biotechnol 44:128–134
Rocha S, Generalov R, Pereira Mdo C, Peres I, Juzenas P, Coelho MA (2011) Epigallocatechin gallate-loaded polysaccharide nanoparticles for prostate cancer chemoprevention. Nanomedicine 6:79–87
Sanna V, Pintus G, Roggio AM, Punzoni S, Posadino AM, Arca A, Marceddu S, Bandiera P, Uzzau S, Sechi M (2011) Targeted biocompatible NPs for the delivery of (−)- epigallocatechin 3-gallate to prostate cancer cells. J Med Chem 54:1321–1332
Sanna V, Siddiqui IA, Sechi M, Mukhtar H (2013) Resveratrol-loaded NPs based on poly (epsilon-caprolactone) and poly (D,L-lactic-co-glycolic acid)–poly (ethylene glycol) blend for prostate cancer treatment. Mol Pharm 10:3871–3881
Shirode AB, Bharali DJ, Nallanthighal S, Coon JK, Mousa SA, Reliene R (2015) Nanoencapsulation of pomegranate bioactive compounds for breast cancer chemoprevention. Int J Nanomedicine 10:475–484
Siddiqui IA, Bharali DJ, Nihal M, Adhami VM, Khan N, Chamcheu JC, Khan MI, Shabana S, Mousa SA, Mukhtar H (2014) Excellent anti-proliferative and pro-apoptotic effects of (−)-epigallocatechin-3-gallate encapsulated in chitosan NPs on human melanoma cell growth both *in vitro* and *in vivo*. Nanomedicine 10:1619–1626
Singh M, Bhatnagar P, Mishra S, Kumar P, Shukla Y, Gupta KC (2015) PLGA-encapsulated tea polyphenols enhance the chemotherapeutic efficacy of cisplatin against human cancer cells and mice bearing Ehrlich ascites carcinoma. Int J Nanomedicine 10:6789–6809
Sulfikkarali N, Krishnakumar N, Manoharan S, Nirmal RM (2013) Chemopreventive efficacy of naringenin-loaded nanoparticles in 7,12-dimethylbenz(a)anthracene induced experimental oral carcinogenesis. Pathol Oncol Res 19:287–296
Sun J, Bi C, Chan HM, Sun S, Zhang Q, Zheng Y (2013) Curcumin-loaded solid lipid nanoparticles have prolonged *in vitro* antitumour activity, cellular uptake and improved *in vivo* bioavailability. Colloids Surf B Biointerfaces 111:367–375
Sun M, Nie S, Pan X, Zhang R, Fan Z, Wang S (2014) Quercetin-nanostructured lipid carriers: characteristics and anti-breast cancer activities *in vitro*. Colloids Surf B Biointerfaces 113:15–24
Sung B, Kunnumakkara AB, Sethi G, Anand P, Guha S, Aggarwal BB (2009) Curcumin circumvents chemoresistance *in vitro* and potentiates the effect of thalidomide and bortezomib against human multiple myeloma in nude mice model. Mol Cancer Ther 8:959–970
Swamy MK, Sudipta K, Jayanta K, Balasubramanya S (2015a) The green synthesis, characterization, and evaluation of the biological activities of silver nanoparticles synthesized from *Leptadenia reticulata* leaf extract. Appl Nanosci 5:73–81
Swamy MK, Akhtar MS, Mohanty SK, Sinniah UR (2015b) Synthesis and characterization of silver nanoparticles using fruit extract of *Momordica cymbalaria* and assessment of their *in vitro* antimicrobial, antioxidant and cytotoxicity activities. Spectrochim Acta A 151:939–944

Tan BJ, Liu Y, Chang KL, Lim BK, Chiu GN (2012) Perorally active nanomicellar formulation of quercetin in the treatment of lung cancer. Int J Nanomedicine 7:651–661
Teskac K, Kristl J (2010) The evidence for solid lipid nanoparticles mediated cell uptake of resveratrol. Int J Pharm 390:61–69
Tzeng CW, Tzeng WS, Lin LT, Lee CW, Yen FL, Lin CC (2016) Enhanced autophagic activity of artocarpin in human hepatocellular carcinoma cells through improving its solubility by a nanoparticle system. Phytomedicine 23:528–540
Udompornmongkol P, Chiang BH (2015) Curcumin-loaded polymeric nanoparticles for enhanced anticolorectal cancer applications. J Biomater Appl 30:537–546
Verma P, Meher JG, Asthana S, Pawar VK, Chaurasia M, Chourasia MK (2016) Perspectives of nanoemulsion assisted oral delivery of docetaxel for improved chemotherapy of cancer. Drug Deliv 23:479–488
Wang L, Liu Z, Liu D, Liu C, Juan Z, Zhang N (2011a) Docetaxel-loaded-lipid-based-nanosuspensions (DTX-LNS): preparation, pharmacokinetics, tissue distribution and antitumor activity. Int J Pharm 13:194–201
Wang XX, Li YB, Yao HJ, Ju RJ, Zhang Y, Li RJ, Yu Y, Zhang L, Lu WL (2011b) The use of mitochondrial targeting resveratrol liposomes modified with a dequalinium polyethylene glycol-distearoylphosphatidyl ethanolamine conjugate to induce apoptosis in resistant lung cancer cells. Biomaterials 32:5673–5687
Wang G, Wang JJ, Yang GY, Du SM, Zeng N, Li DS, Li RM, Chen JY, Feng JB, Yuan SH, Ye F (2012) Effects of quercetin nanoliposomes on C6 glioma cells through induction of type III programmed cell death. Int J Nanomedicine 7:271–280
Wang P, Zhang L, Peng H, Li Y, Xiong J, Xu Z (2013) The formulation and delivery of curcumin with solid lipid nanoparticles for the treatment of on non-small cell lung cancer both *in vitro* and *in vivo*. Mater Sci Eng C 33:4802–4808
Wang J, Wang Y, Liu Q, Yang L, Zhu R, Yu C, Wang S (2016) Rational design of multifunctional dendritic mesoporous silica nanoparticles to load curcumin and enhance efficacy for breast cancer therapy. ACS Appl Mater Interfaces 8:26511–26523
WHO (2017) World Health Organization, factsheets. http://www.who.int/mediacentre/factsheets/fs297/en/. Assessed 10 July 2017
Yallapu MM, Khan S, Maher DM, Ebeling MC, Sundram V, Chauhan N, Ganju A, Balakrishna S, Gupta BK, Zafar N, Jaggi M, Chauhan SC (2014) Anticancer activity of curcumin loaded nanoparticles in prostate cancer. Biomaterials 35:8635–8648
Zaman MS, Chauhan N, Yallapu MM, Gara RK, Maher DM, Kumari S, Sikander M, Khan S, Zafar N, Jaggi M, Chauhan SC (2016) Curcumin nanoformulation for cervical cancer treatment. Sci Rep 6:20051
Zampieri AL, Ferreira FS, Resende EC, Gaeti MP, Diniz DG, Taveira SF, Lima EM (2013) Biodegradable polymeric nanocapsules based on poly(DL-lactide) for genistein topical delivery: obtention, characterization and skin permeation studies. J Biomed Nanotechnol 9:527–534
Zhang P, Ling G, Sun J, Zhang T, Yuan Y, Sun Y, Wang Z, He Z (2011) Multifunctional nanoassemblies for vincristine sulfate delivery to overcome multidrug resistance by escaping P-glycoprotein mediated efflux. Biomaterials 32:5524–5533
Zhu R, Wu X, Xiao Y, Gao B, Xie Q, Liu H, Wang S (2013) Synergetic effect of SLN curcumin and LDH-5-Fu on SMMC-7721 liver cancer cell line. Cancer Biother Radiopharm 28:579–587

Chapter 3
Natural Anticancer Compounds and Their Derivatives in Clinical Trials

Veerabhadrappa Bhavana, Setra Janardhana Shetty Sudharshan, and Dyavaiah Madhu

Abstract Cancer continues to be a global challenge to both clinicians and researchers with an increasing mortality rate. Despite the enormous progress made in the anticancer drug discovery, there is a constant demand for novel therapeutic agents, because of the development of resistance to the existing chemotherapeutic drugs and their adverse side effects. The anticancer drugs derived from the natural sources have shown to be effective and safe in the treatment of cancers. Secondary metabolite compounds from plants such as alkaloids, flavonoids, and carotenoids are known for their cancer prevention and antitumor properties. Peptides produced from marine organisms and anthracyclines synthesized by microbes as secondary metabolites are also known for their anticancer properties. Some of these natural compounds are widely used in cancer therapy, and some are under clinical or preclinical trials. Some of the potential anticancer agents from plants (paclitaxel, vincristine, vinblastine, irinotecan, etoposide, topotecan, and camptothecin), marines (dolastatin 10, cytarabine, and aplidine), and microorganisms (bleomycin, doxorubicin, and dactinomycin) have been used in cancer therapy. Cancers are characterized by the alterations in the cell signaling pathways. Most of the current anticancer therapies involve the modulation of altered signaling targets in cancers. The advantage of using natural compounds with antitumor properties for cancer therapy is that the compounds have well-defined signaling targets with a minimal toxicity. Natural anticancer drugs have been categorized based on their target-specific signaling pathways, which include DNA-damaging drugs, methyltransferase inhibitors, mitotic

V. Bhavana
Department of Biochemistry and Molecular Biology, Pondicherry University, Pondicherry, India

S. J. S. Sudharshan · D. Madhu (✉)
Department of Biochemistry and Molecular Biology, Pondicherry University, Pondicherry, India

DBT-IPLS Program, School of Life Sciences, Pondicherry University, Pondicherry, India

M. S. Akhtar, M. K. Swamy (eds.), *Anticancer Plants: Clinical Trials and Nanotechnology*, https://doi.org/10.1007/978-981-10-8216-0_3

Table 3.1 (continued)

Class	Compound	Plant source	Mode of action	Clinical trial stage	References
	Podophyllotoxins	*Podophyllum peltatum*	Topoisomerase inhibition	Leukemia; brain, ovarian, and breast cancer; and SCLC – phase I/II (r) for etoposide	Gordaliza et al. (2000) and Choi et al. (2015)
	(i) Etoposide – ETP			Leukemia and esophageal cancers – phase III/IV (r) for teniposide	
	(ii) Teniposide – TNP				

The clinical trial stage shown in table is mostly from the NIH (https://clinicaltrials.gov/) or the corresponding EU clinical trial database
r recruiting participants, *a* active participants, *c* completed trial

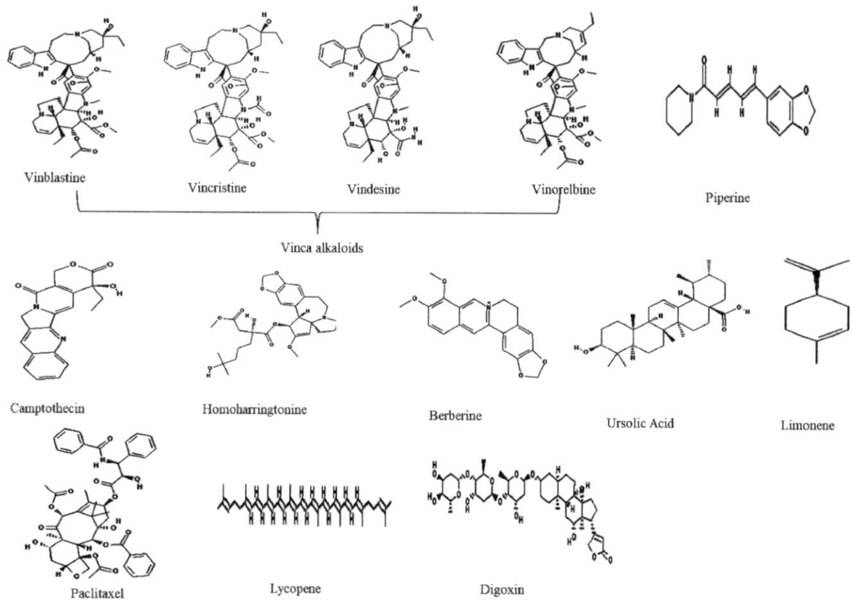

Fig. 3.1a Anticancer compounds from plant under clinical trial

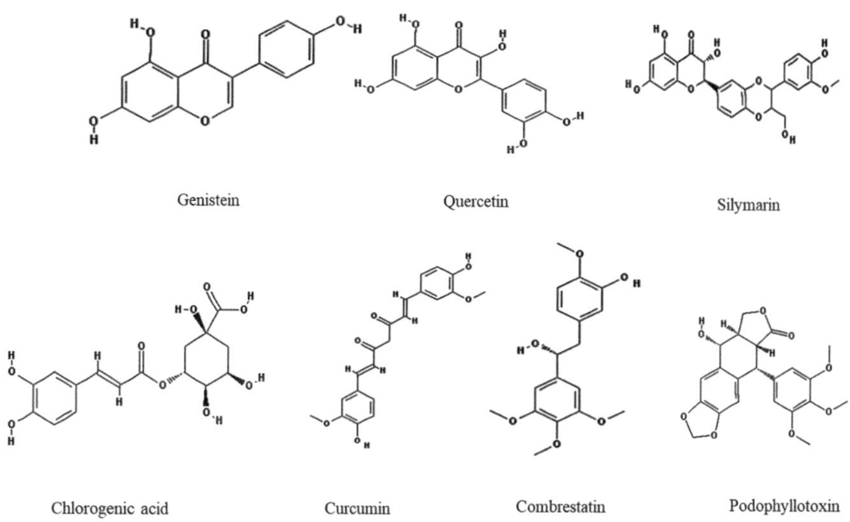

Fig. 3.1b Anticancer compounds from plant under clinical trial

3.2.1 *Alkaloids*

Alkaloids are found mostly in the plant kingdom compared to any other natural resources. They usually exist in higher plants such as Menispermaceae, Leguminosae, and Ranunculaceae. Alkaloids constitute an important class of secondary metabolites having heterocyclic ring containing nitrogen atom (Lu et al. 2012). Alkaloids form around 20% of plant secondary metabolites and have low-molecular-weight structures possessing various medicinal properties with relatively low toxicity (Ziegler and Facchini 2008). Major biological functions of alkaloids include antibacterial, antidiabetic, anti-inflammatory, and anticancer activities (Souto et al. 2011; Tiong et al. 2013; Cushnie et al. 2014; Kaur and Arora 2015). Well-known alkaloids with anticancer potential include vinblastine and vincristine (Jordan and Wilson 2004; Moudi et al. 2013). These alkaloids are the most important active ingredients in the traditional medicines and have also been approved for cancer therapy in recent times.

3.2.1.1 Vinca Alkaloids

Vinca alkaloids are isolated from *Vinca rosea* (*Catharanthus roseus*) belonging to the family Apocynaceae and better known as "Madagascar periwinkle." The plant grows throughout the world and is known for centuries as a traditional medicinal source for diabetes, high blood pressure, ulcer, and scurvy diseases. Vinca alkaloids are one of the oldest classes of plant alkaloids used to treat the cancer, and it was first found out by the Canadian scientists Robert Noble and Charles Beer in the 1950s (Moudi et al. 2013; Anitha Sri 2016). There are four major vinca alkaloids used for cancer treatment, which include vincristine, vinblastine, vinorelbine, and vindesine (Anitha Sri 2016). The mechanism of action of vinca alkaloids is by arresting the cell cycle at M phase by the interaction and disruption of microtubule which consists of mitotic spindles (Jordan and Wilson 2004; Moudi et al. 2013). Vinblastine blocks the endothelial proliferation and spreading of fibronectin and inhibits the malignant angiogenesis by binding to microtubule (Mukhtar et al. 2014). Vincristine destabilizes the microtubules by binding tubulin and blocking polymerization leading to cell cycle arrest and apoptosis (Moudi et al. 2013). Vinorelbine has demonstrated better anticancer property in patients suffering from breast and small cell lung cancer compared to others (Brunello et al. 2010). Vinorelbine and vindesine inhibit tubulin polymerization by disrupting the spindle formation and thereby arrest the cell cycle (Moudi et al. 2013). In 1989 and 1991, France approved vinorelbine for the treatment of small cell lung carcinoma (SCLC) and breast cancer. The USA approved vinorelbine for the treatment of SCLC in 1994 (Bennouna et al. 2008; Schutz et al. 2011). Currently, vinorelbine and vindesine in combination with other drugs such as lapatinib, epirubicin, bevacizumab, gefitinib, cetuximab, and gemcitabine are under the clinical trial phase I, II, and III for the treatment of breast, lung, colon, and skin cancers. Patients are being recruited for phase II clinical trials for the treatment of metastatic breast cancer and non-small cell lung cancer (NSCLC) using vinorelbine. Vincristine has been approved for the treatment of leukemia, Hodgkin's disease, and SCLC. Vinblastine is used to treat a number of cancers such as testicular, brain, and bladder cancers, Hodgkin's lymphoma, and lung cancer. Vinblastine acts by two mechanisms; at low concentrations, it suppresses the microtubule action, whereas, at higher concentrations, it reduces the microtubule mass, which is currently in the recruitment stage of phase II trials for the treatment of low-grade glioma. Currently, vincristine and vinblastine with other compounds such as sirolimus, alisertib, bryostatin I, volasertib, nilotinib, cytarabine, bevacizumab, and doxorubicin are under the clinical trial phase I/II/III for the treatment of acute lymphoblastic leukemia (ALL), sarcoma, and lymphoma.

3.2.1.2 Camptothecin

Camptothecin is a quinoline alkaloid isolated from the stem and bark of *Camptotheca acuminata* and *Mappia foetida* trees commonly available in China. These trees were used for the treatment of cancer in traditional Chinese medicine. Camptothecin was first isolated by Monroe Wall and Mansukh Wani in 1966 from *C. acuminata* (Efferth et al. 2007). The use of camptothecin has gained its importance in the 1980s when they found its molecular target as topoisomerase I. Camptothecin binds to topoisomerase I and forms a DNA complex, which prohibits the DNA re-ligation and finally results in apoptosis. The use of camptothecin suffers from many restrictions, because of its low stability and solubility. To overcome this issue of camptothecin, several derivatives such as irinotecan, topotecan, silotecan, cositecan, exatecan, lurtotecan, and belotecan have been developed. Among them, the Food and Drug Administration (FDA) agency approved topotecan for the treatment of small cell lung cancer and ovarian cancer and irinotecan in combination with 5-fluorouracil for metastatic colorectal cancer (Pommier 2006; Venditto and Simanek 2010; Wahid and Bano 2014). Camptothecin has completed phase I/II/III clinical trials for the treatment of pancreatic, colorectal, and ovarian cancers, and patients are being recruited for phase I/II trials for the treatment of relapsed NSCLC.

3.2.1.3 Piperine

The alkaloid piperine is isolated from the fruits of *Piper nigrum* (black pepper) and from the plant *P. longum* (long pepper). Piperine exhibits antioxidant, anti-inflammatory, antidiarrheal, anticonvulsant, and antimutagenic activities (Vasavirama and Upender 2014). Piperine induces apoptosis by inhibiting the transcription factors and cofactors such as NF-κB, CREB, ATF-2, and ap-1function in various cancer cells. These factors play an important role in the production of pro-inflammatory cytokine and matrix metalloproteinase that promote tumor growth and metastasis. It inhibits proliferation and induces apoptosis in androgen-dependent prostate cancer by activating capsase-3 and cleavage of PARP-1 proteins. Piperine arrests the cell cycle at different phases such as G0/G1, G1/S, and G2/M in various cancer cells (Singh and Duggal 2009; Chinta et al. 2015). Currently, piperine in combination with the curcumin is in phase I clinical trial for the treatment of bladder cancer.

3.2.1.4 Homoharringtonine

Homoharringtonine was isolated from the tree *Cephalotaxus harringtonia* generally known as Japanese/Chinese plum yew. The Earl of Harrington was the first person to grow plum yew plant in European garden centuries ago. To honor his effort, plum yew plant was named after him as *C. harringtonia*. In 1963, Paudler and his team were the first to isolate harringtonine and cephalotaxine from this plant. Later, some alkaloid esters known as harringtonine, isoharringtonine, deoxyharringtonine, and homoharringtonine were also isolated from the same plant. Among these esters, homoharringtonine showed better anticancer property (Bhanot et al. 2011; Badgujar et al. 2015). The mechanism of action of homoharringtonine is inhibition of protein translation. It inhibits translation in the elongation step by interaction with ribosomal A-site and prevents exact positioning of amino acid side chains of incoming aminoacyl-tRNAs. Homoharringtonine decreases the protein efficiency especially proteins with short half-lives such as c-Myc, Mcl-1, and cyclin D1 which are related to cell survival and proliferation, resulting in cell apoptosis. In 2012, homoharringtonine was approved for the treatment of chronic myeloid leukemia (CML) (Lu and Wang 2014; Cao et al. 2015). Currently, homoharringtonine in combination with other drugs such as quizartinib, sorafenib, and cytarabine is under phase I/II clinical trials for the treatment of acute myeloid leukemia (AML) and myelodysplastic syndrome.

3.2.1.5 Berberine

The isoquinoline alkaloid, berberine, is usually isolated from *Berberis vulgaris*, and rhizomes of coptidis (*Rhizoma coptidis* RC). It possesses various biological activities such as antibacterial, antidiabetes, antiulcer, and anti-inflammation (Bhanot et al. 2011; Lu et al. 2012; Kaur and Arora 2015). Several studies have shown the anticancer property of berberine in *in vitro* and *in vivo* experiments. Berberine interacts with DNA and RNA thereby inhibiting the cell proliferation by inducing cell cycle arrest at G1 or G2/M phase in various cancer cell lines. Berberine induces apoptosis in cancer cells by expressing proteins such as Bax, Bcl-2, and Bcl-xl and caspases and regulation of cyclin-dependent kinase proteins. Berberine inhibits COX-2 and telomerase and induces autophagy and endoplasmic reticulum stress. The effect of berberine on metastasis and angiogenesis is through the reduction of

COX-2 and prostaglandin receptors, inhibition of matrix metalloproteinases 2 and 9, and downregulation of pro-inflammatory mediators and vascular endothelial growth factor (Sun et al. 2009; Lu et al. 2012; Raza et al. 2015). Currently, berberine is undergoing phase II/III clinical trials to evaluate its safety and adverse effects in patients suffering from colorectal cancer.

3.2.2 *Terpenes and Sterols*

Terpenes are secondary metabolites produced by plants, known to have various functions in industrial and medical applications. Majority of the flavors, fragrances, and spices are isolated from the plant terpenes that usually have a strong odor to help the plant in pollination and to fight against parasites. The oxidized products of terpenes are usually called terpenoids. However, more research efforts have led to the identification of various terpenes with anticancer potential which act by inhibiting the cell proliferation and inducing metastasis by various mechanisms (Martin 2003; Pichersky 2006; Huang et al. 2012).

3.2.2.1 Ursolic Acid

Ursolic acid is a pentacyclic triterpenoid isolated from *Mirabilis jalapa*, and it is widely found in the peels of fruits like apple, basils, bilberries, and cranberries and in flowers like peppermint, rosemary, lavender, and others (Bhanot et al. 2011; Huang et al. 2012; Singh and Sharma 2015). Ursolic acid was designated as a multitasking agent as it induces several cell signaling pathways simultaneously to protect cells against carcinogenic agents. Novel pharmacological strategies of ursolic acid are not only on the destruction of cancer cells but also modulation of their metabolism to prevent angiogenesis and metastasis to protect healthy tissues against inflammation and oxidative stress that may lead to neoplasm formation. The ability to suppress communication through MAPK/ERK and PI3K/AKT/mTOR signaling pathways is one of the most important anticancer activities of ursolic acid. Apoptosis induction is the most important anticancer property of ursolic acid and has been reported in several cancer types both *in vitro* and *in vivo*. Ursolic acid suppresses the activity of anti-apoptotic protein Bcl-2 and Bcl-xl expression while enhancing the activity of pro-apoptotic protein BAX. Ursolic acid is proved to be an efficient COX-2 inhibitor which is able to suppress progression of inflammation (Tu et al. 2009; Shan et al. 2011; Prasad et al. 2012; Wang et al. 2012; Li et al. 2014b; Wozniak et al. 2015). Ursolic acid has entered phase I/II clinical trials for the treatment of advanced solid tumors (Yan et al. 2013; Wang et al. 2013).

3.2.2.2 Limonene

The monoterpene, limonene, is usually isolated from the citrus fruits and other plants. Limonene occurs in two forms, i.e., L-limonene and D-limonene, which are mirror images to each other. D-limonene is the major constituent of citrus fruits like orange, lemon, mandarin, and grapefruit. It is known for its anti-inflammatory, anti-invasive, and antiangiogenic properties (Sun 2007; Bayala et al. 2014). Although D-limonene is a well-studied monoterpene against different types of cancer, its exact mechanisms of actions are still not understood. However, it is reported to protect tumor growth mainly by pro-apoptotic, antioxidant, and antiangiogenic properties. D-limonene cleaves caspase-3 and caspase-9 by upregulation of Bax protein and helps to release cytochrome c from mitochondria. D-Limonene binds to 5-hydroxy-3-methylglutaryl-coenzyme A (HMG-CoA) reductase and inhibits its activity, which leads to the inhibition of small G proteins which helps in the chemoprevention and cancer therapy. D-Limonene initiates the induction of hepatic detoxification enzymes such as glutathione-s-transferase and uridine diphospho-glucuronosyltransferase in order to protect from carcinogens (Bayala et al. 2014; Sobral et al. 2014). Currently, D-limonene along with gemcitabine or paclitaxel is under phase I clinical trial for the treatment of pancreatic cancer.

3.2.2.3 Lycopene

Lycopene is a red-colored carotenoid terpene, majorly found in tomato (*Solanum lypersicum*) and other red fruits/vegetables such as watermelons, red carrots, and papaya. Lycopene is also present in foods which are not red colored such as asparagus and parsley. Because of its red color, lycopene is approved for food coloring agent in the USA, Australia, New Zealand, and Europe. Lycopene showed very good antioxidant activity *in vitro*. Lycopene is majorly used for cardiovascular and prostate cancer therapy (Bhuvaneswari and Nagini 2005; Wei and Giovannucci 2012). The important mechanisms underlying the anticancer property of lycopene are by ROS scavenging, upregulation of detoxification enzymes, cell proliferation interference, cell cycle arrest, and alteration in signal transduction pathways (Bhuvaneswari and Nagini 2005; Lippi and Targher 2011; Wei and Giovannucci 2012; Chen et al. 2013). Currently, lycopene is in phase II clinical trial for the treatment of hormone-resistant prostate cancer in combination with docetaxel and also for treating colorectal cancer.

3.2.2.4 Paclitaxel

Paclitaxel, a diterpenoid, was first isolated from the bark of Pacific yew tree, *Taxus brevifolia*, in 1971. In 1993, paclitaxel was approved for the cancer treatment, commonly used for the treatment of breast, ovarian, pancreatic, and cervical cancer. The World Health Organization (WHO) declared that paclitaxel is an essential lifesaving drug. The mechanism of action of paclitaxel is mainly tubulin binding related. Upon binding to tubulin, paclitaxel disrupts the spindle assembly, chromosome segregation, and cell divisions in cancer cells, resulting in cell cycle arrest and apoptosis. Apart from tubulin binding, paclitaxel induces gene expressions which are responsible for apoptosis and also induces antiangiogenesis by regulating vascular endothelial growth factor (Jordan and Wilson 2004; Ganguly et al. 2010; Fauzee et al. 2011; Priyadarshini and Keerthi 2012). Commercially, paclitaxel has been sold under the name Taxol. Several paclitaxel analogs have been synthesized and tested for their anticancer properties. Paclitaxel and its derivatives such as nanopaclitaxel (Abraxane) and docetaxel with other compounds such as carboplatin, rituximab, doxorubicin, ibrutinib, durvalumab, neratinib, gemcitabine hydrochloride, cisplatin, and others are under clinical trials of phases I/II/III and IV for the treatment of various cancers, mainly breast, ovarian, and endometrial cancers. Currently, patients are being recruited for phase I/II clinical trials for the treatment of pediatric solid tumors using paclitaxel.

3.2.2.5 Digoxin

Digoxin was first isolated from the plant *Digitalis lantana*, and it belongs to the class cardiac glycosides, a sterol. The key mechanism of action of digoxin is the inhibition of sodium potassium ATPase. It is currently used for the treatment of cardiac disease, and WHO declared digoxin as an essential medicinal drug. The anticancer property of digoxin mainly involves cell proliferation and induces apoptosis in cancer cells by mitochondrial pathway which involves cytochrome c, caspase 8, and caspase 3 (Lopez-Lazaro 2009; Biggar et al. 2012; Kepp et al. 2012; Biggar et al. 2013; Calderón-Montaño et al. 2014; Newman and Cragg 2016a). Currently, digoxin is under phase II clinical trial for patients suffering from stage I–III breast cancer and phase I/II trials for treating head and neck cancer. Digoxin has completed phase I clinical trials for the treatment of metastatic melanoma which cannot be removed by surgery and also for metastatic breast cancer.

3.2.3 Phenols

Phenols and phenolic compounds are the largest classes of plant secondary metabolites, usually present in flowers, fruits, vegetables, and seeds. Phenols are derived from pentose phosphate, shikimate, and phenylpropanoid pathway in plants. Based on the number of phenolic groups, phenols are classified as simple phenol and polyphenol. The major biological activity of polyphenol is the antioxidant property as they have the ability to interact with proteins, lipids, and other molecules. The phenolic compounds are broadly classified as hydroxycinnamic acid, hydroxybenzoic acid (tannins), flavonoids, stilbenes, xanthones, chalcones, lignins, and lignans. Majority of the phenolic compounds have anticancer property as they induce apoptosis, cell cycle arrest, and DNA damages (Dai and Mumper 2010; Ozcan et al. 2014).

3.2.3.1 Genistein

Genistein is a flavonoid usually found in plants such as beans, kudzu, psoralea, and lupin and also in some medicinal plants such as *Flemingia vestita* and *F. macrophylla*. Genistein is known for its antioxidant, antihelmintic, antiproliferative, anti-angiogenesis, immunosuppressive, and phytoestrogen properties. The anticancer property of genistein is by inhibition of tyrosine kinase and topoisomerase, stimulation of autophagy, activation of Nrf2 antioxidant and peroxisome proliferator-activated receptors (PPARs), DNA fragmentation, and cell cycle arrest (Pavese et al. 2010; Spagnuolu et al. 2015; Russo et al. 2016). Currently, genistein is under phase II clinical trial for the treatment of bladder and prostate cancers and also under phase I/II clinical trials for the treatment of pediatric malignancies. Genistein has completed phase I/II clinical trials for the treatment of metastatic colorectal cancer.

3.2.3.2 Quercetin

Quercetin is a natural polyphenolic flavonoid, usually found in fruits and vegetables such as capers, radish, coriander, red onions, radicchio, cranberry, Hungarian wax pepper, fennel leaves, cow peas, sweet potato, broccoli, bilberry, and red kidney beans and also in green tea. Quercetin is a powerful antioxidant and also possesses anti-inflammatory activity. It generates pro-inflammatory cytokines in patients suffering from chronic inflammatory diseases. The anticancer property of quercetin is by the induction of apoptosis, cell cycle arrest, tyrosine kinase inhibition, inhibition of heat shock proteins, and expression of Ras proteins in various cancers (Gibellini et al. 2011; Baghel et al. 2016; Srivastava et al. 2016). At present, quercetin with other compounds such as green tea and genistein is in phase I, active recruiting phase for the treatment of prostate cancer, and also in phase I/II trials for the treatment of renal cancer.

3.2.3.3 Silymarin

Silymarin is a complex mixture of flavolignans and flavonoids, usually isolated from the seeds of *Silybum marianum* L. The mixture of flavolignans includes silybin A, silybin B, isosilybin A, isosilybin B, silychristin, isosilychristin, silydianin, and the flavonoid, taxifolin. It is extensively used for the treatment of chronic viral hepatitis and alcoholic cirrhosis. The anticancer property of silymarin is by inhibiting the cell proliferation and inducing apoptosis through the expression of cell cycle regulators. In most of the cancers, silymarin induces apoptosis by cell cycle arrest at G_1

and G_2 checkpoint and activates the MAPK/JNK pathway. In prostate cancer, silymarin inhibits the epidermal growth factor receptor (EGFR) pathway through transforming growth factors (TGF). In umbilical cancer, silymarin inhibits the growth and survival of human umbilical vein endothelial cell by inhibiting capillary tube formation and induction of apoptosis by arresting the cell cycle (Ramasamy and Agarwal 2008; Mastron et al. 2015). Currently, silymarin is in recruitment stage for phase IV clinical trials for the treatment of metastatic colorectal cancer. It has completed phase II clinical trials for prostate cancer and phase II/III trials for gastrointestinal cancer treatment.

3.2.3.4 Chlorogenic Acid

Chlorogenic acid is a polyphenol and an ester of caffeic acid and quinic acid, usually isolated from bamboo (*Phyllostachys edulis*) and also from the shoots of common heather (*Calluna vulgaris*). Chlorogenic acid is also found in the leaves of *Hibiscus sabdariffa*, flesh of eggplant, peach, prunes, and green tea. It is majorly used as a food additive in coffee products, chewing gum, and mints. Currently, chlorogenic acid is available in the market as Svetol. Chlorogenic acid inhibits DNA damage by scavenging ROS and protecting against carcinogenesis. Chlorogenic acid inhibits matrix metalloproteinase activity and upregulates the genes involved in the immune system by enhancing the activation of macrophages and natural killer cells (Luthria and Mukhopadhyay 2006; Del Rio et al. 2010; Rocha et al. 2012; Zhen et al. 2016). Currently, chlorogenic acid is in phase I clinical trial, recruiting patients for the treatment of glioblastoma and to study the pharmacokinetic properties in advanced cancers.

3.2.3.5 Combretastatin

Natural phenol combretastatin is formed from the backbone of dihydrostilbene (class stilbenes). Combretastatin is usually isolated from the bark of South African bush willow *Combretum caffrum*, which belongs to the family of Combretaceae. The mechanism of action of combretastatin is the inhibition of tubulin polymerization in cancer cells. There are five combretastatin isoforms found in *C. caffrum,* and they are combretastatin A-4, combretastatin A-1, combretastatin B, combretastatin C, and combretastatin D. Among them, combretastatin A-4 has showed better anticancer activity. It binds to tubulin at colchicine site and inhibits the polymerization of the microtubules leading to the arrest of cell cycle at M phase. Apart from tubulin depolymerization, combretastatin A-4 also shows antiangiogenesis by inducing the morphological changes in endothelial cells by a direct damage to the vasculature (Jordan and Wilson 2004; Siemann et al. 2009; Arora et al. 2013). At present, combretastatin A-4 is under phase II/III clinical trials for the treatment of platinum-resistant ovarian cancer and phase I/II clinical trials for the treatment of AML. It has completed phase II clinical trials for the treatment of advanced solid tumors and NSCLC.

3.2.3.6 Flavopiridol

Flavopiridol is a flavonoid isolated from *Amoora rohituka*. It is being used for the treatment of arthritis, atherosclerosis, and cancer. Flavopiridol is a potent inhibitor of CDK 1, 2, 4, 7, and 9, thereby arresting cell cycle at G1 or G2 phase. It induces apoptosis by downregulation of Bcl-2 in cancer cells. In 2014, flavopiridol was approved for the treatment of acute leukemia. Commercially flavopiridol is known as alvocidib (Tan and Swain 2002; Raju et al. 2003; Mariaule and Belmont 2014). Currently, flavopiridol is undergoing phase II clinical trials for the treatment of AML and has completed phase I and phase II clinical trials for treating metastatic solid tumors and pancreatic cancer, respectively.

3.2.3.7 Curcumin

Curcumin is a polyphenolic compound, isolated from *Curcuma* plant species, which belongs to the family Zingiberaceae. It is extensively used in Ayurveda, Unani, Siddha, and Chinese medicine for treating cough, wound healing, inflammation, hepatitis, and cancer. Anticancer property of curcumin is based on its multiple actions such as cell cycle regulation, regulation of apoptosis, proliferation, and metastasis inhibition. Curcumin downregulates the expression of p53, Egr1, and Bcl-xl in cancer cells which results in inhibition of cell cycle progression and leads to apoptosis. Curcumin induces the expression of NF-κB and IKK and inhibits MMP, PKA, PKC, and STAT3 proteins to suppress angiogenesis, tumor invasion, tumor growth, and proliferation in cancer cells. Curcumin suppresses the TNF expression which results in impaired cell proliferation in cancer cells (Wilken et al. 2011; Mehta et al. 2014; Rahmani et al. 2014; Salem et al. 2014). At present, curcumin is under phase II clinical trials for the treatment of metastatic breast cancer and cervical neoplasms and phase III trials for treating prostate cancer. Also, it has completed phase I trials for treating advanced colorectal cancer.

3.2.3.8 Podophyllotoxins

The naturally occurring lignin podophyllotoxin is usually isolated from *Podophyllum peltatum* L. and *P. emodi* Wall. Podophyllotoxin is known for its immunosuppressive and antiviral activities specifically for treating herpes, measles, influenza, and warts and also for its anticancer activity. The anticancer mechanism of podophyllotoxins is by arresting the cell cycle through the inhibition of microtubule polymerization. Podophyllotoxin was found to be unsuitable for clinical use because of its toxic side effects. However, based on its biological activity, semisynthetic compounds such as etoposide, etopophos, and teniposide were derived from podophyllotoxin. These derivatives showed anticancer property by inhibiting DNA topoisomerase II rather than inhibiting tubulin polymerization (Gordaliza et al. 2000; Choi et al. 2015). Currently etoposide and teniposide derivatives are under clinical trials for various cancer treatments. Etoposide with other drugs such as ciprofloxacin, vorinostat, apatinib, dexamethasone, and methotrexate is currently in phase I/II clinical trials for the treatment of leukemia, brain, ovarian, breast, and SCLC. Teniposide with other drugs such as dexamethasone, fotemustine, vindesine, irinotecan, and cladrizine is currently under phase III/IV clinical trials for the treatment of leukemia and esophageal cancers.

3.3 Anticancer Compounds from Microbial Source

Microorganisms are an important source of several clinically important metabolites, which have substantially contributed for the development of pharmaceutically significant compounds. They have played an important role in finding natural compounds with therapeutic benefits against cancer. Screening of natural compounds has led to the discovery of a wide range of anticancer drugs of microbial origin. The most important compounds used in the cancer chemotherapy are microbial origin antibiotics or their derivatives (Demain and Vaishnav 2011). Anticancer drugs such as doxorubicin, dactinomycin, bleomycin, and mitomycin are approved and widely used in the treatment of a variety of cancers. In addition, several derivatives have been developed based on the scaffolds of these microbial compounds. Some of them have demonstrated potent anticancer activity and have entered clinical trials and also in therapeutic stages. Among these antitumor antimicrobial compounds, anthracyclines such as doxorubicin and daunorubicin and glycopeptide antibiotics such as bleomycin are approved for the treatment of numerous cancers as monotherapy or in combination with other drugs. These compounds which belong to these classes cause cytotoxicity in cancer cells by targeting DNA damage response pathway, of which cancer cells are mostly deficient whereas, normal cells are not (Kathiravan et al. 2013).

It all began in the 1940s with the discovery of streptomycin and many other milestone compounds with antimicrobial activity from a variety of microorganisms (Giddings and Newman 2013). Actinomycetes, an important class of antibiotics, were discovered in the 1950s at Abraham Waksman's laboratory, which subsequently led to the discovery of several antimicrobials which were proven effective against a wide range of microbial infections (Waksman and Woodruff 1941). In the 1980s, screening of antibiotics to explore their additional properties other than targeting the microbial infections that included antitumor property was initiated. In the laboratory screening test, several microbial compounds which were considered as failed antibiotics were identified as potent toxins for tumors, and most of the important compounds of microbial origin and their derivatives were successfully used against a variety of cancers (Demain and Vaishnav 2011). Despite their proven antimicrobial action, some antibiotics such as doxorubicin, daunorubicin, and bleomycin have already been used in the treatment of human cancers (Kumar 2015). In addition, many other microbial origin compounds are also under investigation for their antitumor activity. While most of the compounds target DNA (Bhattacharya and Mukherjee 2015), some target tubulin (Jordan and Wilson 2004) and some other compounds induce apoptotic mode of cell death in cancer cells (Chung et al. 2008). In the last few decades, natural products predominately from the actinomycetes and Eukarya or their derivatives have played extremely important roles in cancer drug discovery (Giddings and Newman 2013). Notably, microbial secondary metabolites are the major source of drugs not only used unchanged for the treatment but served as basic chemical structures which can be selectively modified to control tumor cell growth (Cragg et al. 2009; Bhattacharya and Mukherjee 2015). With the advent of high-throughput screening technology, several programs were initiated to discover the small molecules which have anticancer properties (Florento et al. 2012; Hoelder et al. 2012). These research programs focused mainly on quickening the drug development and reducing costs. Several pharmaceutical companies have introduced many small molecules which resemble the core scaffolds of natural products (Giddings and Newman 2013). Yet, there is an inadequacy of lead compounds making their way into the clinical trials. Approximately 50% of all small molecules that received approval by FDA from 2000 to 2006 were not newly synthesized chemical compounds instead derived from combinatorial chemistry but mostly based upon the natural products (Newman and Cragg 2007). Microbial origin anticancer compounds and their derivatives which are under clinical trial for cancer treatment are discussed in this section (Table 3.2; Figs. 3.2a and 3.2b).

3.3.1 Actinomycins

The actinomycins are a group of chromopeptide antibiotics that exhibit high antibacterial and antitumor activity. Several milestone antimicrobial compounds such as streptomycin, neomycin, actinomycin, and others were first discovered by Waksman's group. Actinomycins were discovered in several soil bacterial species of *Streptomyces*, which showed potent inhibition properties against a wide range of microbial infections (Waksman and Woodruff 1941). In 1963, actinomycin D was approved for the treatment of highly malignant pediatric tumors such as rhabdomyosarcoma and Wilms tumor as a single agent and also in combination with other anticancer compounds (Giddings and Newman 2013). Actinomycin D is a DNA intercalating agent that interferes with the binding of transcription factor to DNA, thereby inhibits RNA synthesis and eventually protein translation (Lo et al. 2013). The efficacy of actinomycin D is restrained by its severe cytotoxicity. Therefore, its derivatives from various soil and marine *Streptomyces* species and several synthetic analogs have been developed to modulate the cytotoxicity (Giddings and Newman 2013). Actinomycin D is FDA approved for the treatment of Wilms tumor, advanced breast cancer, and testicular and ovarian cancers (NCI). It has completed phase IV clinical trials for the treatment of various pediatric cancers. Currently, patients are being recruited for phase II clinical trials involving actinomycin D as a single agent and/or in combination with other cytotoxic drugs for the treatment of germ cell tumors and high-risk rhabdomyosarcoma.

Table 3.2 Anticancer compounds from microbial source

Class	Compound	Source	Mode of action	Clinical trial stage	References
Actinomycins	Actinomycin D	*Streptomyces* sp.	DNA damage	Advanced breast cancer, ovarian and testicular cancer, and Wilms tumor approved	Waksman and Woodruff (1941) and Giddings and Newman (2013)
				Germ cell tumors and high-risk rhabdomyosarcoma – phase II (r)	
				Pediatric cancer phase IV (c)	
Anthracyclines	Doxorubicin	*S. peucetius*	DNA damage	Breast, bladder cancers, Kaposi's sarcoma, lymphoma, ALL, and AML – approved triple-negative breast cancer, Ewing's sarcoma – phase III metastatic sarcoma – phase I/II (r) ovarian cancer – phase IV (c)	Khazir et al. (2014) and Cheung-Ong et al. (2013)
	Daunorubicin	*S. caeruleorubidus*	DNA intercalation and topoisomerase inhibition	AML, ALL, CLL – approved	Khazir et al. (2014)
				AML – phase I/II/III (r)	
	Epirubicin	*S. peucetius*	DNA intercalation	Node-positive breast cancer – approved	Bhanot et al. (2011) and Newman and Cragg (2016a)
				DLBCL – phase III/IV (r)	
				Bladder cancer – phase IV (r)	
	Idarubicin	*S. peucetius*	Topoisomerase II inhibition	AML approved	Bhanot et al. (2011) and Newman and Cragg (2016a)
				Metastatic hepatocellular carcinoma – phase I (c)	
Glycopeptides	Bleomycin	*S. verticillus*	DNA damage	Testicular, cervical, head, and neck cancers, Hodgkin's and non-Hodgkin's lymphoma – approved Kaposi's sarcoma – phase III (c) Rectal cancer – phase II and germ cell tumors – phase III (r)	Hecht (2000) and McKinney et al. (2013)
Enediynes	Calicheamicin (an ADC – gemtuzumab ozogamicin)	*Micromonospora echinospora*	DNA damage	AML – phase I/II (a/c)	Nicolaou et al. (1993) and Shen et al. (2015)
				Non-Hodgkin's lymphoma – phase II (c)	
				ALL – phase III (a)	
	Neocarzinostatin	*S. carzinostaticus*	DNA intercalation	Recurrent hepatocellular carcinoma – phase III (r)	Giddings and Newman (2013)

(continued)

Table 3.2 (continued)

Class	Compound	Source	Mode of action	Clinical trial stage	References
Epothilones	Epothilone B	*Myxobacterium, Sorangium cellulosum*	Microtubule inhibition	Metastatic breast cancer, NSCLC, colorectal and ovarian cancer – phase II (c)	Reichenbach and Höfle (2008), Ferrandina et al. (2012), and Fanale et al. (2015)
				Advanced solid tumors – phase I (c)	
				Ovarian and breast cancer – phase II (r)	
	Epothilone D	*Myxobacterium, Sorangium cellulosum*	Microtubule inhibition	Metastatic breast cancer – phase III (r)	
				Advanced solid tumors – phase I (c)	
				Ovarian cancer and urothelial carcinoma – phase II (r)	
	Ixabepilone	Semisynthetic analog of Epothilone B	Microtubule inhibition	Advanced solid tumors – phase I (c)	
				Ovarian cancer and urothelial carcinoma – phase II (r)	
				Metastatic breast cancer – phase III (r)	
Geldanamycins	Tanespimycin	*S. hygroscopicus*	HSP90 inhibition	Metastatic thyroid cancer – phase II (c)	Fukuyo et al. (2010) and Bhanot et al. (2011)
				AML, ALL, CML – phase I (c)	
				Multiple myeloma – phase I/II/III (c)	
Rapamycin	Rapamycin or sirolimus	*S. hygroscopicus*	Inhibits TOR, induces autophagy/apoptosis	NSCLC, ALL, breast cancer, glioma, AML – phase I/II (r) bladder cancer – phase I (r)	Ballou and Lin (2008), Seto (2012), and Li et al. (2014a)
	Everolimus (RAD001)	Synthetic analog of rapamycin	Inhibits TOR, induces autophagy/apoptosis	Glioma – phase II (c) relapsed osteosarcoma – phase II (u)	
				Triple-negative breast cancer – phase I (r)	
				Metastatic renal cell carcinoma – phase II (r)	
	Temsirolimus (CCI-779)	Synthetic analog of rapamycin	TOR inhibition, autophagy, and apoptosis	Renal carcinoma – approved	
				Glioma – phase II (r), advanced cancers and solid tumors – phase I (r)	
	Ridaforolimus	Synthetic analog of rapamycin	TOR inhibition, autophagy, and apoptosis	Metastatic soft tissue and bone sarcomas – phase III (c), advanced solid tumors – phase I (a), and head and neck cancer and NSCLC – phase I (c)	

Romidepsin, Trichostatin A	Romidepsin	*Chromobacterium violaceum*	HDAC inhibition	Advanced colorectal cancer – phase II (r)	Xu et al. (2007), Butler et al. (2014), and Foss et al. (2016), and Newman and Cragg (2016a)
				T-cell lymphoma, CLL, and solid tumors – phase II (a)	
	Vorinostat	*Streptomyces hygroscopicus*	HDAC inhibition	Lymphoma – approved	
				Metastatic soft tissue sarcoma – phase I/II (r), advanced melanoma – phase II (c), advanced solid tumors, NSCLC – phase I (r), and brain metastases – phase I (c)	
Staurosporines	7-Hydroxy staurosporine	*Streptomyces staurosporeus*	PKC inhibition	Advanced solid tumors, triple-negative breast cancer, advanced lymphoma/leukemia – phase I (c) and relapsed SCLC and metastatic pancreatic cancer – phase II (c)	Giddings and Newman (2013)
	Midostaurin	Analog of staurosporine	PKC inhibition	Mast cell leukemia, AML – phase II (a) advanced rectal cancer – phase I (r)	
	Becatecarin	Analog of rebeccamycin	PKC inhibition	AML, CML – phase I (c) NSCLC, stage IIIB/IV breast cancer, non-Hodgkin's lymphoma – phase II (c)	
Nitrosoureas	Streptozocin	*Streptomyces achromogenes*	DNA alkylation and DNA replication inhibition	Metastatic pancreatic cancer – approved	Cheung-Ong et al. (2013), Blasiak et al. (2004), and Newman and Cragg (2016a)
				Metastatic neuroendocrine tumors, recurrent brain neoplasms – phase II (c)	
				Advanced adrenocortical carcinoma – phase III (c) and pancreatic neuroendocrine tumors – phase III (r)	
Mitomycins	Mitomycin C	*Streptomyces caespitosus*	DNA cross linker	Bladder cancer – phase III (c)	Tomasz (1995)
				Metastatic breast cancer – phase II (c)	
				Advanced colorectal cancer – phase II/III (r) and anal carcinoma – early phase I	
Mithramycin	Mithramycin A	*Streptomyces plicatus*	DNA methylation and TRAIL-mediated apoptosis	Malignancies of lungs and esophagus – phase I/II (a)	Lee et al. (2006) and Lin et al. (2007)

The clinical trial stage shown in table is mostly from the NIH (https://clinicaltrials.gov/) or the corresponding EU clinical trial database
ADC antibody-drug conjugate, *TOR* target of rapamycin, *PKC* protein kinase C, *TRAIL* TNF-related apoptosis-inducing ligand, *c* completed, *r* recruiting participants, *a* active

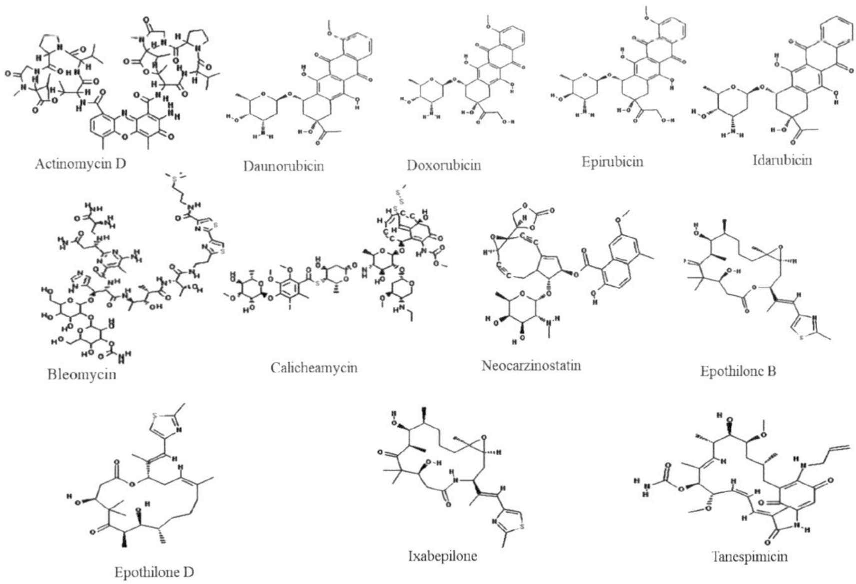

Fig. 3.2a Anticancer compounds from microbes under clinical trial

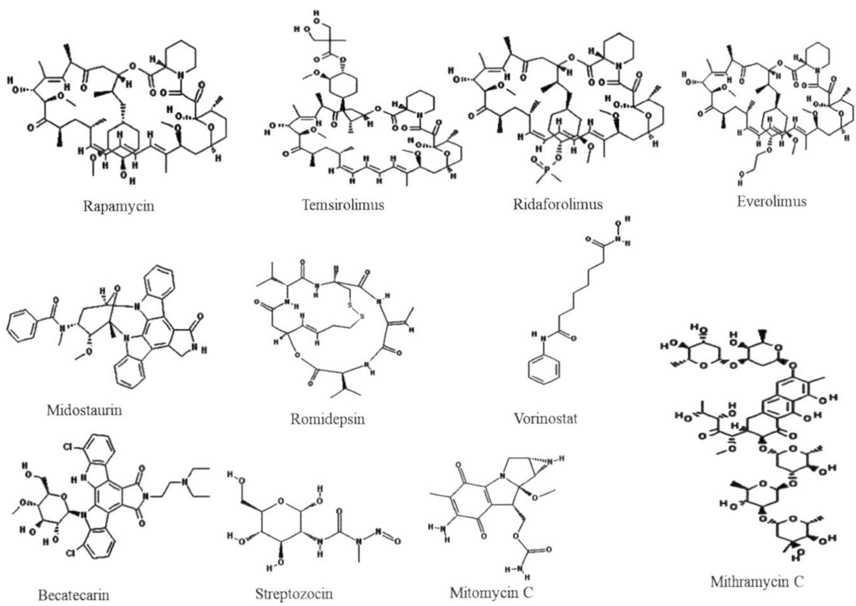

Fig. 3.2b Anticancer compounds from microbes under clinical trial

3.3.2 Anthracyclines

Anthracyclines are a class of antitumor compounds widely used in the treatment of a variety of cancers. The two most useful anthracyclines are doxorubicin (DOX) and daunorubicin (DAN) which were first isolated from *Streptomyces peucetius* and *S. caeruleorubidus*, respectively (Ramasamy and Agarwal 2008; Cheung-Ong et al. 2013; Bhattacharya and Mukherjee 2015). Both DOX and DAN were approved by FDA for cancer therapy in the 1960s (Khazir et al. 2014). DOX is already in clinical use in the treatment of ALL, AML, breast cancer, gastric cancer, Hodgkin's and non-Hodgkin's lymphoma, neuroblastoma, SCLC, soft tissue and bone sarcoma, solid tumors in children, cancers of the bladder and thyroid, and Wilms tumor. Further, it is being studied in combination with other drugs (phase III) against non-metastatic extracranial Ewing's sarcoma, T-cell ALL (phase I), and solid tumors. DOX along with other drugs such as cyclophosphamide and paclitaxel with or without carboplatin is in phase III clinical trial against triple-negative breast cancer. DOX has completed phase IV clinical trials for the treatment of ovarian cancer. DAN is used in the treatment of acute lymphoblastic or myeloblastic lymphoma. DAN is currently undergoing clinical trials ranging from phase I to phase III of combination therapy against AML and ALL.

Both DOX and DAN intercalate between the base pairs of DNA, also inhibit topoisomerase II, and generate free radicals which eventually damage DNA and proteins (Cheung-Ong et al. 2013). Both anthracyclines inhibit DNA and RNA synthesis and cause apoptotic cell death (Kumar 2015). DNA-damaging agents specifically target cancer cells which have impaired DNA damage response and cell cycle checkpoint pathways without causing much harm to the normal cells. The major limitations of these compounds are acute cardiotoxicity and development of resistance. This has led to the investigation of numerous natural, biosynthetic, and synthetic derivatives. Among them are clinically useful analogs such as epirubicin, pirarubicin, idarubicin, amrubicin, valrubicin, aclarubicin, and mitoxantrone hydrochloride (Giddings and Newman 2013). Epirubicin, a semisynthetic anthracycline, is a potent DNA intercalating agent and has been approved by the FDA for the treatment of node-positive breast cancer. Interestingly, epirubicin is reported to cause substantially less cardiotoxicity than DOX (Robert 2007). Patients are being recruited for phase III and IV clinical trials for the treatment of diffuse large B-cell lymphoma (DLBCL) in combination with rituximab and other anticancer drugs which are also in phase IV clinical trials for bladder cancer treatment. Idarubicin, a semisynthetic 4-demethoxy analog of DAN, is an inhibitor of topoisomerase II and has been approved for the treatment of AML (Wellington 2015; Newman and Cragg 2016a). Currently, it is undergoing investigations for the treatment of AML in combination with other cytotoxic agents. It has also completed the phase I clinical trial for the treatment of metastatic hepatocellular carcinoma.

3.3.3 Glycopeptide Antibiotics

Another class of extremely important molecules from actinomycetes is glycopeptide antibiotics, which include bleomycin. Bleomycin A2 is generally used as the representative structure for bleomycin. They all share a core structure but differ in the various functional groups and disaccharides (Cragg and Newman 2013). Bleomycins were originally isolated from *Streptomyces verticillus* and were later developed as antitumor agents (Hecht 2000; Giddings and Newman 2013). Bleomycin is known for its unique mechanism of binding to GC-rich regions of DNA which directly induces strand breaks in DNA. Cleavage of DNA by bleomycin may be due to the chelation of metal ions, particularly iron, resulting in the formation of free radicals which induce single-strand breaks in DNA (McKinney et al. 2013; Shapiro 2015). Nevertheless, bleomycin efficacy has been limited by its acute pulmonary toxicity (Giddings and Newman 2013; Bhattacharya and Mukherjee 2015; Rida et al. 2015). Another similar glycopeptide antibiotic closely related to bleomycin is NC-0604, which was isolated from the fermentation broth of *S. verticillus* var. *pingyangensis* n. sp. and has been reported to possess stronger antitumor action than that of bleomycin and also lower pulmonary cytotoxicity. NC-0604 is currently in preclinical trials (Giddings and Newman 2013).

Bleomycin is known to induce DNA damage in cancer cells with defective DNA repair genes (Simon et al. 2000; Melorose et al. 2008; McKinney et al. 2013). Bleomycin is approved by the FDA for the treatment of Hodgkin's and non-Hodgkin's lymphoma, penile testicular and cervical cancers, vulva carcinoma, and head and neck cancer. It has completed phase III clinical trial for the treatment of AIDS-related Kaposi's sarcoma. It is currently in the phase III clinical trial in combination with other compounds such as taxol and etoposide in treating germ cell cancers. In addition, electrochemotherapy involving administration of bleomycin is being investigated in the phase II clinical trials against locally advanced rectal cancer, inoperable vulva carcinoma, and advanced germ cell cancers.

3.3.4 Enediynes

Enediynes are a class of naturally occurring compounds with potent antitumor property. The limited clinical use of the naturally occurring enediynes is because of their severe cytotoxicity for both tumor and normal cells (Giddings and Newman 2013). Enediynes are structurally unique antitumor antibiotics, whose core structure was first elucidated in 1985, and since then, only 11 enediynes have been discovered, some of which include calicheamicins, dynemicin A, esperamicin, kerdarcidin, and neocarzinostatin. The enediynes contain two acetylenic groups conjugated by a double bond within either a nine- or ten-membered ring based on the structure of the core (Shen et al. 2015). The enediynes are some of the most cytotoxic molecules known today (Giddings and Newman 2013; Shen et al. 2015). Their cytotoxic property is mostly attributed to their ability to cause DNA double-strand breaks upon activation followed by cell death. An alternative mode of action of enediynes has been proposed, which is through the formation of inter-strand cross-links. This property of enediynes has been exploited to target solid tumors or other cancer cells under hypoxic environments which do not respond well to those enediynes that predominantly induce oxygen-dependent DNA double-strand breaks. Due to their very high cytotoxicity, enediynes are used in conjugation with monoclonal antibodies for selectively targeting the cancer cells. This significant mechanism of action of enediynes has made them an excellent payload candidate for antibody-drug conjugates in targeted cancer therapy (Damelin et al. 2015).

Calicheamicin-γ1, the most prominent member of calicheamicin, has been approved for clinical use and is known to possess remarkable anticancer properties (Demain and Vaishnav 2011; Giddings and Newman 2013). In 2000, gemtuzumab ozogamicin, an anti-CD33 humanized antibody linked to a calicheamicin derivative, was approved by the FDA as the first antibody-drug conjugate to use against CML. Later, it was withdrawn in 2010 from the US market because of its cytotoxicity. However, it is still in clinical use in Japan and other parts of world. Calicheamicin is in clinical trials ranging from phase I to III for investigating its cytotoxicity against AML. The antibody conjugated with calicheamicin has completed phase II trial for the treatment of indolent non-Hodgkin's lymphoma and undergoing phase III clinical trial in combination with rituximab and other cytotoxic drugs such as cytarabine and mitoxantrone for the treatment of ALL (Wondrak 2009; Giddings and Newman 2013; Butler et al. 2014; Khazir et al. 2014). Patients are being recruited for phase III clinical trials involving neocarzinostatin for the treatment of recurrent hepatocellular carcinoma. However, research is in progress to produce nontoxic enediyne-based anticancer compounds which are more selective in their action.

3.3.5 *Epothilones*

Epothilones are a class of naturally occurring anticancer compounds which mimic the mechanism of taxol. They are 16-membered polyketide macrolide lactones isolated from myxobacterium *Sorangium cellulosum* (Zhou and Giannakakou 2005; Behrangi et al. 2012). Epothilones were originally used as antifungal agents, but later their antitumor activity was discovered. They promote tubulin polymerization and stabilization which in turn prevent disassembly of microtubules and arrest the cells in G2/M phase and eventually cause apoptosis (Jordan et al. 1993). Though the mechanism of action of epothilones is similar to that of taxanes, they are more soluble in water than taxanes and bind to a distinct region on tubulin (Demain and Vaishnav 2011). Epothilones have been effective against multidrug-resistant cancer cell lines and also those which are resistant to paclitaxel therapy mainly due to specific β-tubulin mutations (Escuin et al. 2005; Stanton et al. 2011). Thus, epothilones have become attractive lead molecules for developing many clinically important derivatives (Ferrandina et al. 2012). Among numerous epothilones identified so far, epothilones A, B, and D are the most important variants (Escuin et al. 2005; Reichenbach and Höfle 2008). Epothilone B has completed phase I and phase II clinical trials for the treatment of advanced breast cancer, advanced colon cancer, advanced melanoma, glioblastoma, metastatic hepatocellular carcinoma, and NSCLC. It has also completed phase III trials for the treatment of a variety of cancers in combination with doxorubicin. Epothilone D has completed phase II clinical trials for the treatment of Her2-positive advanced breast cancer, stage IV NSCLC, and metastatic refractory colorectal cancer in combination with herceptin and phase I trial for the treatment of advanced solid tumors.

Ixabepilone is a semisynthetic analog of epothilone B which is orally bioavailable and effective against taxane-resistant tumors (Harvey 2008; Stanton et al. 2011). Ixabepilone has received FDA approval in 2007 for the treatment of aggressive metastatic or locally advanced breast cancer, which no longer responds to the currently available chemotherapies (Stanton et al. 2011; Scully et al. 2012; Cragg and Newman 2013). Ixabepilone has entered phase I clinical trial for advanced solid tumor treatment. Patients are being recruited for clinical trial phase II for ovarian and phase III for breast cancer treatment using ixabepilone with other cytotoxic agents. In addition, recruitment is being done for the phase II clinical trial based on genomic alteration in advanced urothelial carcinoma involving ixabepilone.

3.3.6 Geldanamycins

Geldanamycins are the naturally occurring benzoquinone ansamycin antibiotics isolated from *Streptomyces hygroscopicus*. Their anticancer property is through the inhibition of HSP90 chaperone function (Li et al. 2009; Fukuyo et al. 2010). Inhibition of HSP90 leads to the proteasomal degradation of cancer-relevant client proteins. HSP90 proteins are expressed in cancer cells two- to tenfold higher than that in normal cells, which makes them a potential target for anticancer therapy (Neckers 2002). Despite of their anticancer properties, the clinical use of geldanamycins is limited by acute hepatotoxicity at clinically relevant doses, poor water solubility, and metabolic instability. Several variants of geldanamycins have been developed with quinone ring modification, which resulted in improved potency, tolerance, metabolic stability, and water solubility (Neckers 2002; Li et al. 2009). Geldanamycin's analog tanespimycin/17-AAG/DMAG has been investigated in a number of clinical trials for the treatment of various cancers as a single agent or in combination with other anticancer drugs (Giddings and Newman 2013). Some of them are in phase III clinical trial for the treatment of relapsed and refractory multiple myeloma; phase I clinical trial for the treatment of AML, ALL, and CML; and phase II clinical trial for the treatment of metastatic thyroid cancer and metastatic breast cancer (with trastuzumab) and stage IV pancreatic and ovarian cancers (with gemcitabine hydrochloride).

3.3.7 Rapamycin

Rapamycin is a macrolide antibiotic isolated from *Streptomyces hygroscopicus* from the soil samples of Rapa Nui Island. It was originally reported to be a potential antifungal agent in 1975, but later, it was found to possess powerful immunosuppressant and antiproliferative activity against cancer cells (Giddings and Newman 2013). Rapamycin is a potent inhibitor of mammalian target of rapamycin (mTOR) in mammalian cells, which is a key regulator of cell growth and metabolism, and increased activation has been implicated in many human cancers (Ballou and Lin 2008; Seto 2012; Li et al. 2014a). In the 1990s, TOR in yeast and its mammalian homolog mTOR were discovered, and it was shown that TOR is involved in the regulation of cell growth by controlling transcription, translation, and ribosome biogenesis (Giddings and Newman 2013). Increased activation of mTORC1 is observed in many human cancers due to mutations in upstream regulators of mTORC1 (Ballou and Lin 2008; Li et al. 2014a). Gain-of-function mutations in oncogenes such as PI3K, AKT, or Ras and/or loss-of-function mutations in tumor suppressor genes such as PTEN, LKB1, or TSC1/2 lead to overexpression of mTORC1. These mutations provide cancer cells with a selective growth advantage in comparison to normal cells. In order to meet the high demands of cell proliferation, cancer cells generally have altered processes such as nutrient uptake and energy metabolism, which are directly controlled by the mTORC1 pathway. Rapamycin and other drugs which selectively target mTORC1 and impair cancer metabolism are considered potential and promising anticancer compounds. Rapamycin was approved as an immunosuppressive agent in 1999, and it has now completed phase I and II clinical trials for the treatment of a wide range of cancers (Giddings and Newman 2013) such as NSCLC, recurrent glioma, advanced ALL, neuroblastoma, advanced sarcoma, and metastatic breast cancer as a single drug and/or in combination with other anticancer compounds. Presently, patients are being recruited for phase IV clinical trials for investigating the efficacy of rapamycin against refractory solid tumors and pancreatic cancer.

Everolimus (RAD001) is a derivative of rapamycin which was initially used as an immunosuppressive agent and was later approved for the treatment of kidney, brain, pancreatic, and breast cancers. It continues to be investigated in combination with other cytotoxic drugs in a huge number of clinical trials for the treatment of various types of cancer (Giddings and Newman 2013). Currently, patient recruitment has been initiated for investigating everolimus in phase II clinical trials for treating metastatic thyroid cancer and renal cell carcinoma and in phase I clinical trials for the treatment of triple-negative breast cancer in combination with other anticancer drugs. Temsirolimus (CCI-779) is another derivative of rapamycin which has been approved by the FDA for the treatment of renal carcinoma. It has completed phase II clinical trials for the treatment of glioma and phase I clinical trials for treating pediatric solid tumors (Cragg and Newman 2013; Kinghorn et al. 2016). At present, it has been investigated in phase I trial for treating advanced cancers and solid tumors. Also, there are phase I clinical trials for the treatment of advanced breast, endometrial, and ovarian cancer and phase II clinical trials for the treatment of hepatocellular carcinoma and Hodgkin's lymphoma using temsirolimus. Ridaforolimus/deforolimus is another analog of rapamycin, which has completed many clinical trial phases for treatment of various cancers (Seto 2012; Giddings and Newman 2013; Li et al. 2014a). Some of the clinical trials include phase III trial for the treatment of metastatic soft tissue and bone sarcomas and phase II trial for the treatment of metastatic endometrial and breast cancers (in combination with trastuzumab). It is also being investigated in phase I clinical trial for the treatment of advanced solid tumors.

3.3.8 Romidepsin and Trichostatin A (HDAC Inhibitors)

Romidepsin, a bicyclic depsipeptide, was first isolated from a gram-negative bacterium, *Chromobacterium violaceum*, isolated from the soil sample of Japan in 1994 (Butler et al. 2014). Romidepsin belongs to class I histone deacetylase (HDAC) inhibitors and forms an unusual disulfide bond between a thiol and D-cysteine. Unlike conventional chemotherapeutic agents, HDAC inhibitors target epigenetic abnormalities responsible for cancer development. In 2009, romidepsin was approved by the FDA for the treatment of cutaneous and peripheral T-cell lymphoma (Mottamal et al. 2015). Romidepsin has completed early stage of clinical trials for the treatment of solid tumors; lung, thyroid, and prostate cancers; and metastatic soft tissue sarcoma (Giddings and Newman 2013; Zwergel et al. 2015). Currently, patients are being recruited for phase I/II clinical trials of romidepsin for the treatment of advanced colorectal cancer and also for phase II trials for the treatment of relapsed or refractory B-cell or T-cell lymphomas, CLL, and solid tumors in combination with other drugs. Similarly, trichostatin A is another potent HDAC inhibitor and was isolated from *S. hygroscopicus* in 1976 and determined to be an antifungal agent at much lower IC_{50}, i.e., less than 10 nM, and showed over 300-fold selectivity for inhibition of class II HDACs. Trichostatin A is currently being investigated for its potential as an anticancer drug, and it has been reported to be effective against hepatocellular carcinoma cells (Mottamal et al. 2015). Suberoylanilide hydroxamic acid (SAHA) also known as vorinostat was synthesized based on the structure of trichostatin A and has been approved by the FDA for the treatment for lymphoma and is currently in clinical trials for the treatment of a variety of cancers (Xu et al. 2007; Giddings and Newman 2013; Foss et al. 2016). Vorinostat has completed phase II clinical trials for treating metastatic soft tissue sarcoma and phase I clinical trials for treating brain metastases. Patients are being recruited for phase I/II clinical trials for the treatment of advanced melanoma and solid tumors using vorinostat.

3.3.9 Staurosporines

Staurosporine is an indolocarbazole alkaloid, which was first isolated from *Streptomyces staurosporeus* in 1977, but its potential as an anticancer compound came to light only a decade after its discovery. It is a potent protein kinase C (PKC) inhibitor with antitumor activity and has led to the development of several analogs including 7-hydroxystaurosporine and midostaurin (Giddings and Newman 2013). 7-Hydroxystaurosporine has completed phase I clinical trials for the treatment of advanced solid tumors and triple-negative breast cancer and phase II clinical trials for treating relapsed SCLC, metastatic pancreatic cancer, and advanced lymphoma/leukemia. Midostaurin is being investigated in the ongoing phase II clinical trials for the treatment of mast cell leukemia and phase II clinical trials for treatment of AML. Patients are being recruited for phase I clinical trial for treating advanced rectal cancer. Indolocarbazole rebeccamycin is another naturally occurring staurosporine. Its derivative, becatecarin, has completed the phase II clinical trials for the treatment of advanced lung cancer, stage IV breast cancer, and non-Hodgkin's lymphoma and phase I clinical trials for treating AML and CML.

3.3.10 Nitrosoureas

Streptozocin (streptozotocin) is a glucosamine nitrosourea compound produced from *Streptomyces achromogenes* (Reusser 1971; Simon et al. 2000). It was originally identified as an antibiotic in the 1950s and approved by the FDA as a therapeutic agent for the treatment of metastatic pancreatic cancer in 1982. It selectively binds to GLUT2 glucose transporter and enters the cells. Like other nitrosoureas, streptozocin induces DNA damage and inhibits DNA synthesis (Blasiak et al. 2004; Cheung-Ong et al. 2013). This could be the possible underlying mechanism by which it causes diabetes in experimental animals by causing DNA alkylation and cell death of pancreatic islet cells (Simon et al. 2000; Lee et al. 2005). However, the DNA alkylating ability of streptozocin has been exploited to test its potential as an anticancer agent against a wide range of cancers including insulinoma, myeloma, glioma, lymphoma, melanoma, and SCLC. Streptozocin has completed phase II clinical trials for the combined chemotherapy against metastatic neuroendocrine tumors and recurrent brain neoplasms and also a phase III clinical trial for advanced adrenocortical carcinoma. Currently, patients are being recruited for the phase III clinical trial in which streptozocin is being evaluated for the treatment of pancreatic neuroendocrine tumors (pNET).

3.3.11 Mitomycin C

Mitomycin C is an aziridine-containing antibiotic isolated from *S. caespitosus* or *S. lavendulae* in 1956 that possesses antitumor properties (Bhattacharya and Mukherjee 2015). It is a strong DNA cross-linking agent, known to cause inhibition of DNA synthesis in mammalian cells (Nijwening et al. 2011; Cheung-Ong et al. 2013). Several synthetic DNA cross-linking agents have been developed based on the structure of mitomycin C. As such, mitomycin C does not react with DNA; however it gets activated upon enzymatic or chemical reduction, which converts mitomycin C into a highly reactive bifunctional alkylating agent that binds to DNA. Thus, it selectively targets and suppresses the growth of the solid tumors which predominantly have hypoxic conditions, while, normal cells do not (Tomasz 1995). Moreover, it has been reported to synergistically work with anticancer drugs such as doxorubicin and 5-fluorouracil in the treatment of gastric adenocarcinoma and a variety of cancers. It has also been reported to be effective in the treatment of lymphoma and leukemia. It has completed phase II trial for metastatic breast cancer. Currently, it is in phase II and phase III clinical trials for the treatment of colorectal cancer and urinary bladder neoplasms respectively.

3.3.12 Mithramycin/Mithramycin A

Mithramycin or plicamycin is an antitumor antibiotic originally isolated from *Streptomyces plicatus*. This anticancer drug has been introduced in the year 1961 for the treatment of bone cancer (Paget's disease), testicular cancer, and tumor-related hypercalcemia. Mithramycin binds to GC-rich region of DNA and inhibits RNA and protein synthesis. A preclinical report suggests that mithramycin may interfere with the DNA methyltransferase activity in CpG hypermethylation of major tumor suppressor genes, which is observed in many human cancers (Lin et al. 2007). It induces TRAIL-mediated apoptosis in cancer cells (Lee et al. 2006). It has been investigated in phase I/II clinical trials for the treatment of Ewing's sarcoma, thoracic malignancies, and germ cell neoplasms. However, its clinical use is limited because of its severe toxicity (Lin et al. 2007).

3.4 Anticancer Compounds from Marine Source

The oceans constitute over 70% of the earth's surface and 95% of its tropical biosphere. Marine organisms cover 50% of the total biodiversity on the earth that represents 34 of the 36 phyla (Thakur et al. 2005; Hussain et al. 2012). Marine flora constitutes most of the oceanic organisms which include marine fungi, bacteria, algae, cyanobacteria, sponges, and seaweeds (Li et al. 2015; Rangel and Falkenberg

2015). With the taxonomical diversity and biological properties, these marine sources have provided an enormous opportunity for the discovery of novel anticancer compounds. Recently, a considerable number of marine compounds with robust anticancer properties have been discovered which have proved their efficiency clinically. Most of them are polyphenols and sulfated polysaccharides, which are known for their potent antioxidant, antitumor, and immunostimulatory activities (Dixit and Suseela 2013; Pangestuti and Kim 2017). Marine origin compounds have also demonstrated other biologically and clinically important properties such as antimicrobial, anti-inflammatory, anticoagulant, photoprotective, and others. Some of the marine organisms such as sponges symbiotically associate with actinomycetes and have also been investigated for anticancer potential. For the past three decades, marine pharmaceutics has developed a number of antitumor drugs (Schwartsmann et al. 2001; Hussain et al. 2012). However, the marine environmental sources have been underexplored for the production of novel antitumor compounds. It is mainly because of the shortfalls in this field (Bhatnagar and Kim 2010). Although there are only a few marine-derived products currently in the market, most of them are in the clinical trials and drug development stage (Newman and Cragg 2014). The drugs from marine sources such as marine alkaloid trabectedin were introduced in 2007 for the treatment of cancer. About 2500 metabolites with antiproliferative activity have been reported during the last decade (Jimeno et al. 2004). Anticancer compounds from marine source and their derivatives under clinical trial and for their clinical status are discussed below (Table 3.3; Figs. 3.3a and 3.3b).

3.4.1 Bryostatins

Bryostatins are a class of naturally occurring macrolides with an immense therapeutic value, originally isolated from the marine invertebrate *Bugula neritina*. They generally share a 20-membered macrolactone core and 3 remotely functionalized polyhydropyran rings. Bryostatins differ from each other in substitution at C7 and C20 and placement of the c-lactone at either C19 or to C23 in the polyhydropyran ring. Several bryostatin analogs have been synthesized such as bryostatin 1, 2, 3, and 7 with several features and activities. Among the bryostatin analogs, the most studied member is bryostatin 1 and is a powerful PKC modulator to which its antitumor activity is attributed. Interestingly, bryostatin 1 has minimal tumorigenic properties unlike most of the PKC modulators (Irie et al. 2012; Giddings and Newman 2013; Newman and Cragg 2016b). Bryostatin 1 is being evaluated for its efficacy against a variety of cancers as a single agent and in combination with other cytotoxic compounds such as vincristine and paclitaxel. Bryostatin 1 has completed phase I clinical trials for the treatment of recurrent HIV-related lymphoma in combination with vincristine and metastatic solid tumors with temsirolimus. It has also completed phase II trials for the treatment of advanced pancreatic cancer, metastatic prostate, and kidney cancers.

Table 3.3 Anticancer compounds from marine source

Class	Compound	Source	Mode of action	Clinical trial stage	References
Bryostatins	Bryostatin 1	*Bugula neritina* (bryozoan)	PKC inhibition and apoptosis	Advanced pancreatic, metastatic prostate, and kidney cancer – phase II (c)	Schwartsmann et al. (2001), Singh et al. (2008), and Butler et al. (2014)
				HIV-related lymphoma, metastatic solid tumors – phase I (c)	
Dolastatin	Dolastatin 10	*Dolabella auricularia*/ *Symploca* sp.	Microtubule inhibition	Metastatic pancreatic cancer, kidney and ovarian cancers, lymphoma – phase II (c), metastatic soft tissue sarcoma – phase II (c)	Ray et al. (2007), and Singh et al. (2008)
	Soblidotin/ auristatin (TZT-1027)	Synthetic derivative of dolastatin 10	Microtubule inhibition	Metastatic soft tissue sarcoma and NSCLC – phase II (c)	Ray et al. (2007), Newman and Cragg (2014), and Fanale et al. (2015)
	Synthadotin/ tasidotin	Synthetic derivative of dolastatin 15	Mitosis and topoisomerase inhibition	Malignant melanoma – phase I/II (c)	
				NSCLC and prostate cancer – phase II (c) solid tumors – preclinical (c)	
	Monomethyl auristatin E (MMAE)/Vedotin	Synthetic variant of auristatin (as ADC)	Microtubule inhibition	Lymphoma – phase II melanoma, triple-negative breast cancer – phase III and solid tumors – phase I/II	
	Monomethyl auristatin F (MMAF)	Synthetic variant of auristatin (as ADC)	Microtubule inhibition	Relapsed/refractory multiple myeloma – phase I (r) and hematologic malignancies – phase I	

Ecteinascidin	Trabectedin (ET-743)	*Ecteinascidia turbinate*	Double-strand DNA damage by minor groove binding	Solid tumors, pediatric cancers – phase II (r)	D'Incalci and Galmarini (2010), Romano et al. (2013), and Petek and Jones (2014)
				Breast and advanced prostate cancer – phase II (c)	
				Soft tissue sarcoma and ovarian cancer – approved	
	Lurbinectedin (PM01183)	Structural variant of trabectedin	Double-strand DNA damage by minor groove binding	Osteosarcoma – phase II (c) NSCLC – phase III (r) and advanced solid tumors, metastatic breast cancer – phase II (r)	
	Zalypsis (PM 00104)	Structural variant of trabectedin	DNA minor groove binding	Metastatic Ewing's family of tumors – phase II (c)	
Kahalalide	Kahalalide F	*Elysia rufescens/ Bryopsis* sp.	Lysosome target and promotion of oncosis	NSCLC – stage IIIB/IV – phase II	Jimeno et al. (2004), Hussain et al. (2012), and Cragg and Newman (2013)
	Elisidepsin (PM02734)	Synthetic derivative of kahalalide F	Oncosis and necrosis promotion	Advanced malignant solid tumors – phase I (c)	
Aplidine	Aplidine	*Aplidium albicans*	Apoptosis and angiogenesis inhibition	Refractory/relapsed multiple myeloma – phase III (r), leukemia, lymphoma – phase II (c) advanced solid tumors – phase I (r)	Schwartsmann et al. (2001), Jimeno et al. (2004), Munoz-Alonso et al. (2008), and Butler et al. (2014)
Epoxomicin	Carfilzomib	Synthetic derivative of epoxomicin	Proteasome inhibition	Multiple myeloma – approved NSCLC; ovarian, renal, and prostate cancers; and hematological malignancies – phase I/II (r)	Moore et al. (2008) and Butler et al. (2014)
Halichondrin	Eribulin mesylate (E7389)	Synthetic derivative of halichondrin B	Microtubule inhibition	Metastatic breast cancer, liposarcoma – approved NSCLC, advanced solid tumors, prostate cancer – phase II (c), soft tissue sarcomas, pediatric, cervical cancers – phase I/ II (r)	Towle et al. (2001) and Newman and Cragg (2014)

(continued)

Table 3.3 (continued)

Class	Compound	Source	Mode of action	Clinical trial stage	References
Salinosporamides	Salinosporamide A (NPI-0052)	*Salinispora* sp. (bacterium)	20 S proteasome inhibition	Grade IV malignant glioma – phase I/II (r)	Moore et al. (2008) and Butler et al. (2014)
				Solid tumors, lymphomas, leukemia, multiple myeloma, pancreatic cancer, melanoma, NSCLC – phase I (c)	
Cytarabine	Ara-C	*Cryptotheca crypta*	DNA synthesis inhibition	AML, ALL and CML – approved meningeal metastasis of breast cancer – phase III (r), medulloblastoma and metastatic solid tumors – phase II (r)	Schwartsmann et al. (2001), Jimeno et al. (2004), and Hussain et al. (2012)
	Gemcitabine	Derivative of cytarabine	DNA synthesis inhibition	Metastatic breast, ovarian, pancreatic cancers, and NSCLC – approved	
				Urothelial cancer – phase III (r), T-cell lymphoma, biliary tract cancers, and metastatic colorectal cancer – phase II (r), solid tumors and bladder cancer – phase I (r)	
Halimide	Plinabulin (NPI-2358)	*Aspergillus* sp.	Microtubule inhibition	Metastatic NSCLC – phase III (r) and advanced solid tumors – phase II/III (r)	Gomes et al. (2015)
Hemiasterlin	E7974	*Hemiasterellaminor* sponge	Microtubule inhibitor	Solid tumors – phase I (c)	Singh et al. (2008 and Kuznetsov et al. (2009)

The clinical trial stage shown in table is mostly from the NIH (https://clinicaltrials.gov/) or the corresponding EU clinical trial database
PKC protein kinase C, *ADC* antibody-drug conjugate, *DDR* DNA damage response, *c* completed, *r* recruiting participants, *a* active

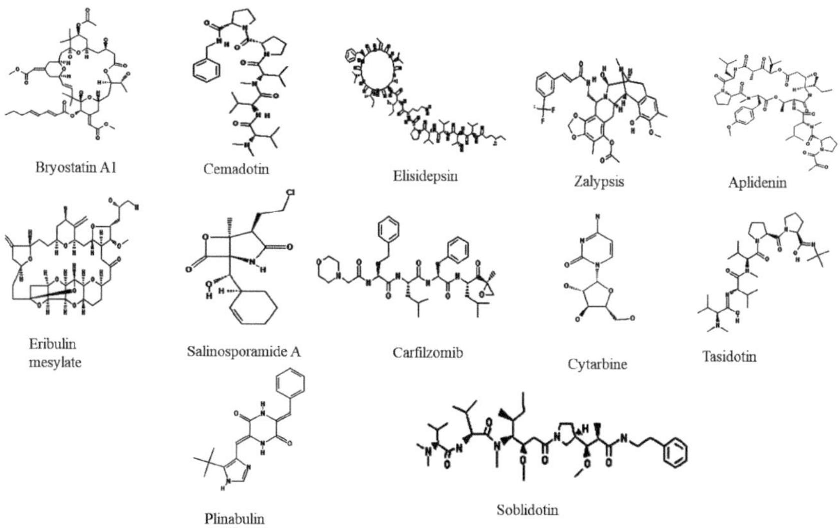

Fig. 3.3a Anticancer compounds from marine under clinical trial

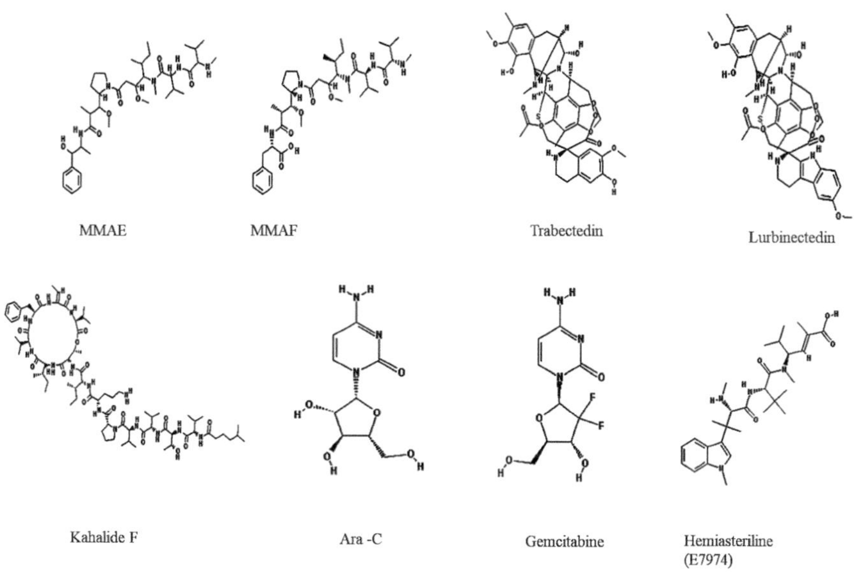

Fig. 3.3b Anticancer compounds from marine under clinical trial

3.4.2 *Dolastatins*

Dolastatins are pentapeptides originally isolated from the sea hare *Dolabella auricularia* in the 1970s from the Indian Ocean and later from cyanobacterium of the genus *Symploca*, which has been recently renamed as *Caldora penicillata* (Newman and Cragg 2016b). Dolastatin 10 is the most active compound of this class with its ED_{50} being 4.6×10^{-5} μg/ml against murine PS leukemia cells (Singh et al. 2008). Dolastatin 10 is a potent antimitotic agent which is similar to the mechanism of action of taxanes and vinca alkaloids (Simmons et al. 2005). Reports suggest that it binds to β-subunit of tubulin, forms tubulin bundles, and inhibits the formation of microtubules which arrests the cells in G_2/M phase and causes Bcl2 mediated apoptotic cell death (Pangestuti and Kim 2017). Also, it acts synergistically with vinca alkaloids and bryostatin 1 against certain cancers (Singh et al. 2008). Dolastatin 10 was evaluated in several phase I and phase II clinical trials for the treatment of solid tumors of the pancreas and kidney and also a multi-institutional phase II clinical trial for the treatment of metastatic soft tissue sarcoma. However, due to minimal responses and side effects like peripheral neuropathy in cancer patients, it was withdrawn from further clinical trials. Nevertheless, because of its antimitotic action against tumors, dolastatin 10 has prompted the development of its synthetic derivatives (Simmons et al. 2005).

3.4.2.1 TZT-1027 (Soblidotin and Auristatin PE)

Soblidotin is a synthetic analog of dolastatin 10, which has been developed by retaining the anticancer property of the lead compound and reducing its toxicity. The only difference between soblidotin and dolastatin 10 is the replacement of the terminal dolaphenine amino acid residue unit of dolastatin with a phenylethylamine group in soblidotin (Simmons et al. 2005). Soblidotin is a microtubule-disrupting agent and has demonstrated to have potency against p53 mutant cell lines, against which DNA-damaging agents are generally less effective. Soblidotin is also known to be a potent tumor-vasculature-disrupting agent due to its ability to interact with VEGF. It entered phase II trials for the treatment of soft tissue sarcoma and lung cancer (Fanale et al. 2015). Due to extreme hematological toxicities and lack of expected cytotoxicity against cancer cells, soblidotin has not progressed beyond phase II clinical trials (Singh et al. 2008; Newman and Cragg 2016b).

3.4.3 *Dolastatin 15*

Dolastatin 15 is a linear depsipeptide, closely related to dolastatin 10, and was also isolated from the sea hare *D. auricularia*. The initial investigations on dolastatin 15 demonstrated that its ED_{50} was 2.4×10^{-3} μg/ml and it binds directly to vinca domain of tubulin. However, due to poor water solubility and low yield, it could not enter clinical trials. Nevertheless, it has encouraged the development of its synthetic analogs such as synthadotin and cematodin with enhanced pharmaceutical properties (Simmons et al. 2005).

3.4.3.1 Synthadotin (Tasidotin)

Tasidotin is a metabolically stable third-generation synthetic analog of dolastatin 15 with a terminal tertiary butyl moiety. It has demonstrated to have potent anticancer activity on cancer cell lines even at very low concentrations. Tasidotin C-carboxylate, a functionally active metabolite of tasidotin, is reported to be 10–30 times more effective with a reduced toxicity profile (Fanale et al. 2015). Its mechanism of action is more similar to that of taxol which binds along the length of microtubules and stabilizes microtubule plus ends at low concentrations and increases the catastrophe frequency at minus ends (Ray et al. 2007). Preclinical studies have suggested that tasidotin has a potent antitumor activity against pediatric sarcomas in xenograft models. This orally active microtubule inhibitor has completed three phase II trials for the treatment of metastatic melanoma, metastatic NSCLC, and hormone refractory prostate cancer which is previously treated with docetaxel. These trials suggested that the toxicity profiles were quite favorable (Simmons et al. 2005; Khazir et al. 2014).

3.4.3.2 Cemadotin

Cemadotin is another synthetic analog of dolastatin 15, which was developed with better chemical properties retaining the high cytotoxic profile of the lead compound with IC_{50} being 0.1 μM. It underwent various phase I clinical trials and phase II clinical trials against malignant melanoma, metastatic breast cancer, and NSCLC. However, the clinical trials were discontinued because of the severe cardiovascular toxicity, hypertension, acute myocardial infarction, and neutropenia caused by cemadotin in phase I clinical evaluation and inconsiderable cytotoxicity in phase II clinical trials (Simmons et al. 2005; Ray et al. 2007).

3.4.4 Monomethylauristatin E and Monomethylauristatin F

Monomethylauristatin E which is also known as vedotin and its phenylalanine variant monomethylauristatin F are the synthetic variants of soblidotin/auristatin E. Due to high cytotoxic properties, these two compounds have been used as warheads to link monoclonal antibodies and are currently in several clinical trials for the treatment of a wide variety of cancers. Currently, there are 68 clinical trials with the status "recruiting" of the total of 139 trials that involve these two potent warheads listed (https://clinicaltrials.gov). Eventually many antibody-drug conjugates received FDA approval (Giddings and Newman 2013).

Vedotin linked with brentuximab has completed phase II clinical trials for the treatment of primary mediastinal large B-cell lymphoma and in combination with rituximab for the treatment of DLBCL. Patients are being recruited for the phase II clinical trials for treating Hodgkin's lymphoma, malignant mesothelioma, and non-Hodgkin's lymphoma. Many active clinical trials are listed in the US clinical trial registry involving brentuximab linked to vedotin in combination with other monoclonal antibodies for the treatment of lymphoma (NIH). Glembatumumab vedotin has completed phase II clinical trials for treating breast cancer, metastatic triple-negative breast cancer, and stage III/IV melanoma. Three phase II clinical trials in which patients are being recruited are for the treatment of SCC of lung, advanced melanoma, and triple-negative breast cancer. Polatuzumab vedotin in combination with other monoclonal antibodies is in three active phase I/II clinical trials for the treatment of non-Hodgkin's lymphoma and DLBCL, and recruitment is going on for phase I/II trials for treating DLBCL in combination with other cytotoxic drugs. Tisotumab vedotin is another antibody-drug conjugate, for which patients are being recruited for phase I/II trials for safety assessment during the treatment of solid tumors. GSK2857916 is the only antibody conjugate of MMEF under clinical trial. There is only one phase I clinical trial listed in which patients are being recruited for the treatment of relapsed/refractory multiple myeloma and other advanced hematologic malignancies expressing B-cell maturation antigen (BCMA) (Newman and Cragg 2016b).

3.4.5 Trabectedin (ET-743)

Trabectedin is a tetrahydroisoquinoline alkaloid, which was originally isolated from the Caribbean tunicate *Ecteinascidia turbinata*. Because of its stability and high natural availability, ET-743 was considered more suitable for clinical development than its analog ET-749, although it showed high cytotoxic potential *in vivo*. Trabectedin received FDA approval in 2015 for the treatment of metastatic liposarcoma and leiomyosarcoma, particularly which are previously treated with anthracyclines (Giddings and Newman 2013; Newman and Cragg 2016b). Trabectedin has completed phase II clinical trials for the treatment of metastatic breast cancer; ovarian, prostate, and pancreatic cancers; metastatic osteosarcoma; Ewing's family of sarcomas; and advanced malignancies. It has completed phase III trials for the treatment of advanced relapsed ovarian cancer in combination with DOXIL and dexamethasone, soft tissue sarcomas with dacarbazine, and advanced translocation related sarcomas with doxorubicin and ifosfamide. Currently, there are many phase I/II/III trials listed with the status "recruiting" for treating ovarian cancer, breast cancer, and soft tissue sarcomas as monotherapy and in combination with other therapies involving cytotoxic drugs/monoclonal antibodies/radiation.

The antitumor activity of ET-743 (IC_{50} of which is well below 1 nM) is attributed to its multifaceted mechanism of action in targeting cancer cells (Singh et al. 2008). It is reported that ET-743 binds to the minor groove of DNA and induces a unique bend in DNA toward the major groove. It inhibits active transcription by directly interacting with RNA polymerase II, causing its ubiquitination and proteasomal degradation. Further, it interferes with the transcription-coupled nucleotide excision repair and homologous recombination machineries and induces double-strand breaks in DNA. It causes cell cycle arrests in G2 phase and induces cell death. The cytotoxicity caused by trabectedin is independent of p53 status (Takebayashi et al. 2001; D'Incalci and Galmarini 2010). ET-743 has demonstrated antiproliferative properties against advanced ovarian and breast cancers, metastatic osteosarcoma, and mesenchymal tumors which have been previously treated with platinum drugs or taxanes in phase I clinical trials. In phase II clinical evaluation, soft tissue sarcoma and ovarian and breast cancer patients responded well to ET-743 treatment (Simmons et al. 2005; Giddings and Newman 2013; Newman and Cragg 2016b). Two structural variants and close relatives of trabectedin, namely, lurbinectedin (PM01183) and Zalypsis, underwent clinical development and entered clinical trials.

3.4.5.1 Lurbinectedin (PM01183)

Lurbinectedin is an isoquinoline alkaloid with the replacement of tetrahydroisoquinoline present in ring C of trabectedin with a tetrahydro- β-carboline moiety (Newman and Cragg 2016b). Like ET-743, PM01183 also binds to the minor groove of DNA that results in delayed progression through S-phase and eventually causes G_2/M phase arrest and cell death. Reports suggest that PM01183 has demonstrated to act in synergy with platinum drugs and is effective against platinum drug-resistant cell lines (Imperatore et al. 2014; Newman and Cragg 2016b). PM01183 has completed its phase II clinical trials for the treatment of NSCLC with docetaxel and gemcitabine and as a single agent against osteosarcoma. Currently, patients are being recruited for phase III clinical trials for the treatment of NSCLC and phase II trials for treating advanced solid tumors and metastatic breast cancer.

3.4.5.2 PM00104 (Zalypsis®)

Zalypsis is a synthetic tetrahydroisoquinoline alkaloid closely related to trabectedin and mimics many natural compounds of marine origin. It is structurally very similar to jorumycin which is isolated from the skin and mucin of the *nudibranch Jorunna funebris*, the family of renieramycins isolated from sponges, and the ecteinascidins, from a Caribbean tunicate (Romano et al. 2013; Petek and Jones 2014). Like ET-743 and PM01183, PM00104 also binds to the minor groove of the DNA through its reactive carbinolamine group. Surprisingly, according to some reports, although it binds to DNA and is cytotoxic, it does not activate DNA damage response. Zalypsis initially demonstrated to be a manageable and reversible toxicity profile in preclinical trials (Romano et al. 2013). It also showed potent antiproliferative effects against a wide range of solid and hematological cancer cell lines *in vitro* (Romano et al. 2013) and human transplantable breast, renal, gastric, and prostate xenograft models in *in vitro* preclinical evaluation (Petek and Jones 2014). A phase II trial using Zalypsis for the treatment of metastatic Ewing's family of tumors has been completed. Currently, there is no active clinical trial listed in the registry.

3.4.6 Kahalalide F

Kahalalides are antitumor depsipeptides, originally isolated from Sacoglossan mollusk, *Elysia rufescens* (Newman and Cragg 2014). Among the kahalalides that were isolated, only kahalalide F demonstrated antitumor activity (Singh et al. 2008). Kahalalide F has a unique mechanism of action which is the induction of oncosis by which it causes cell death in cancer cells (Singh et al. 2008). Reports also suggest that it induces apoptotic mode of cell death by disrupting lysosomal membrane in tumor cells. Also, it targets the proteins involved in the Erb3 and PI3 kinase-Akt/PKB signaling pathways (Giddings and Newman 2013). It may also interfere with the gene expression of the proteins involved in DNA replication and cell proliferation (Mans 2016). Kahalalide F demonstrated to have potent antitumor activity *in vitro* and *in vivo* in various solid tumor models of breast cancer, colon cancer, NSCLC, and, exclusively, prostate cancer. It entered clinical trials for the treatment of prostate cancer, malignant melanoma, and NSCLC (Hussain et al. 2012). Currently, no clinical trials are listed that involve kahalalide in the US clinical trial registry. However, there is one ongoing phase II clinical trial in Europe to treat stage IIIB or stage IV NSCLC using kahalalide F after the first-line chemotherapy.

3.4.6.1 Elisidepsin (PM02734)

Elisidepsin (isokahalalide F, PM-02734) is a synthetic cyclic depsipeptide (Bhatnagar and Kim 2010), a derivative and an isomer of kahalalide F (Giddings and Newman 2013). Like the parent compound, elisidepsin appears to induce oncolytic rather than apoptotic cell death in cancer cells. It has completed phase I clinical trial for the treatment of advanced malignant solid tumors in combination with erlotinib. A study suggests that the trial was closed because of the inadequate antitumor activity of elisipepsin exhibited in combination with erlotinib (Giddings and Newman 2013).

3.4.7 Aplidine

Aplidine is a marine depsipeptide originally isolated from *Aplidium albicans*. It is structurally closely related to didemnin B, an antitumor compound isolated from a Caribbean tunicate *Tridideminum solidum*. Structurally, aplidine is dehydrodidemnin B, and this small structural difference is the reason for its stronger cytotoxicity and lower cardiotoxicity profile (Giddings and Newman 2013). It is reported that the antiproliferative property of aplidine is because of its ability to interfere with multiple regulatory pathways. Aplidine induces activation of Rac1 which leads to a strong activation of JNK pathway that rapidly induces apoptosis in breast, renal, and cervical cancer and melanoma cells (Munoz-Alonso et al. 2008).Currently, aplidine is in phase III clinical trial in combination with dexamethasone for the treatment of refractory/relapsed multiple myeloma (NIH). Aplidine is PharmaMar's second most advanced compound after trabectedin as it has completed phase II trials for the treatment of multiple myeloma and progressed to phase III trials (Khazir et al. 2014). It has completed phase II clinical trials for the treatment of leukemia, lymphoma, and phase I trial for advanced solid tumors treatment. Patients are being recruited for phase I/II trials for the treatment of multiple myeloma in combination with bortezomib and dexamethasone and as a monotherapy in phase II trial for treating lymphoma. The European clinical trial registry has one ongoing multicenter phase II clinical trial against relapsed or refractory indolent non-Hodgkin's lymphoid neoplasm using aplidine.

3.4.8 Epoxomicin and Derivatives/Carfilzomib

Epoxomicin is a tetrapeptide with an epoxy-b-aminoketone moiety isolated from an unidentified strain of actinomycetes. Its antitumor properties were first reported in 1992. Epoxomicin entered preclinical trials conducted by Bristol-Myers Squibb, but was withdrawn because of its unknown mechanism of action. Later, it was discovered that epoxomicin is a proteasome inhibitor by Crews' team at Yale University. With its mechanism of action discovered, several derivatives of epoxomicin were eventually developed (Giddings and Newman 2013).

Carfilzomib is a synthetic derivative of epoxomicin which was designed with a morpholino end group in order to increase its water solubility, oral availability, and other pharmaceutical properties (Giddings and Newman 2013). Carfilzomib is an irreversible proteasomal inhibitor and more selective in its action than bortezomib, which is a well-known proteasomal inhibitor (Crawford et al. 2011). Through the proteasomal inhibition, carfilzomib causes cell cycle arrest, promotes apoptosis, and thus inhibits the growth of the tumors. Unlike bortezomib, carfilzomib does not cause neurotoxicity which is a common side effect caused by bortezomib in multiple myeloma patients. It also synergizes with dexamethasone to induce potent cytotoxicity against multiple myeloma (Orlowski and Kuhn 2008; Arastu-Kapur et al. 2011; Crawford et al. 2011). It received FDA approval in 2012 for the treatment of multiple myeloma and currently is in clinical investigations for the treatment of other cancers. Presently, there are active phase I–II clinical trials with carfilzomib for the treatment of NSCLC; ovarian, renal, and prostate cancers; and hematological malignancies in combination with other cytotoxic drugs. Carfilzomib is the second proteasomal inhibitor to be approved after bortezomib and the first not to cause adverse side effects like painful peripheral neuropathy (Arastu-Kapur et al. 2011).

3.4.9 Halichondrin B

Halichondrins are the complex structured polyether macrolides originally isolated from Japanese marine sponge *Halichondria okadai* in 1986. Halichondrin B exhibited robust antitumor activity with IC_{50} being 0.09 nM in B-16 melanoma cell lines. In 1991, halichondrin B was identified as tubulin-destabilizing agent found to strongly bind to tubulin at a site nearby but different from that of vinca. One of the derivatives which showed very high potency against tumor cells is the truncated analog of halichondrin B against human colon cancer cell lines, and that compound was named eribulin which was then progressed to further clinical evaluation (Singh et al. 2008; Giddings and Newman 2013). Halichondrins prevent microtubule assembly and cause delay in S-phase progression and G2/M phase cell cycle arrest which eventually results in apoptosis in cancer cells (Towle et al. 2001; Kinghorn et al. 2016).

3.4.9.1 Eribulin

Eribulin is one of the synthetic derivatives of the natural marine product halichondrin B, used in the treatment of metastatic breast cancer in patients who have previously been treated with an anthracycline or a taxane and metastasized liposarcoma. Eribulin received FDA approval for treating metastatic breast cancer in 2010 (Khazir et al. 2014). It effectively inhibits microtubule dynamics at low concentrations, causes prolonged cell cycle arrest, and eventually induces apoptosis in tumor cells, whereas at relatively much higher concentrations, it promotes tubulin depolymerization. Reports suggest that eribulin also hampers centromere dynamics (Newman and Cragg 2016b). As of now, clinical trials ranging from phase I to III are listed using eribulin as a monotherapy and/or in combination with other cytotoxic drugs for the treatment of different cancer types. Phase II trials for the treatment of NSCLC, advanced solid tumors, and prostate cancer using eribulin have been completed. Patients have been recruited currently for phase I–II trials for the treatment of soft tissue sarcomas, pediatric cancers, and cervical cancer, and a pilot study for the treatment of brain tumor has also been listed. There are several active clinical trials ranging from phase I to III for the treatment of breast cancer using eribulin in combination with various other anticancer drugs.

3.4.10 Salinosporamide A (NPI-0052 and Marizomib)

Salinosporamide A or NPI-0052 was first isolated from *Salinispora tropica*, a marine streptomycete collected from the Bahamas. Salinosporamide A is structurally closely related to the terrestrial natural compound lactacystin, isolated from bacteria, and its derivative omuralide (Crawford et al. 2011; Giddings and Newman 2013). Salinosporamide A is a potent proteasome inhibitor binds to 20 S proteasome more selective than lactacystin, which is also a proteasome inhibitor (Singh et al. 2008). The antitumor activity of salinosporamide A is attributed to its irreversible binding to 20 S catalytic core subunit proteasome by covalently modifying its threonine residues in the active site (Moore et al. 2008). By inhibiting proteasome, salinosporamide also inhibits NF-kB pathway (Giddings and Newman 2013). Inhibition of proteasome results in an accumulation of poly-ubiquitinated proteins, which leads to cell cycle arrest, promotes apoptosis, and inhibits tumor growth and angiogenesis.

Marizomib has completed phase I clinical trials for the treatment of advanced malignancies including solid tumors, lymphomas, leukemias, and multiple myeloma as a monotherapy and NSCLC, pancreatic cancer, melanoma, or lymphoma in combination with vorinostat. It has completed a phase I trial for treating refractory/relapsed multiple myeloma with marizomib in combination with pomalidomide and low dose of dexamethasone. Currently, patients are being recruited for the multicenter phase I/II clinical trial for the treatment of grade IV malignant glioma using marizomib in combination with bevacizumab and another phase I trial using marizomib and an adjuvant temozolomide in combination with radiotherapy. The status of one more phase I/II clinical trial is unknown which is listed in the US clinical trial registry for the treatment of relapsed/refractory multiple myeloma for patients who have previously been treated with carfilzomib.

3.4.11 Cytarabine

Cytarabine is an FDA-approved anticancer drug for the treatment of AML, ALL, and CML in combination with other drugs (Priyankashukla 2014). It has also been approved as a monotherapy to treat and prevent meningeal leukemia (NCI). Although it is not available naturally in the form Ara-C from the marine environment, initially its development was inspired by a series of C-nucleoside-derived bioactive compounds which were originally isolated from the Caribbean sponge, *Cryptotheca crypta*/*Tethya crypta*. It is a pyrimidine nucleoside analog that contains arabinose rather than ribose or deoxyribose (Schwartsmann et al. 2001; Newman and Cragg 2014, 2016b). Cytarabine is metabolized into cytosine arabinoside triphosphate which is its active triphosphate form and damages DNA by multiple mechanisms (Priyankashukla 2014). It inhibits α-DNA polymerase DNA replication and β-DNA polymerase, and cytosine arabinoside triphosphate incorporates into DNA. It inhibits not only DNA synthesis but even RNA polymerase and nucleotide reductase needed for the synthesis of DNA. It specifically causes S-phase cell cycle arrest (Miura and Izuta 2004).

As of now, there are a total of 1153 clinical trials ranging from phase I to IV have been listed on NIH clinical trial database for the treatment of various cancers using cytarabine as a single therapeutic agent and in combination with other cytotoxic drugs. Other than lymphoma and leukemia treatment, patient recruitment is going on for phase III clinical trial for the treatment of meningeal metastasis of breast cancer, phase II clinical trial for the treatment of medulloblastoma using liposomal cytarabine, and phase II clinical trial for the treatment of leptomeningeal metastasis from metastatic solid tumors using cytarabine, and several active trials have been listed in the registry which involve cytarabine or its derivatives.

3.4.11.1 Gemcitabine

Gemcitabine is a fluorinated derivative of cytarabine which is a broad-spectrum antimetabolite and deoxycytidine analog with anticancer property (Jimeno et al. 2004; Priyankashukla 2014). Upon administration, gemcitabine is converted into the active metabolites which get incorporated into DNA. Like cytarabine, gemcitabine also inhibits DNA polymerase and ribonucleotide reductase activity resulting in diminished deoxynucleotide pool available for DNA synthesis (Miura and Izuta 2004). It has been approved by FDA for the treatment of metastasized breast, ovarian, and pancreatic cancers and NSCLC (Schwartsmann et al. 2001) in combination with other cancer chemotherapeutics such as paclitaxel, cisplatin, carboplatin, and paclitaxel albumin-stabilized nanoparticle formulation (NCI). Currently, numerous active clinical trials involving treatment with gemcitabine as a monotherapy or in combination with other chemotherapeutics are going on. Some of them are a multicenter phase I trial for the treatment of solid tumors and bladder cancer; phase II trials for the treatment of T-cell lymphoma, biliary tract cancers, and metastatic colorectal cancer; and a phase III trial for the treatment of urothelial cancer for which patients are being recruited.

3.4.12 Plinabulin (NPI-2358)

Plinabulin (NPI-2358) is a potent orally active antineoplastic compound isolated from the cultures of a marine fungus *Aspergillus ustus* (Khazir et al. 2014). It is structurally a close relative of the fungal metabolite halimide/phenylahistin which exhibited potent antitumor activity against various *in vitro* and *in vivo* cancer models (Newman and Cragg 2014). Plinabulin interferes with microtubule dynamics by selectively binding to colchicine domain of tubulin, disrupts mitotic spindle assembly, and brings about G_2/M phase cell cycle arrest. It also causes disruption of tumor vasculature and inhibits tumor cell proliferation. Plinabulin has demonstrated to have strong antitumor activity against several human cancer cell lines including multidrug-resistant cell lines *in vitro* (Gomes et al. 2015).Currently, patient recruitment is in progress for phase II/III clinical trials for the treatment of solid tumors by plinabulin in combination with docetaxel/pegfilgrastim, a phase III trial for the treatment of advanced NSCLC in combination with plinabulin and docetaxel, and phase I/II trials for the treatment of metastatic NSCLC with plinabulin plus nivolumab.

3.4.13 E7974

E7974 is a synthetic analog of hemiasterlin which was originally isolated from the marine sponge *Hemiasterella minor* in 1990 (Khazir et al. 2014). Like the parent compound, it also inhibits tubulin polymerization and has demonstrated similar IC_{50} values as those of vinblastine. It causes G_2/M phase cell cycle arrest followed by cell apoptosis (Khazir et al. 2014). A report suggests that it targets α-tubulin than β-tubulin unlike most of the β-tubulin interactive agents such as taxanes and vinca alkaloids although a slight binding to β-subunit of tubulin was also observed. E7974 has demonstrated potency against cancer cells with mutations in β-tubulin gene, which are resistant to the treatment with taxanes (Kuznetsov et al. 2009). E7974 seems to be a promising drug candidate to treat solid tumors. As of now, E7974 has completed three phase I clinical trials for the treatment of solid tumor malignancies.

3.5 Conclusions and Future Prospects

A huge number of natural compounds and their derivatives with anticancer properties have been used in the clinical trials for cancer therapy. The current status of natural compounds in clinical trials and their natural sources was documented in accordance with NIH or the corresponding EU clinical trial database. From the information presented in this chapter, researchers or clinicians get the knowledge of the natural compounds, their anticancer properties, their mechanisms of action, and their clinical trial status so that they can select the compounds of their interest for further study or clinical use. The major advantage of using natural compounds in anticancer therapy is that they are safe and less toxic unlike synthetic compounds which are rather unsafe and toxic to normal cells also, in addition to cancer cells. Selected natural compounds with anticancer properties of plant, microbial, and marine origin have been explored. The most potent natural anticancer drugs include vinblastine, vincristine, and taxol from plants; dolastatin 10, cytarabine, and aplidine from marines; and bleomycin, doxorubicin, and dactinomycin from microorganisms, which have been used in cancer therapy. Plants have contributed numerous bioactive compounds which are clinically approved. Despite their contribution, for more than a decade, screening for bioactive lead molecules in plants was considerably reduced by the leading pharmaceutical companies, which were instead in favor of combinatorial chemistry by which libraries of millions of compounds can be synthesized. Moreover, characterization of the lead molecules in plant extractions was rather a time-consuming and laborious process. However, emerging biotechnology companies have taken up the drug discovery programs for identifying the lead molecules from the natural sources and developing clinically relevant drug molecules. Many of these natural compounds have been undergoing clinical trials. Previously, identification of the lead molecules used to take several months and, in some cases, years. But the remarkable progress made in the instrumentation and technology such as HPLC-MS (high-performance liquid chromatography-mass spectrometry), LC-MS (liquid chromatography-mass spectrometry), NMR (nuclear magnetic resonance spectrometry), and LC-SPE-NMR (liquid chromatography-solid-phase extraction (SPE)-nuclear magnetic resonance-mass spectrometry) coupled with high-throughput technology and also capillary NMR enables the rapid screening, identification, and characterization of the molecules which are available in the trace amounts in their source organisms. In the future, NMR coupled with HPLC and SPE techniques performed with high-throughput screening will be routinely used in anticancer drug discovery. Biotechnology and genetic engineering play a major role in bulk production of the lead molecules from microbes and marine sources with potent anticancer properties which are much needed to meet the increasing demand. Plant cell fermentation is another technology by which the plant-derived compounds can be mass produced, and the same technology has been employed to produce paclitaxel, one of the most successful anticancer drugs, which yields larger quantities of the desired compounds. Plant tissue culture/cell culture can also be used to produce the plant-derived bioactive compounds in bulk. Nature offers the scope for discovery of novel compounds from natural sources with new mechanisms of action, which is simpler now with the aid of advances made in the realm of drug discovery research. Discovery of novel anticancer compounds especially from natural origin has a constant demand as cancer poses new challenges such as detrimental adverse effects by existing synthetic drugs, development of multiple drug resistance, and emergence of new cancer variants. Future challenges in anticancer drug discovery from natural sources focus mainly on recompensing the shortcomings of the existing anticancer chemotherapeutic drugs with the natural compounds and their derivatives with the better understanding of their mechanisms of action. The best possible drug-target combinations are based on the understanding of the cancer-specific environment of mutated oncogenes, tumor suppressor genes, and their regulatory pathways. It would be ideal to design the targeted cancer therapeutic regimens with the natural compounds and their derivatives with relatively lower toxicity profiles in combination or as monotherapeutic agents. The clinical effectiveness of combined drugs has not been robustly demonstrated.

Therefore additional research using *in vivo* systems and better clinical trials is needed to determine the safety and clinical usefulness of nature-based drugs. The future success of anticancer drug discovery from natural source depends on the combinatorial chemistry based on the natural lead compounds and high-throughput drug screening technology. By cataloging the natural anticancer compounds in the clinical trials, we have attempted to create knowledge-based awareness to the scientific community for further exploring the natural sources to discover new lead anticancer molecules and, also, to clinically use existing molecules in cancer therapy.

References

Anitha Sri S (2016) Pharmacological activity of vinca alkaloids. J Pharmacogn Phytochem 4:27–34

Arastu-Kapur S, Anderl JL, Kraus M, Parlati F, Shenk KD, Lee SJ, Muchamuel T, Bennett MK, Driessen C, Ball AJ, Kirk CJ (2011) Nonproteasomal targets of the proteasome inhibitors bortezomib and carfilzomib: a link to clinical adverse events. Clin Cancer Res 17:2734–2743

Arora S, Gonzalez AF, Solanki K (2013) Combretastatin A-4 and its analogs in cancer therapy. Int J Pharm Sci Rev Res 22:168–174

Badgujar VB, Ansari MT, Abdullah MS (2015) Homoharringtonine: a nascent phytochemical for cancer treatment. World J Pharm Pharm Sci 4:1380–1391

Baghel SS, Shrivastava N, Baghel RS (2016) A review of quercetin: antioxidant and anticancer properties. World J Pharm Pharm Sci 1:146–160

Ballou LM, Lin RZ (2008) Rapamycin and mTOR kinase inhibitors. J Chem Biol 1:27–36

Bayala B, Bassole IH, Scifo R, Gnoula C, Morel L, Lobaccaro JMA, Simpore J (2014) Anticancer activity of essential oils and their chemical components-a review. Am J Cancer Res 4:591–607

Behrangi N, Hashemi M, Borna H (2012) Microtubules and tubulins as target for some natural anticancer agents extracted from marines, bacteruim, and fungus. Adv Stud Biol 4:1–9

Bennouna J, Delord JP, Campone M, Nguyen L (2008) Vinflunine: a new microtubule inhibitor agent. Clin Cancer Res 14:1625–1632

Bhanot A, Sharma R, Noolvi MN, Bhanot SR, Noolvi MN (2011) Natural sources as potential anticancer agents: a review. Int J Phytomed 3:9–26

Bhatnagar I, Kim SK (2010) Marine antitumor drugs: status, shortfalls and strategies. Mar Drugs 8:2702–2720

Bhattacharya B, Mukherjee S (2015) Cancer therapy using antibiotics. J Cancer Ther 6:849–858

Bhuvaneswari V, Nagini S (2005) Lycopene: a review of its potential as an anticancer agent. Curr Med Chem 5:627–635

Biggar RJ, Wohlfahrt J, Melbye M (2012) Digoxin use and the risk of cancers of the corpus uteri, ovary and cervix. Int J Cancer 131:716–721

Biggar RJ, Andersen EW, Kroman N, Wohlfahrt J, Melbye M (2013) Breast cancer in women using digoxin: tumor characteristics and relapse risk. Breast Cancer Res 15:R13. https://doi.org/10.1186/bcr3386

Blasiak J, Sikora A, Wozniak K, Drzewoski J (2004) Genotoxicity of streptozotocin in normal and cancer cells and its modulation by free radical scavengers. Cell Biol Toxicol 20:83–96

Brunello A, Roma A, Basso U, Jirillo A (2010) A review of Vinorelbine in the treatment of advanced breast cancer. Clin Med Ther 1:1715–1726

Butler MS, Robertson AAB, Cooper MA (2014) Natural product and natural product derived drugs in clinical trials. Nat Prod Rep 31:1612–1661

Calderón-Montaño JM, Burgos-Morón E, Orta ML, Maldonado-Navas D, García-Domínguez I, López-Lázaro M (2014) Evaluating the cancer therapeutic potential of cardiac glycosides. Biomed Res Int 2014:794930. https://doi.org/10.1155/2014/794930

Cao W, Liu Y, Zhang R, Zhang B, Wang T, Zhu X, Mei L, Chen H, Zhang H, Ming P, Huang L (2015) Homoharringtonine induces apoptosis and inhibits STAT3 via IL-6/JAK1/STAT3 signal pathway in Gefitinib-resistant lung cancer cells. Sci Rep 5:8477. https://doi.org/10.1038/srep08477

Chen J, Song Y, Zhang L (2013) Lycopene/tomato consumption and the risk of prostate cancer: a systematic review and meta-analysis of prospective studies. J Nutr Sci Vitaminol 59:213–223

Cheung-Ong K, Giaever G, Nislow C (2013) DNA-damaging agents in cancer chemotherapy: serendipity and chemical biology. Chem Biol 20:648–659

Chinta G, Syed SB, MSC C, Periyasamy L (2015) Piperine: a comprehensive review of pre-clinical and clinical investigations. Curr Bioact Compd 11:156–169

Choi JY, Hong WG, Cho JH, Kim EM, Kim J, Jung CH, Hwang SG, Um HD, Park JK (2015) Podophyllotoxin acetate triggers anticancer effects against non-small cell lung cancer cells by promoting cell death via cell cycle arrest, ER stress and autophagy. Int J Oncol 47:1257–1265

Chung KS, Yim N-H, Lee SH, Choi SJ, Hur KS, Hoe KL, Kim DU, Goehle S, Kim HB, Song KB, Yoo HS, Bae KH, Simon J, Won M (2008) Identification of small molecules inducing apoptosis by cell-based assay using fission yeast deletion mutants. Investig New Drugs 26:299–307

Cragg GM, Newman DJ (2005) Plants as a source of anti-cancer agents. J Ethnopharmacol 100:72–79

Cragg GM, Newman DJ (2013) Natural products: a continuing source of novel drug leads. Biochim Biophys Acta 1830:3670–3695

Cragg GM, Grothaus PG, Newman DJ (2009) Impact of natural products on developing new anticancer agents. Chem Rev 109:3012–3043

Crawford LJ, Walker B, Irvine AE (2011) Proteasome inhibitors in cancer therapy. J Cell Commun Signal 5:101–110

Cushnie TPT, Cushnie B, Lamb AJ (2014) Alkaloids: an overview of their antibacterial, antibiotic-enhancing and antivirulence activities. Int J Antimicrob Agents 44:377–386

D'Incalci M, Galmarini CM (2010) A review of trabectedin (ET-743): a unique mechanism of action. Mol Cancer Ther 9:2157–2163

Dai J, Mumper RJ (2010) Plant phenolics: extraction, analysis and their antioxidant and anticancer properties. Molecules 15:7313–7352

Damelin M, Bankovich A, Park A, Aguilar J, Anderson W, Santaguida M, Aujay M, Fong S, Khandke K, Pulito V, Ernstoff E, Escarpe P, Bernstein J, Pysz M, Zhong W, Upeslacis E, Lucas JJ, Lucas JJ, Nichols T, Loving K, Foord O, Hampl J, Stull R, Barletta F, Falahatpisheh H, Sapra P, Gerber HP, Dylla SJ (2015) Anti-EFNA4 calicheamicin conjugates effectively target triple-negative breast and ovarian tumor-initiating cells to result in sustained tumor regressions. Clin Cancer Res 21:4165–4173

Del Rio D, Stalmach A, Calani L, Crozier A (2010) Bioavailability of coffee chlorogenic acids and green tea flavan-3-ols. Forum Nutr 2:820–833

Demain AL, Vaishnav P (2011) Natural products for cancer chemotherapy. Microb Biotechnol 4:687–699

Dixit RB, Suseela MR (2013) Cyanobacteria: potential candidates for drug discovery. Antonie Van Leeuwenhoek 103:947–961

Dizon DS, Krilov L, Cohen E, Gangadhar T, Ganz PA, Hensing TA, Hunger S, Krishnamurthi SS, Lassman AB, Markham MJ, Mayer E, Neuss M, Pal SK, Richardson LC, Schilsky R, Schwartz GK, Spriggs DR, Villalona-Calero MA, Villani G, Masters G (2016) Clinical cancer advances 2016: annual report on progress against cancer from the American Society of Clinical Oncology. J Clin Oncol 34:987–1011

Efferth T, Fu YJ, Zu YG, Schwarz G, Konkimalla VSB, Wink M (2007) Molecular target-guided tumor therapy with natural products derived from traditional Chinese medicine. Curr Med Chem 14:2024–2032

Escuin D, Kline ER, Giannakakou P (2005) Both microtubule-stabilizing and microtubule-destabilizing drugs inhibit hypoxia-inducible factor-1. Accumulation and activity by disrupting microtubule function. Cancer Res 65:9021–9028

Fanale D, Bronte G, Passiglia F, Calo V, Castiglia M, Di Piazza F, Barraco N, Cangemi A, Catarella MT, Insalaco L, Listi A, Maragliano R, Massihnia D, Perez A, Toia F, Cicero G, Bazan V (2015) Stabilizing versus destabilizing the microtubules: a double-edge sword for an effective cancer treatment option. Anal Cell Pathol 2015:690916. https://doi.org/10.1155/2015/690916
Fauzee NJS, Dong Z, Wang YL (2011) Taxanes: promising anti-cancer drugs. Asian Pac J Cancer Prev 12:837–851
Ferrandina G, Mariani M, Andreoli M, Shahabi S, Scambia G, Ferlini C (2012) Novel drugs targeting microtubules: the role of epothilones. Curr Pharm Des 18:2793–2803
Florento L, Matias R, Tuaño E, Santiago K, Dela Cruz F, Tuazon A (2012) Comparison of cytotoxic activity of anticancer drugs against various human tumor cell lines using *in vitro* cell-based approach. Int J Biomed Sci 8:76–80
Foss F, Horwitz S, Pro B, Prince HM, Sokol L, Balser B, Wolfson J, Coiffier B (2016) Romidepsin for the treatment of relapsed/refractory peripheral T cell lymphoma: prolonged stable disease provides clinical benefits for patients in the pivotal trial. J Hematol Oncol 9:22. https://doi.org/10.1186/s13045-016-0243-8
Fukuyo Y, Hunt CR, Horikoshi N (2010) Geldanamycin and its anti-cancer activities. Cancer Lett 290:24–35
Ganguly A, Yang H, Cabral F (2010) Paclitaxel-dependent cell lines reveal a novel drug activity. Mol Cancer Ther 9:2914–2923
Gibellini L, Pinti M, Nasi M, Montagna JP, De Biasi S, Roat E, Bertoncelli L, Cooper EL, Cossarizza A (2011) Quercetin and cancer chemoprevention. Evid Based Compl Altern Med 2011:591356. https://doi.org/10.1093/ecam/neq053
Giddings LA, Newman DJ (2013) Microbial natural products: molecular blueprints for antitumor drugs. J Ind Microbiol Biotechnol 40:1181–1210
Gomes NGM, Lefranc F, Kijjoa A, Kiss R (2015) Can some marine-derived fungal metabolites become actual anticancer agents. Mar Drugs 13:3950–3991
Gordaliza M, Castro MA, del Corral JM, Feliciano AS (2000) Antitumor properties of podophyllotoxin and related compounds. Curr Pharm Des 6:1811–1839
Greenwell M, Rahman PKSM (2015) Medicinal plants: their use in anticancer treatment. Int J Pharm Sci Res 6:4103–4112
Harvey AL (2008) Natural products in drug discovery. Drug Discov Today 13:894–901
Hecht SM (2000) Bleomycin: new perspectives on the mechanism of action. J Nat Prod 63:158–168
Hoelder S, Clarke PA, Workman P (2012) Discovery of small molecule cancer drugs: successes, challenges and opportunities. Mol Oncol 6:155–176
Huang M, Lu JJ, Huang MQ, Bao JL, Chen XP, Wang YT (2012) Terpenoids: natural products for cancer therapy. Expert Opin Investig Drugs 21:1801–1818
Hussain MS, Fareed S, Ansari S, Khan MS (2012) Marine natural products: a lead for anti-cancer. Indian J Mar Sci 41:891–903
Imperatore C, Aiello A, D'Aniello F, Senese M, Menna M (2014) Alkaloids from marine invertebrates as important leads for anticancer drugs discovery and development. Molecules 19:20391–20423
Irie K, Yanagita RC, Nakagawa Y (2012) Challenges to the development of bryostatin-type anticancer drugs based on the activation mechanism of protein kinase Cδ. Med Res Rev 32:518–535
Jimeno J, Faircloth G, Sousa-Faro JF, Scheuer P, Rinehart K (2004) New marine derived anticancer therapeutics – a journey from the sea to clinical trials. Mar Drugs 2:14–29
Jordan MA, Wilson L (2004) Microtubules as a target for anticancer drugs. Nat Rev Cancer 4:253–265
Jordan M, Toso RJ, Thrower D, Wilson L (1993) Mechanism of mitotic block and inhibition of cell proliferation by taxol at low concentrations. Proc Natl Acad Sci U S A 90:9552–9556
Kathiravan MK, Khilare MM, Nikoomanesh K, Chothe AS, Jain KS (2013) Topoisomerase as target for antibacterial and anticancer drug discovery. J Enzyme Inhib Med Chem 28:419–435
Kaur R, Arora S (2015) Alkaloids-important therapeutic secondary metabolites of plant origin. J Crit Rev 2:1–8

Kepp O, Menger L, Vacchelli E, Adjemian S, Martins I, Ma Y, Sukkurwala AQ, Michaud M, Galluzzi L, Zitvogel L, Kroemer G (2012) Anticancer activity of cardiac glycosides: at the frontier between cell-autonomous and immunological effects. Oncoimmunology 1:1640–1642
Khazir J, Riley DL, Pilcher LA, De-Maayer P, Mir BA (2014) Anticancer agents from diverse natural sources. Nat Prod Commun 9:1655–1669
Kinghorn AD (2015) Review of anticancer agents from natural products. J Nat Prod 78:2315–2315
Kinghorn AD, Carcache De Blanco EJ, Lucas DM, Rakotondraibe HL, Orjala J, Soejarto DD, Oberlies NH, Pearce CJ, Wani MC, Stockwell BR, Burdette JE, Swanson SM, Fuchs JR, Phelps MA, Xu L, Zhang X, Shen YY (2016) Discovery of anticancer agents of diverse natural origin. Anticancer Res 36:5623–5637
Kumar S (2015) Drug targets for cancer treatment: an overview. Med Chem 5:115–123
Kuznetsov G, TenDyke K, Towle MJ, Cheng H, Liu J, Marsh JP, Schiller SER, Spyvee MR, Yang H, Seletsky BM, Shaffer CJ, Marceau V, Yao Y, Suh EM, Campagna S, Fang FG, Kowalczyk JJ, Littlefield BA (2009) Tubulin-based antimitotic mechanism of E7974, a novel analogue of the marine sponge natural product hemiasterlin. Mol Cancer Ther 8:2852–2860
Lee W, St.Onge RP, Proctor M, Flaherty P, Jordan MI, Arkin AP, Davis RW, Nislow C, Giaever G (2005) Genome-wide requirements for resistance to functionally distinct DNA-damaging agents. PLoS Genet 1:e24. https://doi.org/10.1371/journal.pgen.0010024
Lee TJ, Jung EM, Lee JT, Kim S, Park JW, Choi KS, Kwon TK (2006) Mithramycin A sensitizes cancer cells to TRAIL-mediated apoptosis by down-regulation of XIAP gene promoter through Sp1 sites. Mol Cancer Ther 5:2737–2746
Li Y, Zhang T, Schwartz SJ, Sun D (2009) New developments in Hsp90 inhibitors as anti-cancer therapeutics: mechanisms, clinical perspective and more potential. Drug Resist Updat 12:17–27
Li J, Kim SG, Blenis J (2014a) Rapamycin: one drug, many effects. Cell Metab 19:373–379
Li Y, Lu X, Qi H, Li X, Xiao X, Gao J (2014b) Ursolic acid induces apoptosis through mitochondrial intrinsic pathway and suppression of ERK1/2 MAPK in HeLa cells. J Pharmacol Sci 125:202–210
Li K, Chung-Davidson YW, Bussy U, Li W (2015) Recent advances and applications of experimental technologies in marine natural product research. Mar Drugs 13:2694–2713
Lin RK, Hsu CH, Wang YC (2007) Mithramycin A inhibits DNA methyltransferase and metastasis potential of lung cancer cells. Anti-Cancer Drugs 18:1157–1164
Lippi G, Targher G (2011) Tomatoes, lycopene-containing foods and cancer risk. Br J Cancer 104:1234–1235
Lo YS, Tseng WH, Chuang CY, Hou MH (2013) The structural basis of actinomycin D-binding induces nucleotide flipping out, a sharp bend and a left-handed twist in CGG triplet repeats. Nucleic Acids Res 41:4284–4294
Lopez-Lazaro M (2009) Digoxin, HIF-1, and cancer. Proc Natl Acad Sci U S A 106:E26–E27
Lü S, Wang J (2014) Homoharringtonine and omacetaxine for myeloid hematological malignancies. J Hematol Oncol 7:2. https://doi.org/10.1186/1756-8722-7-2
Lu JJ, Bao JL, Chen XP, Huang M, Wang YT (2012) Alkaloids isolated from natural herbs as the anticancer agents. Evid Based Compl Altern Med 2012:485042. https://doi.org/10.1155/2012/485042
Luthria DL, Mukhopadhyay S (2006) Influence of sample preparation on assay of phenolic acids from eggplant. J Agric Food Chem 54:41–47
Mans DRA (2016) Exploring the global animal biodiversity in the search for new drugs-marine invertebrates exploring the global animal biodiversity in the search for new drugs – marine invertebrates. J Transl Sci 2:170–179
Mariaule G, Belmont P (2014) Cyclin-dependent kinase inhibitors as marketed anticancer drugs: where are we now. A short survey. Molecules 19:14366–14382
Martin DM (2003) Induction of volatile terpene biosynthesis and diurnal emission by methyl jasmonate in foliage of Norway spruce. Plant Physiol 132:1586–1599
Mastron JK, Siveen KS, Sethi G, Bishayee A (2015) Silymarin and hepatocellular carcinoma: a systematic, comprehensive, and critical review. Anti-Cancer Drugs 26:475–486

McKinney JS, Sethi S, Tripp JD, Nguyen TN, Sanderson BA, Westmoreland JW, Resnick MA, Lewis LK (2013) A multistep genomic screen identifies new genes required for repair of DNA double-strand breaks in Saccharomyces cerevisiae. BMC Genomics 14:251

Mehta HJ, Patel V, Sadikot RT (2014) Curcumin and lung cancer-a review. Target Oncol 9:295–310

Melorose J, Perroy R, Careas S, Martin LP, Hamilton TC, Schilder RJ, Helleday T, Petermann E, Lundin C, Hodgson B, Sharma RA (2008) DNA repair pathways as targets for cancer therapy. Nat Rev Cancer 14:1291–1295

Miura S, Izuta S (2004) DNA polymerases as targets of anticancer nucleosides. Curr Drug Targets 5:191–195

Moore BS, Eustáquio AS, McGlinchey RP (2008) Advances in and applications of proteasome inhibitors. Curr Opin Chem Biol 12:434–440

Mottamal M, Zheng S, Huang T, Wang G (2015) Histone deacetylase inhibitors in clinical studies as templates for new anticancer agents. Molecules 20:3898–3941

Moudi M, Go R, Yien CYS, Nazre M (2013) Vinca alkaloids. Int J Prev Med 4:1131–1135

Mukhtar E, Adhami VM, Mukhtar H (2014) Targeting microtubules by natural agents for cancer therapy. Mol Cancer Ther 13:275–284

Munoz-Alonso M, Gonza L, Zarich N, Martı T, Alvarez E (2008) Plitidepsin has a dual effect inhibiting cell cycle and inducing apoptosis via Rac1/c-Jun NH2-terminal kinase activation in human melanoma cells. J Pharmacol Exp Ther 324:1093–1101

Neckers L (2002) Hsp90 inhibitors as novel cancer chemotherapeutic agents. Trends Mol Med 8:55–61

Newman DJ, Cragg GM (2007) Natural products as sources of new drugs over the last 25 years. J Nat Prod 70:461–477

Newman DJ, Cragg GM (2014) Marine-sourced anti-cancer and cancer pain control agents in clinical and late preclinical development. Mar Drugs 12:255–278

Newman DJ, Cragg GM (2016a) Natural products as sources of new drugs from 1981 to 2014. J Nat Prod 79:629–661

Newman DJ, Cragg GM (2016b) Drugs and drug candidates from marine sources: an assessment of the current "State of play". Planta Med 82:775–789

Newman DJ, Cragg GM, Snader KM (2003) Natural products as sources of new drugs over the period 1981–2002. J Nat Prod 66:1022–1037

Nicolaou KC, Smitht AL, Yue EW (1993) Review chemistry and biology of natural and designed enediynes. Proc Natl Acad Sci 90:5881–5888

Nijwening JH, Kuiken HJ, Beijersbergen RL (2011) Screening for modulators of cisplatin sensitivity: unbiased screens reveal common themes. Cell Cycle 10:380–386

Orlowski RZ, Kuhn DJ (2008) Proteasome inhibitors in cancer therapy: lessons from the first decade. Clin Cancer Res 14:1649–1657

Ozcan T, Akpinar-Bayizit A, Yilmaz-Ersan L, Delikanli B (2014) Phenolics in human health. Int J Chem Eng Appl 5:393–396

Pangestuti R, Kim SK (2017) Bioactive peptide of marine origin for the prevention and treatment of non-communicable diseases. Mar Drugs 15:1–23

Pavese JM, Farmer RL, Bergan RC (2010) Inhibition of cancer cell invasion and metastasis by genistein. Cancer Metastasis Rev 29:465–482

Petek BJ, Jones RL (2014) PM00104 (Zalypsis®): a marine derived alkylating agent. Molecules 19:12328–12335

Pichersky E (2006) Biosynthesis of plant volatiles: nature's diversity and ingenuity. Science 311:808–811

Pommier Y (2006) Topoisomerase I inhibitors: camptothecins and beyond. Nat Rev Cancer 6:789–802

Prasad S, Yadav VR, Sung B, Reuter S, Kannappan R, Deorukhkar A, Diagaradjane P, Wei C, Baladandayuthapani V, Krishnan S, Guha S, Aggarwal BB (2012) Ursolic acid inhibits growth and metastasis of human colorectal cancer in an orthotopic nude mouse model by targeting multiple cell signaling pathways: chemosensitization with capecitabine. Clin Cancer Res 18:4942–4953

Priyadarshini K, Keerthi A (2012) Paclitaxel against cancer: a short review. Med Chem 2:142–146
Priyankashukla XX (2014) Marine natural products as anticancer agents. J Pharm Biol Sci 9:60–64
Rahib L, Smith BD, Aizenberg R, Rosenzweig AB, Fleshman JM, Matrisian LM (2014) Projecting cancer incidence and deaths to 2030: the unexpected burden of thyroid, liver, and pancreas cancers in the united states. Cancer Res 74:2913–2921
Rahmani AH, Al Zohairy MA, Aly SM, Khan MA (2014) Curcumin: a potential candidate in prevention of cancer via modulation of molecular pathways. Biomed Res Int 2014:761608. https://doi.org/10.1155/2014/761608
Raju U, Nakata E, Mason KA, Ang KK, Milas L (2003) Flavopiridol, a cyclin-dependent kinase inhibitor, enhances radiosensitivity of ovarian carcinoma cells. Cancer Res 63:3263–3267
Ramasamy K, Agarwal R (2008) Multitargeted therapy of cancer by silymarin. Cancer Lett 269:352–362
Rangel M, Falkenberg M (2015) An overview of the marine natural products in clinical trials and on the market. J Coast Life Med 3:421–428
Ray A, Okouneva T, Manna T, Miller HP, Schmid S, Arthaud L, Luduena R, Jordan MA, Wilson L (2007) Mechanism of action of the microtubule-targeted antimitotic depsipeptide tasidotin (formerly ILX651) and its major metabolite tasidotin C-carboxylate. Cancer Res 67:3767–3776
Raza A, Aslam B, Naseer MU, Ali A, Majeed W, Hassan SU (2015) Antitumor activity of berberine against breast cancer: a review. Int Res J Pharm 6:81–85
Reichenbach H, Höfle G (2008) Discovery and development of the epothilones: a novel class of antineoplastic drugs. Drugs R D 9:1–10
Reusser F (1971) Mode of action of streptozotocin. J Bacteriol 105:580–588
Rida PCG, Livecche D, Ogden A, Zhou J, Aneja R (2015) The noscapine chronicle: a pharmaco-historic biography of the opiate alkaloid family and its clinical applications. Med Res Rev 35:1072–1096
Robert J (2007) Preclinical assessment of anthracycline cardiotoxicity in laboratory animals: predictiveness and pitfalls. Cell Biol Toxicol 23:27–37
Rocha LD, Monteiro MC, Teodoro AJ (2012) Anticancer properties of hydroxycinnamic acids-a review. Cancer Clin Oncol 1:109–121
Romano M, Frapolli R, Zangarini M, Bello E, Porcu L, Galmarini CM, García-Fernández LF, Cuevas C, Allavena P, Erba E, D'Incalci M (2013) Comparison of *in vitro* and *in vivo* biological effects of trabectedin, lurbinectedin (PM01183) and Zalypsis® (PM00104). Int J Cancer 133:2024–2033
Russo M, Russo GL, Daglia M, Kasi PD, Ravi S, Nabavi SF, Nabavi SM (2016) Understanding genistein in cancer: the "good" and the "bad" effects: a review. Food Chem 196:589–600
Salem M, Rohani S, Gillies ER (2014) Curcumin, a promising anti-cancer therapeutic: a review of its chemical properties, bioactivity and approaches to cancer cell delivery. RSC Adv 4:10815
Schutz FA, Bellmunt J, Rosenberg JE, Choueiri TK (2011) Vinflunine: drug safety evaluation of this novel synthetic vinca alkaloid. Expert Opin Drug Saf 10:645–653
Schwartsmann G, da Rocha AB, Berlinck RG, Jimeno J (2001) Marine organisms as a source of new anticancer agents. Lancet Oncol 2:221–225
Scully OJ, Bay BH, Yip G, Yu Y (2012) Breast cancer metastasis. Cancer Genomics Proteomics 9:311–320
Seto B (2012) Rapamycin and mTOR: a serendipitous discovery and implications for breast cancer. Clin Transl Med 1:29. https://doi.org/10.1186/2001-1326-1-29
Shan J, Xuan Y, Ruan S, Sun M (2011) Proliferation-inhibiting and apoptosis-inducing effects of ursolic acid and oleanolic acid on multi-drug resistance cancer cells *in vitro*. Chin J Integr Med 17:607–611
Shapiro RS (2015) Antimicrobial induced DNA damage and genomic instability in microbial pathogens. PLoS Pathog 11:e1004678. https://doi.org/10.1371/journal.ppat.1004678
Shen B, Hindra YX, Huang T, Ge H, Yang D, Teng Q, Rudolf JD, Lohman JR (2015) Enediynes: exploration of microbial genomics to discover new anticancer drug leads. Bioorg Med Chem Lett 25:9–15

Siegel RL, Miller KDJA (2017) Cancer statistics, 2017. CA Cancer J Clin 67:7–30
Siemann DW, Chaplin DJ, Walicke PA (2009) A review and update of the current status of the vasculature-disabling agent combretastatin-A4 phosphate (CA4P). Expert Opin Investig Drugs 18:189–197
Simmons TL, Andrianasolo E, Mcphail K, Flatt P, Gerwick WH (2005) Minireview marine natural products as anticancer drugs. Mol Cancer Ther 4:333–342
Simon J, Szankasi P, Nguyen DK, Ludlow C, Dunstan HM, Roberts CJ, Jensen EL, Hartwell LH, Friend SH (2000) Differential toxicities of anticancer agents among DNA repair and checkpoint mutants of *Saccharomyces cerevisiae* differential toxicities of anticancer agents among DNA repair and checkpoint mutants of *Saccharomyces cerevisiae* 1. Cancer Res 60:328–333
Singh A, Duggal S (2009) Piperine-review of advances in pharmacology. Int J Pharm Sci Nanotechnol 2:615–620
Singh B, Sharma RA (2015) Plant terpenes: defense responses, phylogenetic analysis, regulation and clinical applications. 3 Biotech 5:129–151
Singh R, Sharma M, Joshi P, Rawat DS (2008) Clinical status of anti-cancer agents derived from marine sources. Anti Cancer Agents Med Chem 8:603–617
Sobral MV, Xavier AL, Lima TC, De Sousa DP (2014) Antitumor activity of monoterpenes found in essential oils. Sci World J 2014:953451. https://doi.org/10.1155/2014/953451
Souto AL, Tavares JF, Da Silva MS, De Diniz MFFM, De Athayde-Filho PF, Barbosa Filho JM (2011) Anti-inflammatory activity of alkaloids: an update from 2000 to 2010. Molecules 16:8515–8534
Spagnuolu C, Russo GL, Orhan Llkay E, Habtemariam S, Daglia M, Sureda A, Nabavi SF, Devi KP, Loizzo MR, Tundis R, Nabavi SM (2015) Genistein and cancer: current status, challenges, and future directions. Adv Nutr An Int Rev J 6:408–419
Srivastava S, Somasagara RR, Hegde M, Nishana M, Tadi SK, Srivastava M, Choudhary B, Raghavan SC (2016) Quercetin, a natural flavonoid interacts with DNA, arrests cell cycle and causes tumor regression by activating mitochondrial pathway of apoptosis. Sci Rep 6:24049. https://doi.org/10.1038/srep24049
Stanton RA, Gernert KM, Nettles JH, Aneja R (2011) Drugs that target dynamic microtubules: a new molecular perspective. Med Res Rev 31:443–481
Sun J (2007) D-limonene: safety and clinical applications. Altern Med Rev 12:259–264
Sun Y, Xun K, Wang Y, Chen X (2009) A systematic review of the anticancer properties of berberine, a natural product from Chinese herbs. Anti-Cancer Drugs 20:757–769
Takebayashi Y, Pourquier P, Zimonjic DB, Nakayama K, Emmert S, Ueda T, Urasaki Y, Kanzaki A, Akiyama S, Popescu N, Kraemer KH, Pommier Y (2001) Antiproliferative activity of ecteinascidin 743 is dependent upon transcription-coupled nucleotide excision repair. Nat Med 7:961–966
Tan AR, Swain SM (2002) Review of flavopiridol, a cyclin-dependent kinase inhibitor, as breast cancer therapy. Semin Oncol 29:77–85
Thakur NL, Thakur AN, Müller WEG (2005) Marine natural products in drug discovery. Nat Prod Radiance 4:471–477
Tiong SH, Looi CY, Hazni H, Arya A, Paydar M, Wong WF, Cheah SC, Mustafa MR, Awang K (2013) Antidiabetic and antioxidant properties of alkaloids from *Catharanthus roseus* (L.) G. Don. Molecules 18:9770–9784
Tomasz M (1995) Mitomycin C: small, fast and deadly (but very selective). Chem Biol 2:575–579
Towle MJ, Salvato KA, Budrow J, Wels BF, Kuznetsov G, Aalfs KK, Welsh S, Zheng W, Seletsky BM, Palme MH, Habgood GJ, Singer LA, Dipietro LV, Wang Y, Chen JJ, Quincy DA, Davis A, Yoshimatsu K, Kishi Y, Yu MJ, Littlefield BA (2001) *In vitro* and *in vivo* anticancer activities of synthetic macrocyclic ketone analogues of halichondrin B. Cancer Res 61:1013–1021
Tu HY, Huang AM, Wei BL, Gan KH, Hour TC, Yang SC, Pu YS, Lin CN (2009) Ursolic acid derivatives induce cell cycle arrest and apoptosis in NTUB1 cells associated with reactive oxygen species. Bioorg Med Chem 17:7265–7274

Vasavirama K, Upender M (2014) Piperine: a valuable alkaloid from piper species. Int J Pharm Pharm Sci 6:34–38
Venditto VJ, Simanek EE (2010) Cancer therapies utilizing the camptothecins: a review of the *in vivo* literature. Mol Pharm 7:307–349
Wahid M, Bano Q (2014) Structure activity relationship and clinical development perspective of analogs. J Appl Pharmacol 6:286–295
Waksman SA, Woodruff HB (1941) Actinomyces antibioticus, a new soil organism antagonistic to pathogenic and non-pathogenic bacteria 1. J Bacteriol 42:231–249
Wang J, Ren T, Xi T (2012) Ursolic acid induces apoptosis by suppressing the expression of FoxM1 in MCF-7 human breast cancer cells. Med Oncol 29:10–15
Wang XH, Zhou SY, Qian ZZ, Zhang HL, Qiu LH, Song Z, Zhao J, Wang P, Hao XS, Wang HQ (2013) Evaluation of toxicity and single-dose pharmacokinetics of intravenous ursolic acid liposomes in healthy adult volunteers and patients with advanced solid tumors. Expert Opin Drug Metab Toxicol 9:117–125
Wei MY, Giovannucci EL (2012) Lycopene, tomato products, and prostate cancer incidence: a review and reassessment in the PSA screening era. J Oncol 2012:271063. https://doi.org/10.1155/2012/271063
Wellington KW (2015) Understanding cancer and the anticancer activities of naphthoquinones – a review. RSC Adv 5:20309–20338
Wilken R, Veena MS, Wang MB, Srivatsan ES (2011) Curcumin: a review of anti-cancer properties and therapeutic activity in head and neck squamous cell carcinoma. Mol Cancer 10:12. https://doi.org/10.1186/1476-4598-10-12
Wondrak GT (2009) Redox-directed cancer therapeutics: molecular mechanisms and opportunities. Antioxid Redox Signal 11:3013–3069
Wozniak L, Skapska S, Marszalek K (2015) Ursolic acid – a pentacyclic triterpenoid with a wide spectrum of pharmacological activities. Molecules 20:20614–20641
Xu WS, Parmigiani RB, Marks PA (2007) Histone deacetylase inhibitors: molecular mechanisms of action. Oncogene 26:5541–5552
Yan Z, Zhu Z, Qian Z, Zhao C, Wang H, Ying G (2013) A phase I pharmacokinetic study of ursolic acid nanoliposomes in healthy volunteers and patients with advanced solid tumors. Int J Nanomedicine 8:129–136
Zhen J, Villani TS, Guo Y, Qi Y, Chin K, Pan MH, Ho CT, Simon JE, Wu Q (2016) Phytochemistry, antioxidant capacity, total phenolic content and anti-inflammatory activity of *Hibiscus sabdariffa* leaves. Food Chem 190:673–680
Zhou J, Giannakakou P (2005) Targeting microtubules for cancer chemotherapy. Curr Med Chem Anticancer Agents 5:65–71
Ziegler J, Facchini PJ (2008) Alkaloid biosynthesis: metabolism and trafficking. Annu Rev Plant Biol 59:735–769
Zwergel C, Valente S, Mai A (2015) DNA methyltransferases inhibitors from natural sources. Curr Top Med Chem 16:680–696

Chapter 4
Insight Approaches of Medicinal Plants for the Discovery of Anticancer Drugs

Diana Cláudia Gouveia Alves Pinto, Ana Maria Loureiro Seca, and Artur Manuel Soares Silva

Abstract Cancer is one of the most serious illnesses of our civilization. The International Agency for Research on Cancer estimated that 14.1 million new cancer cases were diagnosed and that 8.2 million patients died from this disease worldwide in 2012 alone. Therefore, there is an urgent need for novel anticancer agents as well as new treatment strategies. Natural products have been valuable sources of new therapeutic agents. Thus, the aim of present chapter is to provide an overview of the anticancer natural compounds highlighting the successful cases with clinical application such as Taxol, vincristine and omacetaxine mepesuccinate; those involved in clinical trials such as parthenolide, betulinic acid, ingenol mebutate and curcumin; and the ones with high degree of activity and safety profile such as resveratrol, ursolic acid, tetrandrine and triptolide. For each one of these compounds, it will be presented and discussed its natural origin, cell target, mechanism of action, pharmacologic aspects as well as the structural modifications that improve its anticancer properties in order to summarize the recent medicinal advances of natural anticancer compounds and their analogues. All these aspects aim to draw our readers' attention to the plants and/or their secondary metabolites therapeutic potential and in consequence value the plants' role in anticancer drug discovery.

Keywords Cancer · Clinical trials · Natural compounds · Medicinal plants · Secondary metabolites

---

D. C. G. A. Pinto (✉) · A. M. S. Silva
Department of Chemistry and QOPNA, University of Aveiro, Aveiro, Portugal
e-mail: diana@ua.pt

A. M. L. Seca
Department of Chemistry and QOPNA, University of Aveiro, Aveiro, Portugal

Centre for Ecology, Evolution and Environmental Changes/Azorean Biodiversity Group & University of Azores, Ponta Delgada, Portugal

M. S. Akhtar, M. K. Swamy (eds.), *Anticancer Plants: Clinical Trials and Nanotechnology*, https://doi.org/10.1007/978-981-10-8216-0_4

4.1 Introduction

Cancer is a generic term for a large group of diseases that can affect any part of the body. In fact, the types and subtypes of cancer that can be found in a specific organ are more than 100 (Liu et al. 2016a). Cancer is a multistage process resulted into an uncontrolled and abrupt division of body cells and is considered one of the leading causes of mortality. The reported numbers of cases per year as well as the predictions for the near future are unthinkable (Gali-Muhtasib et al. 2015; Turek et al. 2016). The fact that tumour development involves physiological capabilities such as limitless replicative potential, tissue invasion and metastasis (Liu et al. 2016a), joined with difficult diagnosis, infers that the therapeutic options are limited and involve mainly the use of synthetic anticancer molecules. Moreover, the approval of these molecule uses involves several stages of security that enhance the treatment budget; they are toxic to normal cells, that is, their side effects are huge, and usually their mechanism of action is through cell apoptosis. To cure this deadly disease, various treatment strategies and therapeutic targets are being developed (Ogbourne et al. 2004; Turek et al. 2016), but still we are so far from seeing the light at the end of the tunnel.

According to WHO (2017), the economic impact of cancer is significantly increasing, and approximately 70% of deaths from cancer occur in low- and middle-income countries. These facts, together with the undesired effects of some synthetic molecules and the knowledge that traditional medicine is the primary health care in several countries of the world, are used as an alternative to cure or improve the health of the countrymen. Thus, many researchers are focused to investigate the potential of medicinal plants or their derived natural products as an anticancer agent (Dias et al. 2012; Katz and Baltz 2016; Fabricant and Farnsworth 2001). The data available from FDA showed that 40% of the approved molecules for drug administration are pure natural compounds or inspired on them (Kinch et al. 2014). And from those, nearly 74% are used in cancer chemotherapy (Bhanot et al. 2011; Giddings and Newman 2013). Natural products have garnered increasing attention in cancer chemotherapy because they are viewed more biologically friendly and consequently more co-evolved with their target sites and less toxic to normal cells (Mishra and Tiwari 2011). Moreover, there are evidences that natural product-derived anticancer drugs have alternative modes of promoting cell death (Gali-Muhtasib et al. 2015; Khalid et al. 2016). Considering the current botanical nomenclature and according to the database "The Plant List" (www.theplantlist.org), several plant species mentioned in this chapter have an alternative name. Therefore, the name cited in the original paper is accepted. Further, taxonomic corrections regarding the author or correct names were rectified following the same database recommendations. Thus, the aim of this chapter is to provide an overview of the recent and most successful cases of medicinal plants and their secondary metabolites proven their anticancer potential in various clinical trials.

4.2 Medicinal Plants with Antitumoural Activity

According to the World Health Organization, around 80% of the global population still relies on botanical drugs, and in several countries, the herbal medicines are used with some level of integration into the traditional health-care system (Allam et al. 2014; Aziato and Odai 2016), while, in Western countries, the herbal medicine use is growing but following tight legislation and under surveillance (Enioutina et al. 2017). Regarding the antitumoural activity, plants constitute a common alternative for cancer treatment in many countries around the world, and more than 3000 plants worldwide have been reported to have anticancer properties (Graham et al. 2000; Alonso-Castro et al. 2011; Ochwangi et al. 2014; Tariq et al. 2017). However, many of the claims for the efficacy of such treatment should be viewed with some critical spirit once cancer, as a disease with specific mechanisms, was not known or was poorly defined in traditional medicine. In fact, in traditional medicine, the cancer term has been applied to describe general conditions such as inflammations, ulcers or dermatological conditions such as "hard swellings", abscesses, calluses, corns, warts and polyps, whereas currently it corresponds to a specific disease entity. The traditional medicine knowledge in the treatment of cancer has promoted investigations into natural sources mainly plants and marine life, resulting in the discovery of lead structures that are useful templates for the pharmaceutical industry.

4.2.1 Medicinal Plants Used in Traditional Medicine to Treat Cancer Diseases

An excellent work was published by Tariq et al. (2017), where the utmost information on ethnomedical, pharmacological and phytochemical aspects of anticancer plants from different countries was reviewed. However, some examples of plants used in traditional medicine to treat cancer, which are not cited by Tariq et al. (2017), are presented in tabular form (Table 4.1). These medicinal plants were chosen on the basis of their anticancer, cytotoxic and antitumour properties, for instance, *Annona crassiflora* (Formagio et al. 2015), *Clinacanthus nutans* (Teoh et al. 2017; Zulkipli et al. 2017), *Solanum lyratum* (Lin et al. 2013; Chiu et al. 2015), *Solanum aculeastrum* (Omosa et al. 2016), *Triumfetta rhomboidea* (Sivakumar et al. 2010), *Sesbania sesban* (Dianhar et al. 2014) and *Prunus africana* (Komakech et al. 2017). In addition, isolation of the anticancer and cytotoxic bioactive phytoconstituents has been also performed in some species. For example, ingenol mebutate isolated from *Euphorbia peplus* (Millsop et al. 2013), polysaccharide-peptide complex (Huang et al. 2016) and lupeol (Teoh et al. 2017) from *Clinacanthus nutans*, lyratol G and 1β-hydroxy-1,2-dihydro-α-santonin from *Solanum lyratum* (Nie et al. 2014), tomatidine and solasodine from *S. aculeastrum* (Koduru et al. 2007), 3-hydroxy-4′,7-dimethoxyflavone from *S. sesban* (Dianhar et al. 2014) and amyrin and β-sitosterol-3-*O*-glucoside from *P. africana* (Maiyo et al. 2016) are frequently used

Table 4.1 Medicinal plants used in traditional medicine for the cancer treatment

Species	Part of plant	Medicinal use	References
Rotheca serrata (*Clerodendrum serratum*)	A paste of leaves	Used by Siddis of Uttara Kannada, India, in their traditional medicine for tumour growth prevention	Bhandary et al. (1995)
Anthocleista djalonensis	Leaves	Used by traditional healers in Ogun State, Nigeria, for the management of cancer	Soladoye et al. (2010)
Olax subscorpioidea	Root		
Euphorbia peplus	Latex	As home remedy in Australia for the treatment of solar keratosis, skin cancer and basal cell carcinoma. Used by England native people to treat skin cancer	Drury (1991) and Butler et al. (2014)
Neonotonia wightii (*Glycine wightii*)	Leaves	Recommended by herbalists from Kakamega County, Kenya, to treat breast cancer	Ochwangi et al. (2014)
Prunus africana	Stem bark and roots	Recommended to treat colorectal, breast, skin and prostate cancer	
Sesbania sesban	Whole plant	Used to treat throat, uterine, skin cancer	
Solanum aculeastrum	Fruits and roots	Recommended to treat skin, breast and cervical cancer	
Triumfetta rhomboidea	Leaves	Used to treat gum cancer	
Urtica massaica	Leaves	Recommended to treat skin, uterine and breast cancer	
Zanthoxylum rubescens	Stem bark, leaves and root bark	Used to treat breast, colorectal, skin and oesophageal cancer	
Solanum lyratum	Not indicated	Used in Chinese traditional herbal medicine to treat liver, lung and oesophageal cancers	Chiu et al. (2015)
Annona crassiflora	Leaves and seeds infusion	Used by Cerrado (Brazil region) native people to treat tumours	Formagio et al. (2015)
Clinacanthus nutans	Leaves	Used in Malaysia and Singapore to treat liver, nasal cavity and general cancers	Zulkipli et al. (2017)

in the traditional medicine. However, several other species, for instance, *Anthocleista djalonensis*, *Glycine wightii*, *Olax subscorpioidea*, *T. rhomboidea*, *Urtica massaica* and *Zanthoxylum rubescens*, have not been evaluated for their anticancer activity or subjected to phytochemical and biological activities studies in order to identify the active secondary metabolites and/or determining its action mechanism. Moreover, despite the beneficial effects of plants in the cancer treatment reported and scientifically confirmed, it must be emphasized that their potential adverse effects, either through directly toxic effects or interactions with conventional drugs, should also be assessed. It should not be ignored that there are risks arising from unmonitored use of herbal medicinal with "miraculous effects" by the general population (Ben-Arye et al. 2016). In addition, it must be considered that several of the plants reported in Table 4.1 have other traditional phytotherapy uses. For instance, *U. massaica* is used to treat diabetes (Kamau et al. 2016) and malaria (Muthaura et al. 2015); *Clinacanthus nutans* is used against dysentery and varicella zoster, herpes simplex and herpes genitals lesions (Zulkipli et al. 2017); *A. crassiflora* oil seeds is used against snake bites (Formagio et al. 2015); *O. subscorpioidea* is used to treat wounds (Adetutu et al. 2011); *T. rhomboidea* is used to treat anaemia (Das et al. 2015); *S. sesban* is used to cure digestive disorders (Dianhar et al. 2014); and *C. serratum* is popularly used in indigenous systems of medicine to treat respiratory disorders especially asthma (Patel et al. 2014). Proved or not, these uses may also interfere with the beneficial anticancer effect.

4.2.2 Plant Extracts with Cytotoxic/Antiproliferative Activities

The American National Cancer Institute, relatively to anticancer activity, assigns a significant cytotoxic effect of promising anticancer products for future bioguided studies if their IC_{50} values are lower than 30 μg/ml. Besides, in assessing the anticancer potential of any mixture and/or pure compound, the selectivity index (SI = IC_{50} non-tumour cells/IC_{50} tumour cells) and the use of a clinical drug as positive control are essential. The literature survey highlighted the most recent and promising results on the basis of their anticancer activity (Table 4.2). The *in vivo* assays were emphasized, even though the results show that the extracts are less active than the positive control (Viral et al. 2011) or even if the authors do not show the results of reference compounds (Ahmed et al. 2016), because they are less frequent and show the extract effect in conditions much closer to those found in clinical therapeutics. However, the plant extracts showed the antitumour activity both in *in vitro* and *in vivo* conditions failed to provide the comparative results of anticancer compound used in clinical trials. For example, Réthy et al. (2007) and Iweala et al. (2015) reported that the extracts are more active than positive control (cisplatin); nonetheless, they do not assay the extract selectivity. Conversely, Koňariková et al. (2015) report a very low IC_{50} value for the extract activity with a very high selectivity index (higher than 400), but they do not present the positive control results. The same, very promising extract activity values but no report about positive control activity can be noticed in Jain et al. (2016) work. On the other hand, Boutennoun et al. (2017) showed the percentage of cell death for the extract at 5 μg/ml, while the percentage of cell death to the positive control colchicine is given at 50 μg/ml; naturally, these different concentrations do not allow proper conclusions. We cannot fail to point out how much these aspects reduce the impact of such works and once again appeal to researchers that use the latest best practice guidelines on their anticancer assays. This research area is fundamental and usually the first step in a chain of methodologies that will allow the identification of new plants and/or their metabolites with potential use in cancer treatment.

Table 4.2 Plants whose extracts are cytotoxic towards cancer cell lines

Species	Extract tested	Anticancer model and activity level	References
Achillea collina	Chloroform extract from herbs	Tumour cell inhibitory activity against HeLa (cervical cancer cell line) (IC_{50} 1.74 μg/ml) (cisplatin IC_{50} 3.73 μg/ml; doxorubicin IC_{50} 0.089 μg/ml)	Réthy et al. (2007)
Xanthium orientale	Chloroform extracts from folium and from buds/flowers	Antiproliferative activity against A431 (epidermoid carcinoma cell line) (IC_{50} 0.71–0.75 μg/ml) (cisplatin IC_{50} 0.852 μg/ml; doxorubicin IC_{50} 0.086 μg/ml)	
Euphorbia peplus	Plant's sap	A phase I/II clinical trial showed 50–75% full remission in 15 months for the topical treatment of human nonmelanoma skin cancers	Ramsay et al. (2011)
Adiantum venustum	Ethanolic extract from leaves and stem	*In vivo* assay on Ehrlich ascites carcinoma bearing Swiss albino mice model. At 250 mg/kg (i.p.) increase 31.8% in life span (vincristine at 0.8 mg/kg increase life span 40.9%) without toxicity at 2000 mg/kg. Moreover, it reduces the packed cell volume and viable tumour cell count identical to vincristine at 0.8 mg/kg	Viral et al. (2011)
Annona sylvatica	Methanolic extract from leaves	Antitumour activity against UA251 (glioma cell line) (GI_{50} = 0.05 μg/ml; Vero cell line GI_{50} = 41.60 μg/ml) (doxorubicin GI_{50} = 5.13 μg/ml; Vero cell line GI_{50} = 1.38 μg/ml)	Formagio et al. (2015)
A. coriacea	Methanolic extract from floral capitulum	Antitumour activity against K562 (leukaemia cell line) (GI_{50} = 0.04 μg/ml; Vero cell line GI_{50} = 6.33 μg/ml) (doxorubicin GI_{50} = 0.89 μg/ml; Vero cell line GI_{50} = 1.38 μg/ml)	
Piper guineense	Dichloromethane extract from leaves	Cytotoxic activity against HL-60 (myeloid leukaemia cell line) and MCF-7 (breast cancer cell line) (IC_{50} = 3.62 and 6.54 μg/ml, respectively) (cisplatin IC_{50} = 3.06 and 17.48 μg/ml, respectively)	Iweala et al. (2015)
Camellia sinensis	Black tea extract	Antiproliferative activity against MCF-7 (breast cancer cell line) (IC_{50} = 0.0125 μg/ml; mouse healthy fibroblast cell line NIH-3 T3 IC_{50} > 5 μg/ml)	Koňariková et al. (2015)
Alcea rosea	Ethyl acetate from dried seeds	Extract inhibits tumour growth in HCT116 (colon cancer cell line) tumour xenografts in nude mice treated with 200 mg/kg body weight of extract (i.p.) for 21 days	Ahmed et al. (2016)

(continued)

Table 4.2 (continued)

Species	Extract tested	Anticancer model and activity level	References
Aphanamixis polystachya	Ethyl acetate extract from the stem bark	On mice inoculated with Dalton's lymphoma ascites cells, the extract at doses of 20 mg/kg/day (i.p.) prolonged the median survival time. *In vitro* activity against Dalton's lymphoma ascites cells (IC_{50} = 9 μg/ml)	Jain et al. (2016)
Vitis vinifera	Standardized grape seed extract (GSE) (90% procyanidins)	GSE significantly inhibited the 1198 bronchial premalignant cells (IC_{50} <10 μg/ml), while Leucoselect Phytosome (containing GSE at 56–112 mg/kg/day) administered via gavage inhibits the growth of A549 (human lung carcinoma) xenograft tumour in athymic nude mice model	Mao et al. (2016)
Solanum aculeastrum	Ethanol extract of the berries	Cytotoxic activity against human CCRF-CEM leukaemia cell line (0.8% of cell viability at 10 μg/ml, IC_{50} values of 1.36 μg/ml), more active than doxorubicin (6.5% of cell viability at 10 μg/ml)	Omosa et al. (2016)
A. odorata	Methanol extract from leaves	Cytotoxic activity against Wehi cell line (63.19% cell death at 5 μg/ml) (colchicine 81.01% of cell death at 50 μg/ml)	Boutennoun et al. (2017)
E. terracina	Chloroformic and hydromethanolic extracts from shoots	Cytotoxic activity against human leukaemia monocytes (THP-1) by induction of apoptosis (IC_{50} = 14.43 and 2.08 μg/ml, respectively) while is not toxic against normal cells (CD14+ and IEC6) (IC_{50}>100 μg/ml), both after 48 h of exposure	Jannet et al. (2017)
Sarcocephalus pobeguinii	Methanol extracts from leaves and bark	Cytotoxic activity against cervical cancer cell line (HeLa) (IC_{50} 10.19 and 15.26 μg/ml) with selectivity index of 18 and 14, respectively (doxorubicin IC_{50} 1.04 μg/ml)	Njoya et al. (2017)

4.3 Natural Compounds As Anticancer Agents

An important method of drug discovery is the isolation of natural and bioactive compounds from diverse life forms, medicinal plants being one of the most important sources. Thus, an increasing number of studies are being carried out on plants due to their anticancer ethnopharmacological use; these scientific investigations involved phytochemical studies to isolate and characterize the secondary metabolites and pharmacological evaluations to find the active principle. In these researches, both plant traditional applications are recognized, and new bioactive compounds are obtained. In cancer research, lead compound structures from natural sources can be used to design novel chemotherapies with enhanced biological properties.

Throughout history, plants have been a rich source of affordable natural compounds with diverse chemical structures and bioactivities (Chudzik et al. 2015; Das and Dhanjal 2015; Nwodo et al. 2016; Habli et al. 2017). In fact, several new cytotoxic compounds are isolated from plants each year constituting a source of new possibilities to explore in order to fight against cancer. The significant achievements of plant secondary metabolites for its anticancer potential are well explored. Some of them are used in the clinical or preclinical trials and are discussed below.

4.3.1 Natural Compounds with High Degree of Activity and Safety

The natural compounds have high degree of anticancer and cytotoxic activities. The mechanism of action and levels of safety of some selected natural compounds are well documented (Table 4.3 and Fig. 4.1), but still some of them are under clinical trials for fruitful results. The analysis of data showed that the various compounds from different families have some limitations that may prevent its use based on their low solubility and poor bioavailability. Having in mind the compound structure (Table 4.3 and Fig. 4.1), it is obvious that without chemical transformations just ursolic acid and tetrandrine could have their water solubility increased if transformed in the correspondent salt, especially in the case of tetrandrine, an alkaloid, for which there are few studies and could be easily transformed in a hydrochloric derivative. The rapid metabolization of polyphenols apigenin and resveratrol is certainly a drawback in their utilization, but it can be overcome using proper carriers. Deoxypodophyllotoxin and triptolide seem to be more complicated cases because chemical transformations will be necessary, being the first and obvious suggestions the methyl groups cleavage and the epoxy ring opening, respectively (Table 4.4).

4.3.2 Natural Compounds in Clinical Trials

4.3.2.1 Parthenolide

Parthenolide is the most noticeable germacranolide sesquiterpene lactone, due to its high cytotoxicity and the fact that it is currently being tested in clinical trials as anticancer agent (Orofino et al. 2012). Chemically, parthenolide has a 15-carbon (C_{15}) backbone derived from farnesyl diphosphate through cyclization and oxidative transformation (like all sesquiterpene) and is structurally characterized by a ten-membered ring with a five-membered-fused lactone moiety, being the presence of an α-methylene group and a C_4-C_5 epoxide ring are also characteristic in its structure (Fig. 4.2).

Parthenolide was first isolated from few leaves of *Tanacetum parthenium* (L.) Sch. Bip. an ornamental and medicinal plant used in the traditional medicine to

Table 4.3 Plant secondary metabolites with high antitumour activity and in preclinical trials

Compounds (type)	Activities/ actions	Remarks	References
Apigenin (flavonoid)	Natural sources	*Salvia officinalis*, *Lawsonia inermis*, *Turnera aphrodisiaca*, *Ocimum basilicum*, *Tamarindus indica* but also found in many other fruits and vegetables	Torquato et al. (2017)
	Cytotoxicity	*In vitro* and *in vivo* inhibition of some tumour cells of melanoma, lung, breast, colon, prostate, leukaemia and pancreatic cancers	Bridgeman et al. (2016), Sung et al. (2016), Erdogan et al. (2016), Zhou et al. 2017 and Zhao et al. (2017c)
	Other significant antitumour effects	Inhibitory effects of cisplatin on cell migration via downregulation of SNAIL expression. MAPK activation and subsequent p53 phosphorylation	Erdogan et al. (2017), and Liu et al. (2017)
	Mechanism	Targets leptin/leptin receptor pathway and induces caspase-dependent extrinsic apoptosis in HER-2. Induces apoptosis by downregulation of PI3K/Akt/NF-κB signalling. Induces autophagy and inhibition of UVB-induced mTOR signalling by a not Akt-dependent pathway	Seo et al. (2015), Bridgeman et al. (2016), Duan et al. (2016), Erdogan et al. (2016), and Sung et al. (2016)
	In vivo use limitations	Rapid metabolization in the intestine	Ali et al. (2017a) and Tang et al. (2017)
Deoxypodophyllotoxin (lignan)	Natural sources	*Anthriscus sylvestris*, *Podophyllum peltatum*; *Juniperus* spp.	Khaled et al. (2013), Seca et al. (2015), and Zhao et al. (2016a)
	Cytotoxicity	Inhibition of HL-60 cells (2 ng/ml), H460 cells, H460 xenografts, glioma cell xenograft and SGC-7901 cells *in vivo* and the cell lines H460/Bcl-xL *in vitro*. Inhibition of EphB2-EPD (IC50 1.93 nM), prostate DU-145, U-87 MG and SF126 cell lines	Muto et al. (2008), Wu et al. (2013), Yang et al. (2014a), Guerram et al. (2015), Wang et al. (2015a), Ma et al. (2016), and Hu et al. (2016)
	Other significant antitumour effects	Investigation on Heps tumour-xenografted mice and antitumour capacity. Exhibits higher anticancer activity than etoposide both at 15 mg/kg	Wang et al. (2016a)

(continued)

Table 4.3 (continued)

Compounds (type)	Activities/ actions	Remarks	References
	Mechanisms	Induces apoptosis by G_2/M cell cycle arrest in a caspase-dependent pathway and exerts potent antiangiogenic and vascular disrupting effects via stimulation of LKB1-AMPK. Induces apoptosis through Akt/p53/Bax/PTEN signalling pathway and stimulate upregulation of PARP-1, cytoplasmic accumulation of cytotoxic PAR polymer, nuclear translocation of AIF and overproduction of ROS	Muto et al. (2008, Wang et al. (2015b), Guerram et al. (2015), Hu et al. (2016), Jiang et al. (2013), and Ma et al. (2016)
	In vivo use limitations	Poor water solubility, acquired drug resistance and rapid elimination	Khaled et al. (2013), and Yang et al. (2014b)
Resveratrol (stilbene)	Natural sources	*Polygonum cuspidatum*, *Vitis vinifera*, *Arachis hypogaea*, *Vaccinium* spp., *Prunus* spp.	Burns et al. (2002), Sales and Resurreccion (2014), and Torquato et al. (2017)
	Cytotoxicity	Prevention and therapy of neuroblastomas, lymphoid, myeloma, breast, prostate, stomach, colon, pancreas, thyroid, ovarian and cervical carcinoma	Aggarwal et al. (2004), Peng et al. (2013), Gupta and Pramanik (2016), Sinha et al. (2016), Tsai et al. (2017), and Varoni et al. (2016)
	Other significant antitumour effects	Multidrug resistance reversion and synergetic effect with several approved anticancer drugs	Pavan et al. (2016), Sinha et al. (2016), Varoni et al. (2016), Alamolhodaei et al. (2017), and Redondo-Blanco et al. (2017)
	Mechanism	Acting in molecular targets associated with the modulation of transcription factors, protein kinases, cell cycle regulatory proteins and inhibition of angiogenesis and including signalling pathways involved in cancer metastasis	Pavan et al. (2016), Sinha et al. (2016), and Varoni et al. (2016)
	In vivo use limitations	Low aqueous solubility and rapid metabolization	Cottart et al. (2014)
Tetrandrine (alkaloid)	Natural sources	*Stephania tetrandra*, *Mahonia bealei*	Zhu et al. (2015), and Liu et al. (2016a)

	Cytotoxicity	Antitumour effects on oral, prostate, lung, gastric, breast, hepatic colon, renal, glioma, leukaemia, cervical and nasopharyngeal cancers cell lines at 1–30 μM concentration level	Liu et al. (2016a)
	Other significant antitumour effects	Synergistic cytotoxic effects with cisplatin on A549 and A549/DDP cell lines. Prevents MDR in U-2OS/Taxol by NF-κB inhibition. Nanoscale delivery systems using a Trojan horse strategy for the delivery of tetrandrine and modified liposomes delivery in combination therapy. Enhance a radiation sensitization effect on human glioma	Xu et al. (2014), Lu et al. (2017), Ma et al. (2017), Song et al. (2017a), and Ye et al. (2017)
	Mechanism	Promotes in oral cancer cells G_0/G_1 phase arrest, ROS intracellular accumulation, Ca^{2+} production, ER stress and finally apoptotic death through a caspase-dependent pathway and Beclin-1-induced autophagy. Also the inactivation of PI3K/Akt signalling through TGF-β upregulation to decrease the phosphorylation of PTEN. Triggers in DU145 cells the LC3-independent autophagy. Repress the nuclear translocation and expression of β-catenin	Chen et al. (2017), Lien et al. (2017), Qiu et al. (2017), and Zhang et al. (2017a)
	In vivo use limitations	Water insolubility	Liu et al. (2016a)
Triptolide (diterpene)	Natural sources	*Tripterygium hypoglaucum*, *T. regelii*, *T. forrestii*	Zhou et al. (2012)
	Cytotoxicity	Inhibits the proliferation of all 60 NCI cancer cell lines at IC_{50} values lower than 50 nM. *In vitro* and *in vivo* anti-invasion and anti-tumourigenesis activity against hepatocellular carcinoma cells. Supress tumour growth, invasion, migration and angiogenesis of head and neck squamous cell carcinoma. Inhibits human prostate cancer cells, CT26 and HCT116 colon and colorectal cancer cells and pancreatic cancer cells	Zhou et al. (2012), Gali-Muhtasib et al. (2015), Wang et al. (2016b), Yuan et al. (2016), Zhang et al. (2016a), Zhao et al. (2016b), Xi et al. (2017), and Yang et al. (2017a)
	Other significant antitumour effects	New delivery systems improve efficacy and reduce toxicity. *In vitro* and *in vivo* anti-MDR activity against: MCF-7/ADR. A549/Taxol via modulation of MAPK and PI3K/Akt signalling pathways; KB/VCR via CDK7/RPB1. Synergistic effect in the treatment of nasopharyngeal carcinoma as well as in thyroid cancer and liver cancer	Alsaied et al. (2014), Li et al. (2014, 2016), Kim et al. (2016), Xie et al. (2016), Yi et al. (2016), Xi et al. (2017), and Zhang et al. (2016b, 2017b)

(continued)

Table 4.3 (continued)

Compounds (type)	Activities/ actions	Remarks	References
	Mechanism	Induces apoptosis by extrinsic and intrinsic pathway, causing mitochondrial damage and then activating caspase-dependent pathways involving MAPK activation and NF-kB inhibition. Induces autophagy by inhibition of Akt/mTOR/p70S6 K pathway and activating the ERK1/2 pathway. In pancreatic cells activate the ER stress pathway; show ability to block NF-kB/MMP-9 activation in hepatocellular carcinoma cells. Inhibits the functional centres of Na+/K+-ATPase and Ca^{2+}/Mg^{2+}-ATPase. Inhibits the growth of orthotopically xenografted NSCLC cells by suppression of HA-CD44/RHAMM signalling pathway. Promote T and B cell populations	Meng et al. (2014), Mujumdar et al. (2010, 2014), Wang et al. (2016b), Chan et al. (2017), Xi et al. (2017), and Song et al. (2017b)
	In vivo use limitations	Short $t½$ values after oral administration. Poor water solubility, severe multi-organ toxicity	Li et al. (2014), Gong et al. (2015), and Xi et al. (2017)
Ursolic acid (triterpene)	Natural sources	*Oldenlandia diffusa*, *Vaccinium* spp. In many aromatic plants (*Thymus*, *Lavandula*, *Origanum* and *Eucalyptus*)	Kondo et al. (2011), Woźniak et al. (2015), Jain et al. (2016), and Torquato et al. (2017)
	Cytotoxicity	*In vitro* and *in vivo* activity against more than 60 cell lines belonging to 20 tumour types such as glioma, melanoma, colon, hepatic, lung, breast, gastric and prostate cancers without significant toxicity to normal cells	Kondo et al. (2011), Kim and Moon (2015), Woźniak et al. (2015), Jain et al. (2016), Pattnaik et al. (2017)
	Other significant antitumour effects	Growth inhibition of MDR cell line through apoptosis poly(lactic acid) nanoparticles load ursolic acid decrease cytotoxicity over normal cells, while chitosan-coated ursolic acid liposome exhibited better pharmacokinetic parameters	Jain et al. (2016), Wang et al. (2017a), and Antônio et al. (2017)
	Mechanism	Able to induce apoptosis; inhibit tumour promotion, metastasis and angiogenesis by suppressing the MAPK/ERK and PI3K/AKT/mTOR signalling cascades; decrease NF-B activity, modelling the Bcl-2, Bcl-xL BAX activity, reverse transcriptase inhibition, modelling matrix metalloproteinases activity, G_0/G_1 cell cycle arrest	Woźniak et al. (2015), Jain et al. (2016), Lewinska et al. (2017), and Pattnaik et al. (2017)
	In vivo use limitations	Low aqueous solubility, short plasma half-life and low oral bioavailability	Shanmugam et al. (2013)

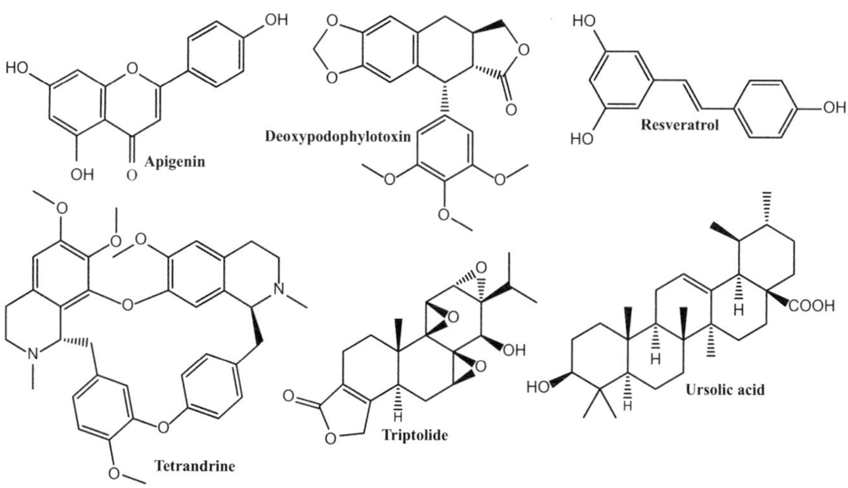

Fig. 4.1 Structures of natural compounds with high antitumour activity and in preclinical trials

relieve intermittent fever, arthritis pain and infant colic and as inducer of uterine contractions and also insect repellent (Ghantous et al. 2013). Although the concentration of parthenolide in native plants is usually quite low, the commercially available parthenolide is still nowadays obtained by extraction from its first-known natural source where it is found in the superficial leaf glands (0.2–0.5%) and comprises up to 85% of the total sesquiterpene content (Pareek et al. 2011). It is noteworthy that the *T. argenteum* (Lam.) Willd. leaves which possess higher parthenolide content (2.261 ± 0.002%) are not the main source of this compound (Orhan et al. 2015). Other new natural sources of parthenolide, including several *Tanacetum* species, were evaluated and subjected to new extraction procedures and to genetic engineering strategies (Seca et al. 2017), aiming to obtain this natural compound in higher yield and in an economically feasible manner. In the last years, there have been also an increasing number of attempts to uncover efficient methodologies towards its synthesis. Although two routes for total synthesis of parthenolide have been recently described (Long et al. 2014; Yang et al. 2015), the number of steps is huge, in some cases isomers were obtained, and the desired 6,7-*trans*-10-membered ring system is achieved with low selectivity. Besides the several steps needed, which contribute to lower yields, the methods are not green and altogether are the foundation for the economic failure of the procedures. One of the parthenolide strengths is its great ability to get involved in different mechanisms that induces apoptosis, and as more is known about its targets, more likely new applications will be exposed, and therefore its pharmacology continues to stimulate great interest in the scientific community. Therefore, in the last 3 years, a significant number of studies on parthenolide cytotoxic mechanism of action were published and showing that parthenolide can induce *in vitro* and *in vivo* inhibition of cell proliferation by distinct mechanisms and in various human cancer cells (Seca et al. 2017).

Table 4.4 Natural compound derivatives in clinical therapeutic and clinical development

Name	Lead compound (source)/ development phase	Major area of application	Other applications	References
Cabazitaxel	Paclitaxel (*Taxus brevifolia*)/in the market (Jevtana®) since 2010	Treatment of metastatic castration-resistant prostate cancer	In testicular germ cell tumours	Yao et al. (2017), and Oing et al. (2016)
Vinflunine	Vinblastine (*Catharanthus roseus*)/in the market (Javlor®) since 2009	Treatment of metastatic transitional-cell carcinoma of the urothelial tract	In non-small cell lung cancer	Genova et al. (2016)
Vintafolide	Vinblastine (*Catharanthus roseus*)/in clinical trial	Promise in the treatment of platinum-resistant ovarian and NSLC cancers (NCT01577654)	Phase II trial to evaluate its activity in higher FR expression solid tumours (NCT01002924)	Vergote and Leamon (2015)
Topotecan	Camptothecin (*Camptotheca acuminata*)/in the market (Hycamtin®) since 2009	To treat metastatic ovarian cancer and relapsed small cell lung cancer (EMA/248288/2015 EMEA/H/C/000123)	In glioma undergoing a clinically indicated surgical resection (NCT02500459)	Hanna et al. (2015)
Phenoxodiol	Daidzein (*Glycine max*)/in clinical trial development	Chemosensitizing agent for carboplatin in resistant ovarian cancer (NCT00382 811)	Effect on castrate and non-castrate prostate cancer cells (NCT005570 37)	Mahoney et al. (2014), and Xiao et al. (2016)
Alvocidib	Rohitukine (*Dysoxylum binectariferum*/ clinical trial development	To treat advanced, fludarabine refractory chronic lymphocytic leukaemia (NCT00464 633)	Biomarker-driven phase II acute myeloid leukaemia (NCT02520011)	Lanasa et al. (2015)

Parthenolide is a nuclear factor kappaB (NF-kB) cells inhibitor with great potential, since this family of proteins and their regulated genes have been correlated with cell proliferation, inflammation, angiogenesis, invasion and metastasis of the cancer cells (Vazquez-Santillan et al. 2015; Li et al. 2015). Parthenolide can prevent the DNA binding of p65/NF-kB subunits causing the alkylation of cysteine sulfhydryl groups in the p65 subunit, effect mainly observed at lower concentrations, while at higher concentrations, it preferentially inhibits the IkB kinase complex (Liu and Ou 2013; Gali-Muhtasib et al. 2015). Furthermore it can also act inhibiting tubulin carboxypeptidase by an independent NF-kB pathway (Fonrose et al. 2007), inhibiting the mitogen-activated protein kinase and causing impact on epigenetic mechanisms (Ghantous et al. 2013; Gali-Muhtasib et al. 2015). Besides it can induce the apoptosis of cancer stem cells in acute myelogenous leukaemia (AML) and blast

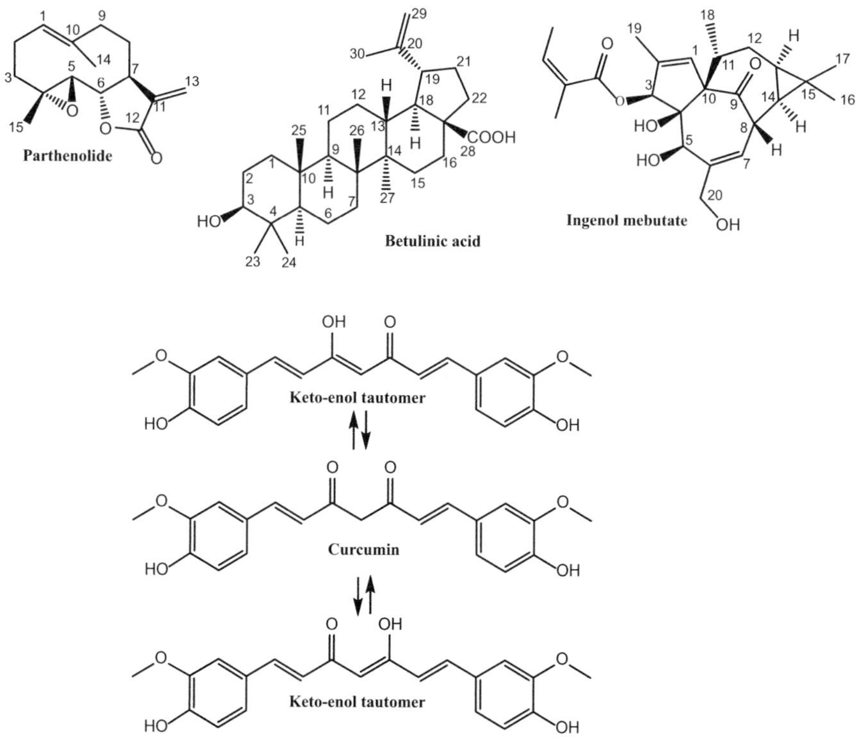

Fig. 4.2 Structures of natural compounds in clinical trials

crisis chronic myelogenous leukaemia not only through the inhibition of NF-kB but also through mechanisms involving proapoptotic activation of p53 and elevating ROS levels (Liu and Wang 2015). New evidence indicates that the antitumour action of parthenolide is also related to its ability to induce a caspase-independent form of death, stimulating oxidative stress and autophagic process (D'Anneo et al. 2013; Jeyamohan et al. 2016; Carlisi et al. 2016).

Parthenolide applied at concentrations lower than 25 μM exhibits a strong cytotoxic effect against several different types of tumour cell, such as non-small lung cancer cells, oral cancer cells, glioma radioresistant cells, cervical cancer cells, breast cancer cells, bone osteosarcoma cells and multiple myeloma cells (Seca et al. 2017). However, the most amazing cytotoxic effect of the parthenolide is to possess selective anticancer stem cell activity in multiple myeloma cells (MMSC) by inducing apoptosis (IC_{50} 1.5–3.5 μM) (Gunn et al. 2011). Cancer stem cells (CSCs), a subset of the heterogeneous cell population of a tumour, are self-renewable, can differentiate into all types of cancer cells and are rather resistant to active drugs which difficult its elimination (Khan et al. 2015; Moselhy et al. 2015). Currently, it is accepted that compounds that eliminate CSCs are expected to confer more durable and potentially curative therapies, and targeting CSCs seems to be a promising strategy for cancer therapy (Gupta and Pramanik 2016; Torquato et al. 2017).

Parthenolide was the first small molecule that selectively target leukaemia stem cells (LSCs) and AML cells, in addition to the aforementioned MMSC, while sparing normal haematopoietic cells (Pareek et al. 2011; Gunn et al. 2011; He et al. 2014). Based on the structure (Fig. 4.2) and anticancer activity results, some structure/activity relationship (SAR) can be drawn: (1) C-6 and C-7 configurations have more impact on the activities against some cancer cell lines than the C-1=C-10 double bond configuration (Long et al. 2014); (2) the replacement of the epoxide moiety by the potentially bioisostere cyclopropyl moiety might provide a more stable parthenolide derivative without changing the cytotoxicity level (Long et al. 2013); (3) replacement of the lactone moiety with lactam moiety greatly decreased the anticancer activity (Long et al. 2016); and (4) the C-14 methyl group of parthenolide might be important for its high anticancer activity (Long et al. 2016).

Parthenolide seems to reduce the frequency of ABCB5-positive cells (ABCB5 transporter is one of the recognized mediators of chemoresistance in melanoma and a marker for melanoma stemlike cells), and the melanoma cells that survive its treatment lose their self-renewing capacity (Czyz et al. 2013). Moreover, the cytotoxicity of parthenolide against stemlike cells derived from triple-negative breast cancers (TNBCs), an aggressive form of breast carcinoma associated with a high rate of recidivism, strongly suggests that it can be used as new therapeutic strategy against TNBCs (Carlisi et al. 2016). Regarding the selectivity demonstrated by parthenolide, it is expected that when used in combination with other drugs, it can significantly contribute to the complete tumour eradication.

The use of a single drug in the cancer treatment has proven to be limited because there are numerous types of cancer cells and resistance to chemotherapy is customary. Therefore, studies to assess synergistic effects involving parthenolide have given a new impulse to its use. This subject was reviewed recently (Wyrebska et al. 2014; Seca et al. 2017) and shows that combined effects of doxorubicin and parthenolide seem to be effective against melanoma cells highly resistant to chemotherapy (Wozniak et al. 2013) and enhance the apoptotic cytotoxicity of DOX in A549 DOX-resistant cells (Xin et al. 2013). The parthenolide cytotoxicity against AML Kasumi-1 cells is enhanced by the antifungal agent with antitumour properties ciclopirox (10–20 μM) (Sen et al. 2013), and the synergistic effect of parthenolide/dacarbazine and parthenolide/5-FU was established for metastatic melanoma or TRAIL-resistant human colorectal cancer cell lines (Koprowska et al. 2013; Kim et al. 2013). Recently it was reported that parthenolide potentiates the efficacy of balsalazide (a colon-specific prodrug used to treat ulcerative colitis) through synergistic inhibition of NF-kB activation and the combination of these dual agents prevents colitis-associated colon cancers (Kim et al. 2017).

Despite the positive effects of parthenolide as cytotoxic agent, in particular its great efficiency to selectively target CSCs and its safety profile established in phase I clinical trials, it exhibits a low bioavailability derived from low solubility in plasma blood, and because of this, it did not pass to the next phase. Moreover, the parthenolide stability is pH dependent, being unstable under both high acid and basic conditions (Adio 2009), and its half-life in mouse plasma is only 0.34 h (Zhang et al. 2012). To address these limitations to the parthenolide application as a clinical drug, very interesting studies of new delivery systems have emerged. Recent advances have shown that nanoparticles used for drug therapy provide some advantages over conventional formulations, including increased solubility, enhanced storage stability, improved permeability and bioavailability, prolonged half-life, tissue targeting and minimal side effects, both *in vitro* and *in vivo* assays (Aljuffali et al. 2016) and some of them are approved by the Food and Drug Administration (Watkins et al. 2015). Also vitamin E-based nanomedicines already provide sufficient preclinical data, for example, to parthenolide delivery, to be applied as anticancer drug delivery system (Duhem et al. 2014). The synthetic $PSMA_{100}$-b-PS_{258} micelles encapsulating parthenolide have demonstrated excellent properties (increase the aqueous solubility, significant stability improvement and attractive physical dimensions and surface characteristics) giving them high likelihood for long circulation times *in vivo* (Baranello et al. 2015). Recently, EGF-conjugated mixed micelles of PEG2000-DSPE and vitamin E-TPGS were developed for targeted combination therapy with paclitaxel and parthenolide against paclitaxel-resistant and paclitaxel-sensitive non-small cell lung (Gill et al. 2016).

As a conclusion, parthenolide itself remains a very promising multiple target drug in terms of combating a wide range of human cancer types, showing a remarkable ability to beneficially interfere with almost all pathways that lead to activation of apoptosis and inhibition of cellular inflammation. Its low bioavailability entails limitations to its clinical use, but this seems to be overcome with innovative forms of parthenolide release. So if parthenolide presents higher bioavailability, its clinical application will be a reality and no doubt that it will be a great success and a stunning medical breakthrough for different cancer diseases.

4.3.2.2 Betulinic Acid

Betulinic acid (3*β*-hydroxy-lup-20(29)-en-28-oic acid), a lupane-type pentacyclic triterpene with a molecular formula of $C_{30}H_{48}O_3$ and biosynthesized from six different isoprene units (Fig. 4.2), was first identified and isolated from *Gratiola officinalis* L. and named "graciolon", then from the bark of *Platanus acerifolia* (Aiton) Willd. (named "platanolic acid") and after from *Cornus florida* L. and called "cornolic acid". Later on, after the confirmation that all have the same structure, was named betulinic acid (Sarek et al. 2011). This triterpene is extensively spread throughout the plant kingdom, and considerable amounts (up to 2.5%) are available in the outer bark of a wide range of plants, for instance, *Betula* spp., *Diospyros* spp., *Syzygium* spp., *Ziziphus* spp., *Paeonia* spp., *Sarracenia flava* L., *Anemone raddeana* Regel and *Lycopodium cernuum* L., among others (Ali-Seyed et al. 2016). However, these sources are not sufficient to meet the growing demand for betulinic acid. So it can be obtained through a selective oxidation of betulin (lup-20(29)-en-3,28-diol) (Pichette et al. 2004; Csuk et al. 2006), far more abundant (up to 30%) in birch bark than the acid (Holonec et al. 2012), a more profitable process. Pisha et al. (1995) were the first who reported the antitumour activity of betulinic acid and found that it killed melanoma cells in mice with low IC_{50} values (IC_{50} 0.5–1.5 μg/ml). Later on,

Singh et al. (2016) have also tested the anticancer activities of betulinic acid with respect to melanoma cells (Singh et al. 2016). However, recent evidence indicates that betulinic acid possesses a broader spectrum of activity against other cancer cell types, and because of that, betulinic acid has been selected by the National Cancer Institute to be added into the Rapid Access to Intervention in Development (RAID) program. Betulinic acid exhibits significant *in vitro* cytotoxicity in a variety of tumour cell lines and also inhibits the growth of solid tumours *in vivo*, comparable to some clinically used drugs and showing a good selectivity index for cancer over normal cells even at doses up to 500 mg/kg body weight (Lee et al. 2015; Gali-Muhtasib et al. 2015; Zhang et al. 2015; Luo et al. 2016). It shows anticancer proprieties against colorectal lung, colon, breast, prostate, hepatocellular, bladder, head and neck, stomach, pancreatic, ovarian and cervical carcinoma, glioblastoma, chronic myeloid leukaemia cells and human melanoma with IC_{50} values mainly between 1.0 and 13.0 μg/ml (Sarek et al. 2011; Gheorgheosu et al. 2014; Lee et al. 2015; Zhang et al. 2015; Ali-Seyed et al. 2016; Singh et al. 2016; Ali et al. 2017b).

Betulinic acid exhibits potent anticancer activity by multiple molecular targets (Luo et al. 2016), the induction of apoptosis by direct regulation of the mitochondrial apoptotic pathway being the best characterized mechanism. This pathway can be associated with mitochondrial collapse through direct opening of the permeability transition pore, decreases of mitochondrial outer membrane potential, downregulation of Bcl-2 family members, release of proapoptotic factors such as cytochrome c, increase of caspase activities, attenuation of both the constitutive and inducible STAT3 phosphorylation and nuclear translocation and its DNA binding (Zhang et al. 2015; Luo et al. 2016; Ali-Seyed et al. 2016). Betulinic acid also may induce apoptosis by stabilizing p53 and downregulating NF-kB-mediated signalling (Ali-Seyed et al. 2016; Shankar et al. 2017), although this does not appear to be a target common to all cells (Zhang et al. 2015). The anti-metastatic effect of betulinic acid seems to be through the prevention of the epithelial-to-mesenchymal transition in highly aggressive melanoma cells (Gheorgheosu et al. 2014), while in breast cancer cells acts by downregulating the matrix metalloproteinases expression (Luo et al. 2016). It also induces an antiangiogenic response (Ali-Seyed et al. 2016) under hypoxia mediated by STAT3/HIF-1α/VEGF signalling pathway (Zhang et al. 2015), can block the cell cycle in the G1 phase through inhibition of Cyclin B1 and Hiwi in mRNA and potently induces autophagy as a survival mechanism in response to permeability transition pore opening and mitochondrial damage (Gali-Muhtasib et al. 2015; Luo et al. 2016). Recently, it was described as a new cell death pathway that depends on cardiolipin modification, whereby betulinic acid selectively kills tumour cells (Potze et al. 2016a, b). It was demonstrated that betulinic acid is able to affect the cancer stem cells directly, inducing cell death via loss of clonogenic capacity by inhibition of the stearoyl-CoA-desaturase (SCD-1), an enzyme that is overexpressed in tumour cells and rapidly impacts on the saturation level of cardiolipin, a mitochondrial lipid that, among other functions, regulates mitochondria-dependent cell death (Potze et al. 2016a, b). Proteasome inhibition assays suggest the proteasome is the main target for betulinic acid (Waechter et al. 2017). However the regulatory effects of betulinic acid on NF-κB pathway and on Bax or Bak expression are not well clarified (Zhang et al. 2015).

Betulinic acid seems to be a very effective chemosensitizer for anticancer drug treatment in chemoresistant cell lines once it promotes the inhibition of multidrug resistance proteins *in vivo* and *in vitro* (Luo et al. 2016). For example, the combination of betulinic acid with 5-FU reverts apoptosis induction in the 5-FU-resistant cells (Jung et al. 2007; Luo et al. 2016). These results clearly demonstrate that in some cases, it is possible to circumvent acquired chemoresistance by combination therapy of anticancer drugs with chemosensitizers as betulinic acid. Moreover, betulinic acid has strong synergy with mithramycin A by inhibition migration and invasion on pancreatic cancer cells at nontoxic concentrations by suppressing Sp1 and uPAR level (Gao et al. 2011). Also a synergistic effect of betulinic acid with tumour necrosis factor-related apoptosis-inducing ligand (TRAIL) to inhibit liver cancer progression *in vitro* and *in vivo* through targeting p53 signalling pathway was reported by Xu et al. (2017). Betulinic acid is slightly soluble in water, and therefore its water solubility is a drawback that should be overcome to improve its absorption and bioavailability. The main targets for SAR studies were the 3-hydroxy, C-20 alkene and 28-carboxylic acid moieties (Fig. 4.2). The 3-OH oxidation increased cytotoxic activity but decreased selectivity, introduction of groups, such as amine or hydroxy, at the C-28 increased activity, while modifications at C-20 did not enhance cytotoxicity (Yogeeswari and Sriram 2005; Gali-Muhtasib et al. 2015; Ali-Seyed et al. 2016). It can be concluded that modifications may improve the cytotoxicity and/or the water solubility but not the selectivity. It seems that the presence of the free 3-hydroxy group and the 28-carboxyl group is the most important feature. Recently, a clearer and more realistic method, 3D-QSAR by CoMFA and CoMSIA, shows the structure-cytotoxicity relationship of betulinic acid derivatives against human ovarian cancer cell A2780, and the main conclusions were an electropositive group at C-2 α-site; an electronegative and hydrogen bond acceptor group at C-2 β-site; bulky groups at the C-3 β-site; bulky and electronegative groups at C-3 α-site; and bulky, electronegative and hydrogen bond donor or acceptor groups at C-28 side chain would be beneficial for its antitumour potency (Zhang et al. 2015). Due to its extraordinary potential as antitumour agent, betulinic acid was involved in phase I/II clinical trials to evaluate its safety and effectiveness. The study involved the topical application of 20% betulinic acid ointment to treat dysplastic nevi that can be transformed in melanoma.

4.3.2.3 Ingenol Mebutate

The phytochemical study of *Euphorbia peplus* latex sap yielded several macrocyclic diterpenes (Rizk et al. 1985), including ingenol mebutate (Fig. 4.2). It was identified as the most active antitumour component (Ogbourne et al. 2004). In fact, the *E. peplus* sap, as referred in Table 4.2, has been shown to be effective against human nonmelanoma skin cancer in a phase I/II clinical study (Ramsay et al. 2011). This ingenene-type diterpene can also be isolated from other *Euphorbia* species such as *E. paralias*, *E. antiquorum*, *E. palustris*, *E. helioscopia* and *E. quadrialata* (Gotta et al. 1984). Ingenol mebutate has been prepared by semisynthesis in three steps, involving the selective esterification of the ingenol 3-OH (Liang et al. 2012), which is isolated from the seeds of *E. lathyris* (yield ~100 mg/kg). A recent report for fairly short and efficient route of the ingenol synthesis was published (Jørgensen et al. 2013), but the total synthesis of ingenol is still unaccomplished. Ingenol mebutate, also known as ingenol angelate and PEP005, is a 3-monoester of the ingenol diterpene with the (*Z*)-2-methylbut-2-enoic acid having the chemical formula $C_{25}H_{34}O_6$ (Fig. 4.2). Ingenol mebutate chemical stability is pH dependent and can undergo facile acyl migration involving the hydroxy groups, mainly the 5- and 20-OH (Fig. 4.2), which are also required for the biological activity. In this regard, the ester moiety at the three-position is also important (Liang et al. 2013).

Ingenol mebutate showed potent antiproliferative effects on different cell lines in a dose- and time-dependent manner (Serova et al. 2008; Benhadji et al. 2008), especially against colon 205 cell lines with IC_{50} 0.01 μM that means more active than staurosporine (IC_{50} 0.029 μM) or doxorubicin (IC_{50} 1.5 μM) (Serova et al. 2008). This compound is also very effective against a wide range of established subcutaneous tumours in mice after three daily topical applications of 42 nmol (Ogbourne et al. 2004). Its effectiveness to damages in the tumour vasculature is related with the fact that ingenol mebutate is transported through the epidermis into the deep dermis via a P-glycoprotein (Collier et al. 2014).

Treatment with this compound, both *in vitro* (230 μM) and *in vivo* (42 nmol), rapidly caused swelling of mitochondria probably by loss of their membrane potential and cell death by primary necrosis and is, therefore, unlikely to have its activity compromised by the development of apoptosis resistance in tumour cells (Ogbourne et al. 2004). There are evidences that this rapid action of ingenol mebutate is due to its dual action combining cytotoxic and immunomodulatory effects in which rapid lesion necrosis and antibody-dependent cellular cytotoxicity mediated by neutrophils occur (Rosen et al. 2012). The mechanism of action of ingenol mebutate is also partially related to the modulation of protein kinase C (PKC) to which it has a potent binding affinity by activating PKCδ and inhibiting PKCα (Benhadji et al. 2008; Matias et al. 2017). *In vitro* low isozyme selectivity was verified with a Ki ranging from 0.105 to 0.376 nM (Kedei 2004). The immunomodulatory effects seem to be at least partially mediated by PKC (Serova et al. 2008; Doan et al. 2012; Liang et al. 2013). All these results support the potential of ingenol mebutate for further development of cancer immunotherapy. In fact, the cutaneous treatment of non-hyperkeratotic, non-hypertrophic actinic keratosis (a premalignant precursor of squamous cell carcinoma) with a gel formulation of ingenol mebutate (formerly PEP005 produced by LEO Pharma and marketed as Picato®) was recently approved by both FDA and EMA agencies (Tzogani et al. 2014). Unfortunately, the current gel based on isopropanol gives rise to some local skin responses, being the majority resolved within 2 weeks after treatment with no scarring observed (Collier et al. 2014). The most frequently reported adverse reactions associated with this application are lower to moderate "local skin responses" and included erythema, flaking/scaling, swelling, crusting, erosion/ulceration and vesiculation/postulation (Tzogani et al. 2014; Berman 2015). Ingenol mebutate shows a favourable safety and tolerability profile exhibiting lack of systemic absorption and photosensitivity, and there is no evidence of drug interactions (Collier et al. 2014).

4.3.2.4 Curcumin

Curcumin or diferuloylmethane is a polyphenol compound that has been extracted from the rhizome of turmeric (*Curcuma longa* L.; Fig. 4.2), a tropical Southeast Asian plant mainly known due to its use as spice. Curcumin constitutes 2–5% of turmeric powder (Chainani-Wu 2003) and has been used for a long time in countries such as China and India as traditional medicines (Hatcher et al. 2008; Kocaadam and Sanlier 2017). This ancient remedy has brought the attention of scientific community for a wide range of curcumin beneficial properties including anti-inflammatory, antioxidant, chemopreventive, chemotherapeutic and chemosensitizing activity (Hatcher et al. 2008; Kocaadam and Sanlier 2017). Research interest on curcumin anticancer properties has been developed based on low rate occurrence of gastrointestinal mucosal cancers in Southeast Asian populations and its association with the regular turmeric use in their dietary (Mohandas and Desai 1999; Sinha et al. 2003). Curcumin (a bis-α,β-unsaturated β-diketone) has a molecular formula of $C_{21}H_{20}O_6$ and is an orange-yellow crystalline lipophilic phenolic substance that, in solution, exists in equilibrium with its keto-enol tautomeric forms (Fig. 4.2). It is not very soluble in water and also not very stable, although its degradation increases in basic medium (Wang et al. 1997).

A large volume of experimental data establishing the therapeutic efficacy of curcumin in *in vitro* cellular level as well as in some ex vivo tumour-derived cancer cells/solid tumours like brain tumours; pancreatic, colorectal, lung, breast, prostate and skin cancers; leukaemia; hepatocellular carcinoma; and head and neck carcinoma and including cytotoxic effects on cancer stem cells and anti-metastatic activity were reviewed recently (Chinembiri et al. 2014; Perrone et al. 2015; Osterman and Wall 2015; Mock et al. 2015; Gupta and Pramanik 2016; Pavan et al. 2016; Chen et al. 2016; Jordan et al. 2016; Niedzwiecki et al. 2016; Deng et al. 2016; Klinger and Mittal 2016; Imran et al. 2016; Redondo-Blanco et al. 2017; Borges et al. 2017). Equally important, free curcumin was shown not to be cytotoxic to normal cells, including hepatocytes, mammary epithelial cells, kidney epithelial cells, lymphocytes, neural myelin sheaths and fibroblasts, at the dosages required for therapeutic efficacy against cancer cell lines (Bisht et al. 2010; Buyuklu et al. 2014; Soliman et al. 2014; Sordillo and Helson 2015; Yu et al. 2016). The scientific interest and pharmacological potential of curcumin anticancer effects become also evident from the number of patents on curcumin-based therapeutics registered in the last 5 years. Several studies have shown that curcumin can modulate a variety of cancer-related targets or pathways (Hatcher et al. 2008; Strimpakos and Sharma 2008; Deogade and Ghate 2015; Osterman and Wall 2015; Shanmugam et al. 2015; Kumar et al. 2016; Chen et al. 2016; Kocaadam and Sanlier 2017; Imran et al. 2016; Pavan et al. 2016; Allegra et al. 2017), which may be responsible for its effectiveness in combating cancer diseases. The mechanism of action includes modulation of CYP enzymes by elevation of transcription factor Nrf2 level via mitogen-activated protein kinase (MAPK) signalling pathway and Akt pathway (Schwertheim et al. 2017); mitotic catastrophe induction due to caspase activation and mitochondrial membrane polarization (Gali-Muhtasib et al. 2015); and promotion of autophagic cell death, an important death inducer in apoptosis-resistant cancer cells by beclin-1-dependent and beclin-1-independent pathways (Gali-Muhtasib et al. 2015; Yang et al. 2017b). Additionally, curcumin-mediated arrest of the cell cycle at the checkpoints G1, S-phase and G2/M phase, modulating the cell cycle regulators including upregulation of cyclin-dependent kinase inhibitors (CDKIs) (Dasiram et al. 2017), promotes the inhibition of transcription factor NF-κB by preventing nuclear translocation of NF-κB and attenuates DNA-binding ability of NF-κB contouring the problem of chemoresistance (Uwagawa and Yanaga 2015) and promotes the inhibition the crucial steps to angiogenesis by downregulation of the PGDF, VEGF and FGF expression and downregulation of MMPs via NF-κB, ERKs, MAPKs, PKC and PI3K inhibition (Fu et al. 2015). Moreover, several *in vitro* studies disclose the ability of curcumin to bind and inhibit tubulin polymerization, to bind DNA, to mediate modulation in expression of sirtuins like SIRT1, to promote the downregulation of the pro-oncogenic and upregulation of anti-oncogenic miRNAs expression, to mediate suppression of HMGB1 followed by protection from inflammation and oxidative stress and to act as a potent immune modulator (Ramasamy et al. 2015; Kumar et al. 2016; Haris et al. 2017; Sri Ramya et al. 2017). Despite the existing knowledge about the multiple action mechanisms of curcumin, it is far from a full understanding of their biological properties. For example, alongside the antiproliferative ability, curcumin may have a proliferative effect on some specific cells. Probably, curcumin survival and proliferative effects depend on its concentration, treatment period and the type of cells.

On the other hand, although curcumin is a natural product used in the diet (granted an acceptable daily intake level of 0.1–3 mg/kg-BW by the Joint FAO/WHO Expert Committee on Food Additives), the doses administered in clinical trials exceed those normally consumed in the diet. An average intake of curcumin of 60–100 mg per day (typical value to Indian diet) and even the systematic *in vivo* (at doses of curcumin up to 300 mg–3.5 g/kg-BW administered for up to 14–90 days) or clinical studies (1.2–12 g daily of oral intake for 6 weeks to 4 months) did not demonstrate any adverse effects at the population, animal and patient level (Strimpakos and Sharma 2008; Gupta et al. 2013). Moreover, curcumin has been reported to act as chemosensitizer to some clinical anticancer drugs (e.g., gemcitabine, paclitaxel and 5-fluorouracil, doxorubicin) and exhibits synergistic effect in combination with other natural products (e.g., resveratrol, honokiol, epigallocatechin-3-gallate, licochalcone and omega-3), aspects that could be used as an effective strategy to overcome tumour resistance and reduce recurrence (Deng et al. 2016; Klippstein et al. 2016; Pimentel-Gutiérrez et al. 2016; Allegra et al. 2017). These observations therefore suggest that a superior therapeutic index may be achieved with curcumin when used in combination and could be advantageous in the treatment of some tumours. Anyway, extensive studies are still needed to assess the exact mechanism of curcumin synergic effect. Nevertheless, the clinical translation of curcumin has been significantly hampered once it is poorly absorbed, improperly metabolized and shows poor systemic bioavailability. Consequently patients would have to consume up to 8–10 g of the free curcumin orally each day, in order for detectable levels in the circulation (Gupta et al. 2013; Kumar et al. 2016). Thus several strategies have been proposed to counter the bioavailability issue of curcumin involving (1) the use of adjuvants like piperine, which interfere with curcumin metabolism by glucuronidation; (2) curcumin formulations based on the nanotechnology with liposomes, micelles and phospholipid, among others; and (3) the use of curcumin analogues (Yallapu et al. 2015; Rahimi et al. 2016; Chen et al. 2016; Liu et al. 2016b; Puneeth et al. 2016; Sri Ramya et al. 2017). As a result of the anticancer potential of curcumin and despite the limitations to its application in clinical therapeutic, there are currently 17 open clinical studies involving curcumin, mainly studies of combined curcumin therapy with other substances for the treatment of several types of cancer.

4.3.3 Natural Compounds in Clinical Use

In addition to the compounds extracted from plants with high anticancer potential described above, several other compounds have successfully run the entire long, selective, expensive and bureaucratic process from their chemical identification to their effectiveness in the cancer treatment, being approved and available in the market. Each of these compounds has its history of success and limitations, which has explained in respective to their historical, chemical, pharmaceutical and clinical trials. The paclitaxel (trademark Taxol™; Fig. 4.3), the diterpene alkaloid, is the focus of several past studies during the last decades because of its complex structure and unique therapeutic mechanism (promoting tubulin assembly and stabilizing microtubules). It is isolated from the bark of *Taxus brevifolia* Nutt., which has become one of the most active cancer chemotherapeutic drugs that is marketed since 1992. It was first used to treat advanced ovarian cancer and then metastatic breast cancer, and from there paclitaxel has been found to be active in a wide variety of cancers in humans and also with veterinary use (Khanna et al. 2015; Bernabeu et al. 2017). Besides its anticancer activity, paclitaxel can exert a variety of positive influences on the immune system (Pawar et al. 2014) and play a potential role against neurodegenerative diseases (King et al. 2013) as well as inhibiting botulinum neurotoxin (Dadgar et al. 2013).

Despite its clinical success, research on this powerful natural anticancer agent does not slow down, looking for new solutions to some of the weaknesses of this drug like found new sources to obtained paclitaxel since it is not abundant in nature and its synthesis is very difficult (Liu et al. 2016c; Gallego et al. 2017; Wang et al. 2017b). Nowadays, the supply of paclitaxel for the production of drugs is met by semisynthesis from the precursor baccatin III (Fig. 4.3), which is present in much higher quantities and readily available from the needles of *T. brevifolia* and can be

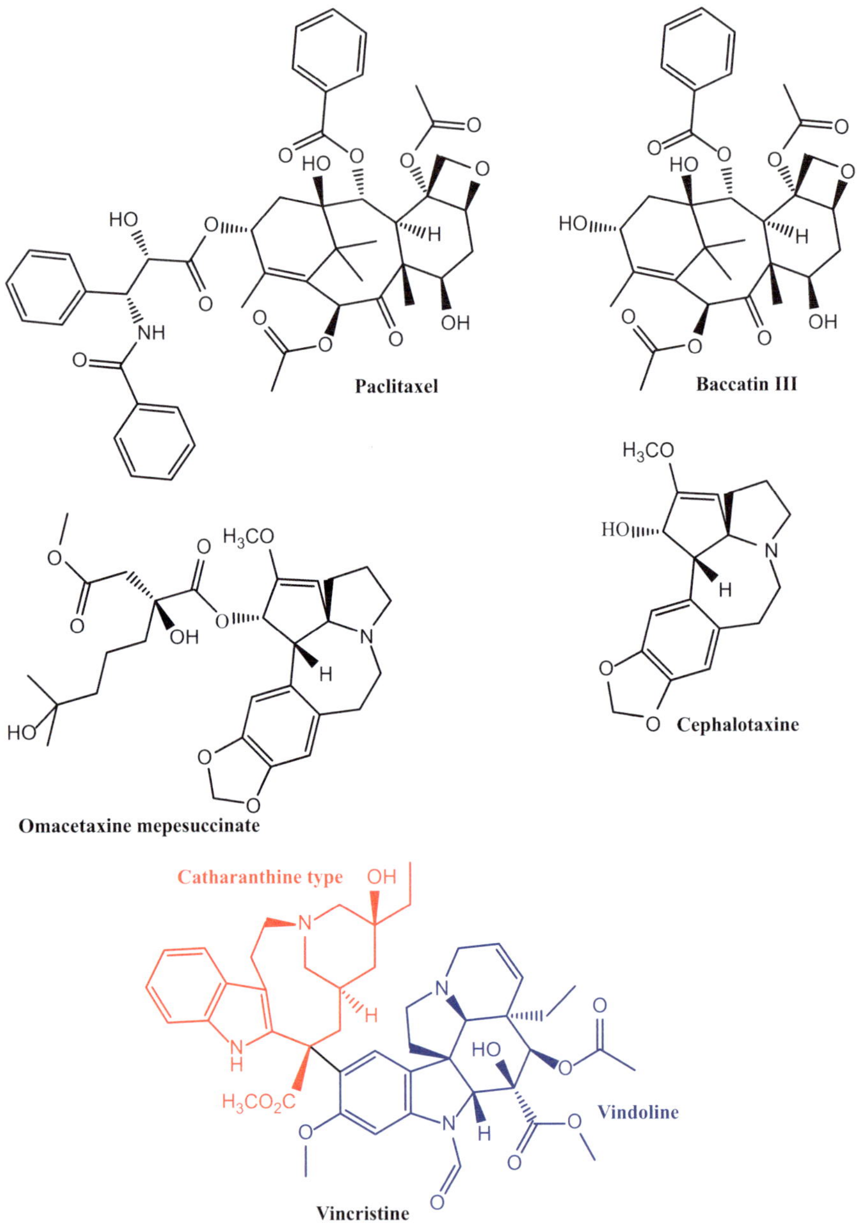

Fig. 4.3 Structure of natural compounds in clinical use and their precursors

efficiently transformed into paclitaxel. Fermentation technology with microbes or plant cell culture followed by extraction, chromatographic purification and crystallization gives paclitaxel (Zhang et al. 2016c; Gallego et al. 2017; Wang et al. 2017b). Recent research on paclitaxel also promotes the development of new delivery systems, new formulations and new derivatives to overcome its poor aqueous solubility, clinical neurotoxicity, neutropenia and multidrug resistance (Khanna et al. 2015; Barbuti and Chen 2015; Chen et al. 2015; Wen et al. 2016; Meng et al. 2016; Kim 2017; Brufsky 2017; Soliman 2017; Zong et al. 2017; Bernabeu et al. 2017).

In 2012, Abraxane®, an injectable nanoparticle of paclitaxel attached to a human protein albumin, was approved by EMA (EMA/99258/2015, EMEA/H/C/000778) as being more effective than conventional paclitaxel-containing medicines. Abraxane® is recommended to treat metastatic breast cancer. All the same, paclitaxel is already a blockbuster of pharmacy industry, and the most recent developments suggest that it will continue to be a success story for many more years. Other example of successful anticancer compound isolated from plants is omacetaxine mepesuccinate (Fig. 4.3), firstly known as homoharringtonine. Omacetaxine mepesuccinate story is one of the ups and downs; in fact it is the compound that has the record for the longest time spent before its approval by the US FDA (40 years). Before the acceptance of the name omacetaxine mepesuccinate, a cephalotaxine alkaloid had already been isolated from the evergreen tree bark *C. harringtonii* Koch and identified as homoharringtonine. This compound aroused great interest in the scientific community because of its potent antiproliferative activity against murine P-388 leukaemia cells with IC_{50} values of 17 nM (Chang et al. 2017). Scientific Chinese community was the first to demonstrate its role as an antileukaemic substance in animal models and also its effectiveness in the treatment of patients with acute myeloid leukaemia, chronic myeloid leukaemia, myelodysplastic syndrome, acute promyelocytic leukaemia and intrathecal therapy for central nervous system leukaemia, but not acute lymphoblastic leukaemia and solid tumours (Lü and Wang 2014; Isah 2016).

The main mechanism of action of homoharringtonine involves inhibition of protein translation and consequently inhibition of elongation step of protein synthesis. Homoharringtonine interacts with the ribosomal A-site, especially of the short-lived oncoproteins (BCR-ABL1) and of the antiapoptotic proteins that are upregulated in leukemic cells, leading to apoptosis (Lü and Wang 2014; Rosshandler et al. 2016). Unfortunately, its development into a useful anticancer drug was hindered by its difficult production. Actually the required *C. harringtonii* trees to isolate this alkaloid in the necessary quantity are unaccomplished. Additionally, the discovery and popularization of the tyrosine kinase inhibitor (TKI), like imatinib (synthetic compound sold under the brand names Gleevec® and Glivec®), to treat chronic myeloid leukaemia have drastically reduced the interest in homoharringtonine. The first semisynthetic homoharringtonine, currently known as omacetaxine mepesuccinate, Ceflatonin, CGX-653 or Myelostat, was obtained by direct esterification of cephalotaxine (Fig. 4.3), an alkaloid extracted from dry leaves of *Cephalotaxus* sp. (instead bark where it exists in lower percentage). This semisynthetic methodology is a supply guaranty and can be used in industrial production (Lü and Wang 2014).

Reports about patients in which the TKI therapy failed, due to resistance or intolerance, prompted the omacetaxine mepesuccinate reappearance. Positive results with these patients, as well as with those carrying the T315I mutation (Chung 2014), associated with the omacetaxine mepesuccinate ability to downregulation of Mcl-1 and effectively kill stem/progenitor cells (Nazha et al. 2013; Damlaj et al. 2016; Rosshandler et al. 2016), increased significantly the interest in omacetaxine mepesuccinate and its approval as commercial drug to treat chronic myeloid leukaemia. After complete clinical development, it was demonstrated that omacetaxine mepesuccinate produced durable haematologic and cytogenetic responses in patients with chronic myeloid leukaemia in chronic and accelerated phases, regardless of mutational status (Lü and Wang 2014; Cortes et al. 2015; Rosshandler et al. 2016; Damlaj et al. 2016). Omacetaxine mepesuccinate was approved in 2012 by FDA to US market under the trademark Synribo® to be used in the treatment of advanced chronic myeloid leukaemia resistant to two or more tyrosine kinase inhibitors (FDA database; Rosshandler et al. 2016). During clinical development, it was also confirmed that omacetaxine mepesuccinate exhibits bioavailability after subcutaneous administration similar to those observed following intravenous administration (Heiblig et al. 2014), and in February 2014, after demonstrating that reconstituted omacetaxine mepesuccinate can remain stable and sterile if refrigerated for up to 6 days, the FDA granted full approval of omacetaxine mepesuccinate for subcutaneous injection. This was indeed an improvement because patients have the opportunity to self-administer their therapy (Shen et al. 2014; Rosshandler et al. 2016; Damlaj et al. 2016).

Myelosuppression (mainly thrombocytopenia, neutropenia and anaemia) is the primary haematologic toxicity observed with omacetaxine mepesuccinate treatment. However, its damage can be limited with frequent parameter control, adequate dose modifications and patient training for signs and symptoms and of appropriate medical support (Chung 2014; Akard et al. 2016). Thus, the myelosuppression events should not prevent omacetaxine mepesuccinate administration. Recent studies showed new improvement for omacetaxine mepesuccinate use (Lam et al. 2016). It is a good candidate to work synergistically with FLT3 inhibitors in the treatment of FLT3-ITD acute myeloid leukaemia subtype and can be used in phase II/III clinical trials. The success of omacetaxine mepesuccinate continues, being involved in 23 clinical studies, 6 (NCT03135054, NCT03170895, NCT02440568, NCT02078960, NCT01873495, NCT02159872) of them are still open, and developments in the near future regarding therapies combined with other drugs in therapeutic use and the application in therapy of other haematologic malignancies.

Similarly, vincristine has also proven success in cancer therapy (Fig. 4.3). It is a natural alkaloid extracted from the leaves of *C. roseus* (L.) G. Don and is one of the first natural compounds to be identified as having oncologic properties although initially it has been studied, along with vinblastine (Figs. 4.4 and 4.5), for their antidiabetic properties (Gidding et al. 1999; Vipasha et al. 2016). In chemical terms, vincristine is asymmetrical and dimeric, composed of a dihydroindole nucleus, vindoline, linked by a carbon-carbon bond to an indole nucleus, catharanthine-type portion (Fig. 4.3). It has been extensively used in chemotherapy for more than

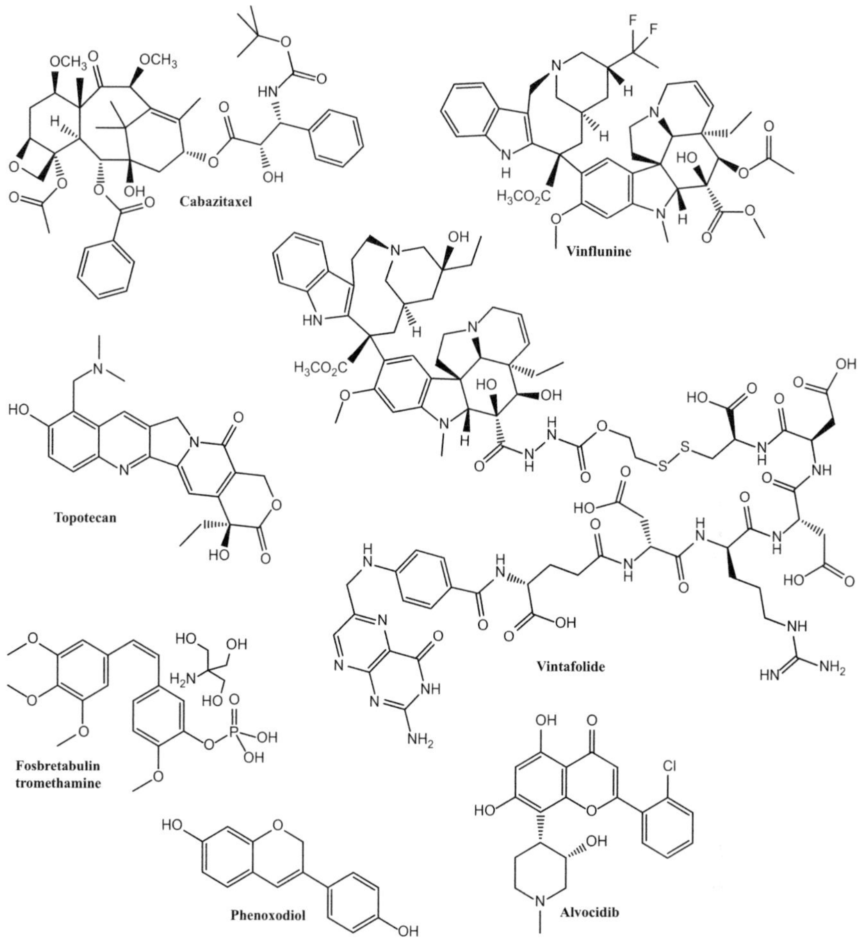

Fig. 4.4 Structure of the derivatives of natural compounds

50 years (mainly under trade name Oncovin®) in both adult and paediatric oncology practice against acute lymphoblastic leukaemia, rhabdomyosarcoma, neuroblastoma, lymphomas and nephroblastoma (Moore and Pinkerton 2009; Basmadjian et al. 2014).

Vincristine exerts its cytotoxic effects via stabilization of the microtubule structure by blocking the polymerization of tubulin molecules into microtubules and thus preventing the formation of the mitotic spindle, disrupt of intracellular transport and decrease of tumour blood flow, with the latter probably as a consequence of anti-angiogenesis (Wang et al. 2016c). At the end of vincristine-mediated cellular disruption, apoptosis occurs and consequently inhibition of cancer cell propagation. Although the applications of vincristine are extensive and with success, the side effects of vincristine cannot be ignored, and the major effect is neurotoxicity, a dose-

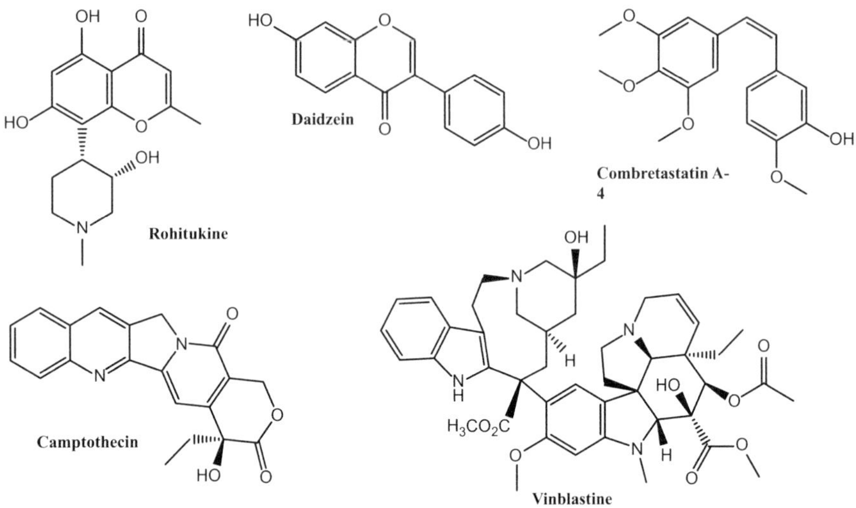

Fig. 4.5 Structure of the lead compounds

dependent side effect causing peripheral neuropathy (Moore and Pinkerton 2009; Wang et al. 2016c). The symptoms appear after a few vincristine administrations and normally disappear a few days after the treatment, although several cases of permanent sequels were reported (Jain et al. 2014; Mora et al. 2016; Velde et al. 2017).

Despite the long history of the clinical application of vincristine, there are three factors that diminish its impact in the fighting cancer: (1) its antitumour mechanism which is cell cycle-specific and the duration of its exposure to tumour cells that can significantly affect its activity, (2) its rapid clearance rate and wide biodistribution (Shah et al. 2016a; Wang et al. 2016c) and (3) vincristine-related neurotoxicity. One way to mitigate these three weaknesses is the encapsulation of vincristine into liposomes, and in 2012 the sphingomyelin/cholesterol liposomal vincristine (Marqibo®) was approved by FDA to treat adults with relapsed acute lymphoblastic leukaemia. Clinical trials involving Marqibo® are underway to paediatric patients with relapsed or chemotherapy-refractory solid tumours and leukaemia (NCT01222780; Shah et al. 2016b). Combined therapy has been the direction followed in the research involving vincristine. In fact, combination chemotherapy cannot only enhance the destruction of tumour cells but also decrease toxicity and drug resistance with drugs exhibiting different mechanisms of action. Therefore, there are in-progress open clinical trials involving combined vincristine therapy (NCT02879643; NCT01527149). Very recently, the treatment of infantile fibrosarcoma by adjuvant therapy after excision, using vincristine and dactinomycin, where the duration of chemotherapy was determined according to tumour response, was reported. At the end, there was no functional impairment and no evidence of recurrence at 18 months after therapy (Yoshihara et al. 2017). Vincristine and some of its encapsulated formulations are involved in clinical studies in which they are tested against other types of cancer such as small cell lung cancer (NCT02566993), advanced cervical cancer (NCT02471027) or liver cancer (NCT00980460). Thus, vincristine alone or combined, encapsulated or not, continues its long history of therapeutic application, increasingly effective and each time with fewer side effects.

4.4 Natural Compound Derivatives with Anticancer Applications

Several natural compounds with unique anticancer effects are not used in clinical practice due to their physico-chemical properties and/or their toxicity. On the other hand, plant occurring secondary metabolites often can be excellent lead compounds for drug development despite sometimes having complex structures and limited oral bioavailability. Thus, modifying the chemical structure of these more promising compounds is one strategic way to increase their anticancer action and selectivity, improve their ADME (absorption, distribution, metabolism and excretion) properties and decrease their toxicity and side effects. The most recent anticancer natural compound-derived drugs as well as their structure and area of application are listed and depicted in Fig. 4.4. Comparing these structures with their precursors, it can be noticed that in some cases, the modifications were slight, such as protection of hydroxy groups (e.g., fosbretabulin tromethamine), addition (e.g., topotecan) or elimination (e.g., phenoxodiol) of a functional group and introduction of bulky groups (e.g., alvocidib), among others. The most noticeable one is the vinblastine transformation into vintafolide (Fig. 4.4) that involves the introduction of a peptide-type chain. The mentioned transformations do not reveal a typical change, quite the opposite. So each case is unique, and it would be beneficial if more SAR studies were available as well as action mechanisms were disclosed. The above-mentioned strategies for the development of new anticancer drugs were explained in detail by using the examples of dimethylaminoparthenolide (DMAPT), an important bioactive compound in preclinical development, and minnelide, a prodrug in clinic trial development.

Dimethylaminoparthenolide (DMAPT) (Fig. 4.6) is a parthenolide derivative obtained by the introduction of an amino group, which can improve the hydrophilic character if a salt form is used. DMAPT showed the ability to selectively eradicate LSCs and thus led to clinical trials for the treatment of AML, ALL and CLL in the United Kingdom (Peese 2010), but unfortunately the conclusions of this study are not known. It is, however, recognized that DMAPT exhibits a LD_{50} of 1.7 μM against AML cell lines and *in vivo* studies, using rats, demonstrated an oral bioavailability of 70% (Peese 2010). *In vitro* and *in vivo* studies with DMAP, alone or in combination with other drugs, have evidenced its beneficial effect on chemoprevention and treatment of pancreatic cancer, mainly by inhibition of NF-κB target genes (Yip-Schneider et al. 2013). It induces cytotoxic effects on hormone-insensitive

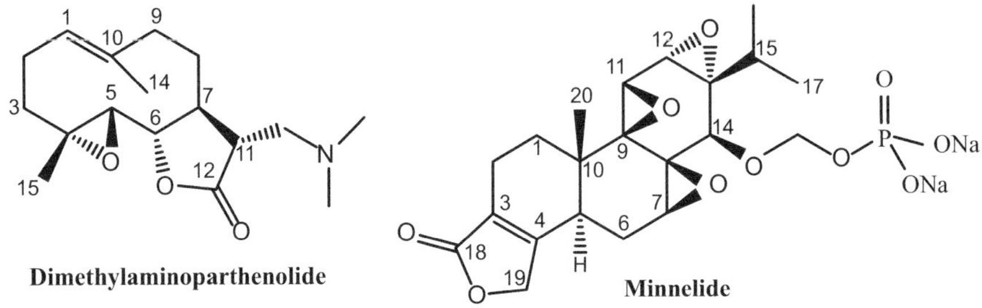

Dimethylaminoparthenolide **Minnelide**

Fig. 4.6 Structures of dimethylaminoparthenolide and minnelide

human breast cancer MDA-MB-231 cells, through a similar mechanism of parthenolide (D'Anneo et al. 2013), being also highly promising for the treatment of glioma. In fact it was tested for antiproliferative activity against glioblastoma multiform cell lines, the most aggressive form of glioma (Ortensi et al. 2013; Ostrom et al. 2015), and the obtained results (Hexum et al. 2015) demonstrate that (a) DMAPT and parthenolide were found to be similarly active against human and murine glioma cell lines; (b) DMAPT can be transported through the blood-brain barrier; (c) DMAPT accumulates in brain tissue with biologically relevant concentrations; (d) DMAPT is nontoxic even following continuous dosing for several weeks (100 mg/kg, 30 days); and finally (e) a statistically significant delay in tumour growth kinetics *in vivo* is observed after treatments with DMAPT. All these results showed that DMAPT is highly promising for the treatment of glioma and may become one of the next-generation multiform drugs for glioblastoma.

An *in vivo* study demonstrates that DMAPT significantly reduce the number and size of lung tumours, mediated, at least in part, via inhibition of STAT3 signalling pathway, suppressing levels of pSTAT3 and MCL1 in lung tissues of NNK-treated mice (Song et al. 2014). Nakshatri et al. (2015) demonstrated that DMAPT is a potent pharmacological agent that can reverse NF-κB-dependent and NF-κB-independent cancer-specific epigenetic abnormalities. DMAPT showed ability to upregulate NSD1 and SETD2, two tumour suppressors, upon NF-κB inhibition. Additionally DMAPT increased KMT5C and H4K20me3 independent of its NF-κB inhibition attribute and adding KMT5C to the list of NF-κB-independent epigenetic targets of parthenolide (Nakshatri et al. 2015). The current knowledge and mechanisms of action of DMAPT are insufficient and its potential application in a variety of biological situations yet to be explored by future investigators.

Similarly, minnelide is a triptolide with low water solubility and serious toxicity. Thus, the researchers have tried to find ways to enhance its bioavailability and reduce its toxicity, by systematic structural modifications, mainly protecting the 14-hydroxy group of triptolide (Fig. 4.6). The details of its synthesis are under patent protection (WO/2010/129918), but it is known that it involves a four-step pathway: (1) react triptolide with acetic acid and acetic anhydride at room temperature for 5 days, (2) transform into a dibenzyl ester derivative by reaction with dibenzylphosphate, (3) obtain the corresponding dihydrogen phosphate by reductive removal of dibenzyl group and (4) obtain the disodium salt by subsequent reaction with sodium carbonate (Chugh et al. 2012). In fact, minnelide or 14-*O*-phosphonooxymethyltriptolide disodium salt (Fig. 4.6) is ineffective against tumour or non-tumour cells, but in the presence of alkaline phosphatase, which is common in all tissues in the body, the phosphate ester group is cleaved, and the unstable *O*-hydroxymethyl intermediate spontaneously will release formaldehyde and triptolide (Rivard et al. 2014). Because of this behaviour in biological systems, minnelide is classified as a prodrug. It is much more water soluble than triptolide (it is a sodium salt), and no overt signs of toxicity were observed when animals received minnelide for 385 days at a concentration of 0.42 mg/kg (Chugh et al. 2012). However, two cases of patients within a phase I multicentre trial exhibit symptoms of minnelide acute toxicity involving the cerebellar cortex that fade with discontinuation of therapy (Roshan et al. 2017).

Minnelide is more effective in preclinical studies in multiple animal models of pancreatic cancer (0.28 mg/kg daily) than gemcitabine (100 mg/kg twice a week), which is one of the first-line chemotherapeutic agents for pancreatic cancer (Chugh et al. 2012). After this first report about the potent minnelide activity against several *in vivo* models of pancreatic cancer (Chugh et al. 2012), the efficacy of minnelide has been increasingly documented. Minnelide exhibited antiproliferative effects and induced apoptosis in non-small cell lung carcinoma and tumours partially mediated by inhibition of NF-κB pathway (Rousalova et al. 2013). Alsaied et al. (2014) report a strong *in vivo* hepatocellular carcinoma tumour growth inhibition rates after 2 weeks of treatment with minnelide [84% at 0.21 mg/kg (i.p.), while sorafenib 10 mg/kg orally exhibits only 59%]. In the same year, the effectiveness of minnelide, in monotherapy or combination therapy, to inhibit *in vitro* and *in vivo* the epithelial ovarian cancer was also demonstrated (Rivard et al. 2014). Banerjee and Saluja (2015) also showed that minnelide was a more effective drug reducing tumour burden and tumour-related morbidity in different and complementary *in vivo* pancreatic cancer models. Moreover, they confirmed that the combination of minnelide with sorafenib is indeed a successful mixture to treat hepatocellular carcinoma (Banerjee and Saluja 2015). Minnelide, at a low dose of 12.5 nM, showed not only significant pancreatic ductal adenocarcinoma tumour regression but also decreases significantly the pancreatic cancer stem cell population (Nomura et al. 2015). Modi et al. (2016) show that combination of low doses of minnelide and oxaliplatin does not have potential serious adverse effects, but this combined therapy is very effective to potentiated apoptotic cell death by suppressing oxaliplatin-induced DNA damage repair pathway in pancreatic cancer. Recently, it was demonstrated that minnelide (at a dose of 0.21 mg/kg) regressed subcutaneous tumours derived from castration-resistant prostate cancer cells more efficacious than the standard docetaxel and enzalutamide and induces apoptosis by downregulation of androgen receptor and its splice variants (Isharwal et al. 2017). Minnelide has been the subject of a phase I clinical trial in the treatment of pancreatic cancer and other gastrointestinal tumours (NCT01927965). Moreover phase I trials to study Minnelide™ capsules in patients with advanced solid tumours (gastric, breast, pancreatic, colon-rectal, stomach and prostate metastatic cancer) (NCT03129139) and phase II trials in patients with refractory pancreatic cancer (NCT03117920) were recently initiated. All these studies show that minnelide, a water-soluble derivative of natural compound extracted from a Chinese herb, is very effective against a number of malignant diseases and points out to its high pharmaceutical potential.

4.5 Conclusions and Future Prospects

Medicinal plants and their anticancer secondary metabolites can serve as an important tool in chemotherapy. Naturally, a long way from the isolation to the treatment utilization is necessary, but the successful cases herein presented are, surely, a motivation to find new one. It is clear that plants, in particular those known as medicinal ones, are an important source of biologically active molecules that may be used as drugs or inspire the synthesis of new and safer compounds. In this regard several bioactive metabolites should be subjected to in-depth studies to fully understand their mode of action and consequently mimic their structures to develop new anticancer drugs. But no less important is the phytochemical study of several plant extracts aiming the isolation and characterization of the active metabolites. New active compounds can be obtained as well as new sources of inspiration to develop anticancer drugs.

Acknowledgement Thanks are due to FCT/MEC for the financial support of the QOPNA research unit (FCT UID/QUI/00062/2013) and the cE3c centre (FCT UID/BIA/00329/2013) through national funds and, where applicable, co-financed by the FEDER, within the PT2020 Partnership Agreement.

References

Adetutu A, Morgan WA, Corcoran O (2011) Ethnopharmacological survey and *in vitro* evaluation of wound-healing plants used in South-western Nigeria. J Ethnopharmacol 137:50–56

Adio AM (2009) Germacrenes A-E and related compounds: thermal, photochemical and acid induced transannular cyclizations. Tetrahedron 65:1533–1552

Aggarwal BB, Bhardwaj A, Aggarwal RS, Seeram NP, Shishodia S, Takada Y (2004) Role of resveratrol in prevention and therapy of cancer: preclinical and clinical studies. Anticancer Res 24:2783–2840

Ahmed I, Roy BC, Subramaniam D, Ganie SA, Kwatra D, Dixon D, Anant S, Zargar MA, Umar S (2016) An ornamental plant targets epigenetic signaling to block cancer stem cell-driven colon carcinogenesis. Carcinogenesis 37:385–396

Akard L, Kantarjian HM, Nicolini FE, Wetzler M, Lipton JH, Baccarani M, Jean Khoury H, Kurtin S, Li E, Munteanu M, Cortes J (2016) Incidence and management of myelosuppression in patients with chronic- and accelerated-phase chronic myeloid leukemia treated with omacetaxine mepesuccinate. Leuk Lymphoma 57:654–665

Alamolhodaei NS, Tsatsakis AM, Ramezani M, Hayes AW, Karimi G (2017) Resveratrol as MDR reversion molecule in breast cancer: an overview. Food Chem Toxicol 103:223–232

Ali F, Rahul, Naz F, Jyoti S, Siddique YH (2017a) Health functionality of apigenin: a review. Int J Food Prop 20:1197–1238

Ali MTM, Zahari H, Aliasak A, Lim SM, Ramasamy K, Macabeoacabeo APG (2017b) Synthesis, characterization and cytotoxic activity of betulinic acid and sec-betulinic acid derivatives against human colorectal carcinoma. Orient J Chem 33:242–248

Ali-Seyed M, Jantan I, Vijayaraghavan K, Bukhari SN (2016) Betulinic acid: recent advances in chemical modifications, effective delivery, and molecular mechanisms of a promising anticancer therapy. Chem Biol Drug Des 87:517–536

Aljuffali IA, Fang CL, Chen CH, Fang JY (2016) Nanomedicine as a strategy for natural compound delivery to prevent and treat cancers. Curr Pharm Des 22:4219–4231

Allam S, Moharam M, Alarfaj G (2014) Assessing patients preference for integrating herbal medicine within primary care services in Saudi Arabia. J Evid-Based Compl Altern Med 19:205–210

Allegra A, Innao V, Russo S, Gerace D, Alonci A, Musolino C (2017) Anticancer activity of curcumin and its analogues: preclinical and clinical studies. Cancer Investig 35:1–22

Alonso-Castro AJ, Villarreal ML, Salazar-Olivo LA, Gomez-Sanchez M, Dominguez F, Garcia-Carranca A (2011) Mexican medicinal plants used for cancer treatment: pharmacological, phytochemical and ethnobotanical studies. J Ethnopharmacol 133:945–972

Alsaied OA, Sangwan V, Banerjee S, Krosch TC, Chugh R, Saluja A, Vickers SM, Jensen EH (2014) Sorafenib and triptolide as combination therapy for hepatocellular carcinoma. Surgery 156:270–279

Antônio E, Junior ORA, de Araújo IS, Khalil NM, Mainardes RM (2017) Poly(lactic acid) nanoparticles loaded with ursolic acid: characterization and *in vitro* evaluation of radical scavenging activity and cytotoxicity. Mater Sci Eng C 71:156–166

Aziato L, Odai PNA (2016) Exploring the safety and clinical use of herbal medicine in the contemporary Ghanaian context: a descriptive qualitative study. J Herb Med 8:62–67

Banerjee S, Saluja A (2015) Minnelide, a novel drug for pancreatic and liver cancer. Pancreatology 15:39–43

Baranello MP, Bauer L, Jordan CT, Benoit DSW (2015) Micelle delivery of parthenolide to acute myeloid leukemia cells. Cell Mol Bioeng 8:455–470

Barbuti AM, Chen ZS (2015) Paclitaxel through the ages of anticancer therapy: exploring its role in chemoresistance and radiation therapy. Cancers 7:2360–2371

Basmadjian C, Zhao Q, Bentouhami E, Djehal A, Nebigil CG, Johnson RA, Serova M, de Gramont A, Faivre S, Raymond E, Désaubry LG (2014) Cancer wars: natural products strike back. Front Chem 2:20. https://doi.org/10.3389/fchem.2014.00020

Ben-Arye E, Mahajna J, Aly R, Ali-Shtayeh MS, Bentur Y, Lev E, Deng G, Samuels N (2016) Exploring an herbal "wonder cure" for cancer: a multidisciplinary approach. J Cancer Res Clin Oncol 142:1499–1508

Benhadji KA, Serova M, Ghoul A, Cvitkovic E, Le Tourneau C, Ogbourne SM, Lokiec F, Calvo F, Hammel P, Faivre S, Raymond E (2008) Antiproliferative activity of PEP005, a novel ingenol angelate that modulates PKC functions, alone and in combination with cytotoxic agents in human colon cancer cells. Br J Cancer 99:1808–1815

Berman B (2015) Safety and tolerability of ingenol mebutate in the treatment of actinic keratosis. Exp Opin Drug Saf 14:1969–1978

Bernabeu E, Cagel M, Lagomarsino E, Moretton M, Chiappetta DA (2017) Paclitaxel: what has been done and the challenges remain ahead. Int J Pharm 526:474–495

Bhandary MJ, Chandrashekar KR, Kaveriappa KM (1995) Medical ethnobotany of the Siddis of Uttara Kannada district, Karnataka, Indian. J Ethnopharm 47:149–158

Bhanot A, Sharma R, Noolvi MN (2011) Natural sources as potential anticancer agents: a review. Int J Phytomed 3:9–26

Bisht K, Wagner KH, Bulmer AC (2010) Curcumin, resveratrol and flavonoids as anti-inflammatory, cyto- and DNA-protective dietary compounds. Toxicology 278:88–100

Borges GÁ, Rêgo DF, Assad DX, Coletta RD, De Luca CG, Guerra EN (2017) *In vivo* and *in vitro* effects of curcumin on head and neck carcinoma: a systematic review. J Oral Pathol Med 46:3–20

Boutennoun H, Boussouf L, Rawashdeh A, Al-Qaoud K, Abdelhafez S, Kebieche M, Madani K (2017) *In vitro* cytotoxic and antioxidant activities of phenolic components of Algerian *Achillea odorata* leaves. Arab J Chem 10:403–409

Bridgeman BB, Wang P, Ye B, Pelling JC, Volpert OV, Tong X (2016) Inhibition of mTOR by apigenin in UVB-irradiated keratinocytes: a new implication of skin cancer prevention. Cell Signal 28:460–468

Brufsky A (2017) *nab*-Paclitaxel for the treatment of breast cancer: an update across treatment settings. Exp Hematol Oncol 6:7. https://doi.org/10.1186/s40164-017-0066-5

Burns J, Yokota T, Ashihara H, Lean MEJ, Crozier A (2002) Plant foods and herbal sources of resveratrol. J Agric Food Chem 50:3337–3340

Buyuklu M, Kandemir FM, Ozkaraca TM, Bakirci EM, Topal E (2014) Protective effect of curcumin against contrast induced nephropathy in rat kidney: what is happening to oxidative stress, inflammation, autophagy and apoptosis. Eur Rev Med Pharmacol Sci 18:461–470

Carlisi D, Buttitta G, Di Fiore R, Scerri C, Drago-Ferrante R, Vento R, Tesoriere G (2016) Parthenolide and DMAPT exert cytotoxic effects on breast cancer stem-like cells by inducing oxidative stress, mitochondrial dysfunction and necrosis. Cell Death Dis 7:e2194. https://doi.org/10.1038/cddis.2016.94

Chainani-Wu N (2003) Safety and anti-inflammatory activity of curcumin: a component of turmeric (*Curcuma longa*). J Altern Complement Med 9:161–168

Chan SF, Chen YY, Lin JJ, Liao CL, Ko YC, Tang NY, Kuo CL, Liu KC, Chung JG (2017) Triptolide induced cell death through apoptosis and autophagy in murine leukemia WEHI-3 cells *in vitro* and promoting immune responses in WEHI-3 generated leukemia mice *in vivo*. Environ Toxicol 32:550–568

Chang Y, Meng FC, Wang R, Wang CM, Lu XY, Zhang QW (2017) Chemistry, bioactivity, and the structure-activity relationship of cephalotaxine-type alkaloids from *Cephalotaxus* sp. In: Atta-ur-Rahman FRS (ed) Studies in natural products chemistry. Elsevier Science Publishers, Amsterdam, pp 339–373

Chen J, Li W, Yao H, Xu J (2015) Insights into drug discovery from natural products through structural modification. Fitoterapia 103:231–241

Chen J, He ZM, Wang FL, Zhang ZS, Liu XZ, Zhai DD, Chen WD (2016) Curcumin and its promise as an anticancer drug: an analysis of its anticancer and antifungal effects in cancer and associated complications from invasive fungal infections. Eur J Pharmacol 772:33–42

Chen QZ, Li Y, Shao Y, Zeng YH, Ren WY, Liu RX, Zhou LY, Hu XL, Huang M, He F, Sun WJ, Wu K, He BC (2017) TGF-β1/PTEN/PI3K signaling plays a critical role in the anti-proliferation effect of tetrandrine in human colon cancer cells. Int J Oncol 50:1011–1021

Chinembiri TN, du Plessis LH, Gerber M, Hamman JH, du Plessis J (2014) Review of natural compounds for potential skin cancer treatment. Molecules 19:11679–11721

Chiu CH, Chou YC, Lin JP, Kuo CL, Lu HF, Huang YP, Yu CC, Lin ML, Chung JG (2015) Chloroform extract of *Solanum lyratum* induced G0/G1 arrest via p21/p16 and induced apoptosis via reactive oxygen species, caspases and mitochondrial pathways in human oral cancer cell lines. Am J Chin Med 43:1453–1469

Chudzik M, Korzonek-Szlacheta I, Król W (2015) Triterpenes as potentially cytotoxic compounds. Molecules 20:1610–1625. https://doi.org/10.3390/molecules20011610

Chugh R, Sangwan V, Patil SP, Dudeja V, Dawra RK, Banerjee S, Schumacher RJ, Blazar BR, Georg GI, Vickers SM, Saluja AK (2012) A preclinical evaluation of minnelide as a therapeutic agent against pancreatic cancer. Sci Transl Med 4:156ra139. https://doi.org/10.1126/scitranslmed.3004334

Chung C (2014) Omacetaxine for treatment-resistant or treatment-intolerant adult chronic myeloid leukemia. Am J Health Syst Pharm 71:279–288

Collier NJ, Ali FR, Lear JT (2014) Ingenol mebutate: a novel treatment for actinic keratosis. Clin Pract 11:295–306

Cortes JE, Kantarjian HM, Rea D, Wetzler M, Lipton JH, Akard L, Khoury HJ, Michallet M, Guerci-Bresler A, Chuah C, Hellmann A, Digumarti R, Parikh PM, Legros L, Warzocha K,

Baccarani M, Li E, Munteanu M, Nicolini FE (2015) Final analysis of the efficacy and safety of omacetaxine mepesuccinate in patients with chronic or accelerated-phase chronic myeloid leukemia: results with 24 months of follow-up. Cancer 121:1637–1644

Cottart CH, Nivet-Antoine V, Beaudeux JL (2014) Review of recent data on the metabolism, biological effects, and toxicity of resveratrol in humans. Mol Nutr Food Res 58:7–21

Csuk R, Schmuck K, Schäfer R (2006) A practical synthesis of betulinic acid. Tetrahedron Lett 47:8769–8770

Czyz M, Koprowska K, Sztiller-Sikorska M (2013) Parthenolide reduces the frequency of ABCB5-positive cells and clonogenic capacity of melanoma cells from anchorage independent melanospheres. Cancer Biol Ther 14:135–145

D'Anneo A, Carlisi D, Lauricella M, Puleio R, Martinez R, Di Bella S, Di Marco P, Emanuele S, Di Fiore R, Guercio A, Vento R, Tesoriere G (2013) Parthenolide generates reactive oxygen species and autophagy in MDA-MB231 cells. A soluble parthenolide analogue inhibits tumour growth and metastasis in a xenograft model of breast cancer. Cell Death Dis 4:e891. https://doi.org/10.1038/cddis.2013.415

Dadgar S, Ramjan Z, Floriano WB (2013) Paclitaxel is an inhibitor and its boron dipyrromethene derivative is a fluorescent recognition agent for botulinum neurotoxin subtype A. J Med Chem 56:2791–2803

Damlaj M, Lipton JH, Assouline SE (2016) A safety evaluation of omacetaxine mepesuccinate for the treatment of chronic myeloid leukemia. Exp Opin Drug Saf 15:1279–1286

Das A, Dhanjal JK (2015) Medicinal plants, a gold mine of anticancer compounds. Am Int J Res Formal Appl Nat Sci 9:14–23

Das C, Mishra P, Das D, Maharana L, Ghosh G (2015) Development of quality control parameters for standardization of *Triumfetta rhomboidea*. Int J Pharmacol Phytochem Res 7:934–942

Dasiram JD, Ganesan R, Kannan J, Kotteeswaran V, Sivalingam N (2017) Curcumin inhibits growth potential by G1 cell cycle arrest and induces apoptosis in p53-mutated COLO 320DM human colon adenocarcinoma cells. Biomed Pharmacother 86:373–380

Deng Y, Verron E, Rohanizadeh R (2016) Molecular mechanisms of anti-metastatic activity of curcumin. Anticancer Res 36:5639–5647

Deogade SC, Ghate S (2015) Curcumin: therapeutic applications in systemic and oral health. Int J Biol Pharma Res 6:281–290

Dianhar H, Syah YM, Mujahidin D, Hakim EH, Juliawaty LD (2014) A flavone derivative from *Sesbania sesban* leaves and its cytotoxicity against murine leukemia P-388 cells. AIP Conf Proc 1589:411–413

Dias DA, Urban S, Roessner U (2012) A historical overview of natural products in drug discovery. Metabolites 2:303–336

Doan HQ, Gulati N, Levis WR (2012) Ingenol mebutate: potential for further development of cancer immunotherapy. J Drugs Dermatol 11:1156–1157

Drury (1991) Plants and wart cures in England from the seventeenth to the nineteenth century: some examples. Folklore 102:97–100

Duan R, Tian F, Sun J (2016) Structural basis and energy landscape of apigenin-induced cancer cell apoptosis mechanism in PI3K/Akt pathway. Mol Simul 42:138–148

Duhem N, Danhier F, Préat V (2014) Vitamin E-based nanomedicines for anticancer drug delivery. J Control Release 182:33–44

Enioutina EY, Salis ER, Job KM, Gubarev MI, Krepkova LV, Sherwin CM (2017) Herbal medicines: challenges in the modern world. Part 5. Status and current directions of complementary and alternative herbal medicine worldwide. Exp Rev Clin Pharmacol 10:327–338

Erdogan S, Doganlar O, Doganlar ZB, Serttas R, Turkekul K, Dibirdik I, Bilir A (2016) The flavonoid apigenin reduces prostate cancer CD44+ stem cell survival and migration through PI3K/Akt/NF-κB signalling. Life Sci 162:77–86

Erdogan S, Turkekul K, Serttas R, Erdogan Z (2017) The natural flavonoid apigenin sensitizes human CD44+ prostate cancer stem cells to cisplatin therapy. Biomed Pharmacother 88:210–217

Fabricant DS, Farnsworth NR (2001) The value of plants used in traditional medicine for drug discovery. Environ Health Perspect 109:69–75
Fonrose X, Ausseil F, Soleilhac E, Masson V, David B, Pouny I, Cintrat J-C, Rousseau B, Barette C, Massiot G, Lafanechère L (2007) Parthenolide inhibits tubulin carboxypeptidase activity. Cancer Res 67:3371–3378
Formagio AS, Vieira MC, Volobuff CR, Silva MS, Matos AI, Cardoso CA, Foglio MA, Carvalho JE (2015) *In vitro* biological screening of the anticholinesterase and antiproliferative activities of medicinal plants belonging to Annonaceae. Braz J Med Biol Res 48:308–315
Fu Z, Chen X, Guan S, Yan Y, Lin H, Hua ZC (2015) Curcumin inhibits angiogenesis and improves defective hematopoiesis induced by tumor-derived VEGF in tumor model through modulating VEGF-VEGFR2 signaling pathway. Oncotarget 6:19469–19482
Gali-Muhtasib H, Hmadi R, Kareh M, Tohme R, Darwiche N (2015) Cell death mechanisms of plant-derived anticancer drugs: beyond apoptosis. Apoptosis 20:1531–1562
Gallego A, Malik S, Yousefzadi M, Makhzoum A, Tremouillaux-Guiller J, Bonfill M (2017) Taxol from *Corylus avellana*: paving the way for a new source of this anticancer drug. Plant Cell Tissue Organ Cult 129:1–16
Gao Y, Jia ZL, Kong XY, Li Q, Chang DZ, Wei DY, Le XD, Huang SD, Huang SY, Wang LW, Xie KP (2011) Combining betulinic acid and mithramycin a effectively suppresses pancreatic cancer by inhibiting proliferation, invasion, and angiogenesis. Cancer Res 71:5182–5193
Genova C, Alama A, Coco S, Rijavec E, Dal Bello MG, Vanni I, Biello F, Barletta G, Rossi G, Grossi F (2016) Vinflunine for the treatment of non-small cell lung cancer. Exp Opin Invest Drugs 25:1447–1455
Ghantous A, Sinjab A, Herceg Z, Darwiche N (2013) Parthenolide: from plant shoots to cancer roots. Drug Discov Today 18:894–905
Gheorgheosu D, Duicu O, Dehelean C, Soica C, Muntean D (2014) Betulinic acid as a potent and complex antitumor phytochemical: a minireview. Anti Cancer Agents Med Chem 14:936–945
Gidding CE, Kellie SJ, Kamps WA, de Graaf SS (1999) Vincristine revisited. Crit Rev Oncol Hematol 29:267–287
Giddings LA, Newman DJ (2013) Microbial natural products: molecular blueprints for antitumor drugs. J Indian Microbiol Biotechnol 40:1181–1210
Gill KK, Kamal MM, Kaddoumi A, Nazzal S (2016) EGFR targeted delivery of paclitaxel and parthenolide co-loaded in PEG-Phospholipid micelles enhance cytotoxicity and cellular uptake in non-small cell lung cancer cells. J Drug Deliv Sci Technol 36:150–155
Gong X, Chen Y, Wu Y (2015) Absorption and metabolism characteristics of triptolide as determined by a sensitive and reliable LC-MS/MS method. Molecules 20:8928–8940
Gotta H, Adolf W, Opferkuch HJ, Hecker E (1984) On the active principles of the Euphorbiaceae IX. Ingenane type diterpene esters from 5 *Euphorbia* species. Z Naturforsch Sect B 39:683–694
Graham JG, Quinn ML, Fabricant DS, Farnsworth NR (2000) Plants used against cancer – an extension of the work of Jonathan Hartwell. J Ethnopharmacol 73:347–377
Guerram M, Jiang ZZ, Sun L, Zhu X, Zhang LY (2015) Antineoplastic effects of deoxypodophyllotoxin, a potent cytotoxic agent of plant origin, on glioblastoma U-87 MG and SF126 cells. Pharmacol Rep 67:245–252
Gunn EJ, Williams JT, Huynh DT, Iannotti MJ, Han C, Barrios FJ, Kendall S, Glackin CA, Colby DA, Kirshner J (2011) The natural products parthenolide and andrographolide exhibit anticancer stem cell activity in multiple myeloma. Leuk Lymph 52:1085–1097
Gupta S, Pramanik D (2016) Phytochemicals and cancer stem cells: a pancreatic cancer overview. Curr Chem Biol 10:98–108
Gupta SC, Patchva S, Aggarwal BB (2013) Therapeutic roles of curcumin: lessons learned from clinical trials. AAPS J 15:195–218
Habli Z, Toumieh G, Fatfat M, Rahal ON, Gali-Muhtasib H (2017) Emerging cytotoxic alkaloids in the battle against cancer: overview of molecular mechanisms. Molecules 22:E250. https://doi.org/10.3390/molecules22020250
Haris P, Mary V, Aparna P, Dileep KV, Sudarsanakumar C (2017) A comprehensive approach to ascertain the binding mode of curcumin with DNA. Spectrochim Acta A 175:155–163

Hatcher H, Planalp R, Cho J, Torti FM, Torti SV (2008) Curcumin: from ancient medicine to current clinical trials. Cell Mol Life Sci 65:1631–1652
He YC, Zhou FL, Shen Y, Liao DF, Cao D (2014) Apoptotic death of cancer stem cells for cancer therapy. Int J Mol Sci 15:8335–8351
Heiblig M, Sobh M, Nicolini FE (2014) Subcutaneous omacetaxine mepesuccinate in patients with chronic myeloid leukemia in tyrosine kinase inhibitor-resistant patients: review and perspectives. Leuk Res 38:1145–1153
Hexum JK, Becker CM, Kempema AM, Ohlfest JR, Largaespada DA, Harki DA (2015) Parthenolide prodrug LC-1 slows growth of intracranial glioma. Bioorg Med Chem Lett 25:2493–2495
Holonec L, Ranga F, Crainic D, Truţa A, Socaciu C (2012) Evaluation of betulin and betulinic acid content in birch bark from different forestry areas of western carpathians. Not Bot Hortic Agrobot 40:99–105
Hu S, Zhou Q, Wu WR, Duan YX, Gao ZY, Li YW, Lu Q (2016) Anticancer effect of deoxypodophyllotoxin induces apoptosis of human prostate cancer cells. Oncol Lett 12:2918–2923
Huang D, Li Y, Cui F, Chen J, Sun J (2016) Purification and characterization of a novel polysaccharide-peptide complex from *Clinacanthus nutans* Lindau leaves. Carbohydr Polym 137:701–708
Imran M, Ullah A, Saeed F, Nadeem M, Arshad MU, Suleria HAR (2016) Cucurmin; anticancer and antitumor perspectives – a comprehensive review. Crit Rev Food Sci Nutr 22:1–23
Isah T (2016) Anticancer alkaloids from trees: development into drugs. Pharmacogn Rev 10:90–99
Isharwal S, Modi S, Arora N, Uhlrich C, Giri B, Barlass U, Soubra A, Chugh R, Dehm SM, Dudeja V, Saluja A, Banerjee S, Konety B (2017) Minnelide inhibits androgen dependent, castration resistant prostate cancer growth by decreasing expression of androgen receptor full length and splice variants. Prostate 77:584–596
Iweala EEJ, Liu FF, Cheng RR, Li Y, Omonhinmin CA, Zhang YJ (2015) Anticancer and free radical scavenging activity of some Nigerian food plants *in vitro*. Int J Cancer Res 11:41–51
Jain P, Gulati S, Seth R, Bakhshi S, Toteja GS, Pandey RM (2014) Vincristine-induced neuropathy in childhood ALL (acute lymphoblastic leukemia) survivors: prevalence and electrophysiological characteristics. J Child Neur 29:932–937
Jain S, Dwivedi J, Jain PK, Satpathy S, Patra A (2016) Medicinal plants for treatment of cancer: a brief review. Pharm J 8:87–101
Jannet SB, Hymery N, Bourgou S, Jdey A, Lachaal M, Magné C, Ksouri R (2017) Antioxidant and selective anticancer activities of two *Euphorbia* species in human acute myeloid leukemia. Biomed Pharmacother 90:375–385
Jeyamohan S, Moorthy RK, Kannan MK, Arockiam AJV (2016) Parthenolide induces apoptosis and autophagy through the suppression of PI3K/Akt signaling pathway in cervical cancer. Biotechnol Lett 38:1251–1260
Jiang Z, Wu M, Miao J, Duan H, Zhang S, Chen M, Sun L, Wang Y, Zhang X, Zhu X, Zhang L (2013) Deoxypodophyllotoxin exerts both anti-angiogenic and vascular disrupting effects. Int J Biochem Cell Biol 45:1710–1719
Jordan BC, Mock CD, Thilagavathi R, Selvam C (2016) Molecular mechanisms of curcumin and its semisynthetic analogues in prostate cancer prevention and treatment. Life Sci 152:135–144
Jørgensen L, McKerrall SJ, Kuttruff CA, Ungeheuer F, Felding J, Baran PS (2013) 14-step synthesis of (+)-ingenol from (+)-3-carene. Science 341:878–882
Jung GR, Kim KJ, Choi CH, Lee TB, Han SI, Han HK, Lim SC (2007) Effect of betulinic acid on anticancer drug-resistant colon cancer cells. Basic Clin Pharmacol Toxicol 101:277–285
Kamau LN, Mbaabu MP, Mbaria JM, Karuri GP, Kiama SG (2016) Knowledge and demand for medicinal plants used in the treatment and management of diabetes in Nyeri County, Kenya. J Ethnopharmacol 189:218–229
Katz L, Baltz RH (2016) Natural product discovery: past, present and future. J Indian Microbiol Biotechnol 43:155–176
Kedei N (2004) Characterization of the interaction of ingenol 3-angelate with protein kinase C. Cancer Res 64:3243–3255

Khaled M, Jiang ZZ, Zhang LY (2013) Deoxypodophyllotoxin: a promising therapeutic agent from herbal medicine. J Ethnopharmacol 149:24–34
Khalid EB, Ayman E-ME-K, Rahman H, Abdelkarim G, Najda A (2016) Natural products against cancer angiogenesis. Tumor Biol 37:14513–14536
Khan IN, Al-Karim S, Bora RS, Chaudhary AG, Saini KS (2015) Cancer stem cells: a challenging paradigm for designing targeted drug therapies. Drug Discov Today 20:1205–1216
Khanna C, Rosenberg M, Vail DM (2015) A review of paclitaxel and novel formulations including those suitable for use in dogs. J Vet Intern Med 29:1006–1012
Kim G (2017) *nab*-paclitaxel for the treatment of pancreatic cancer. Cancer Manag Res 9:85–96
Kim ES, Moon A (2015) Ursolic acid inhibits the invasive phenotype of SNU-484 human gastric cancer cells. Oncol Lett 9:897–902
Kim SL, Kim SH, Trang KTT, Kim IH, Lee SO, Lee ST, Kim DG, Kang SB, Kim SW (2013) Synergistic antitumor effect of 5-fluorouracil in combination with parthenolide in human colorectal cancer. Cancer Lett 335:479–486
Kim SH, Kang JG, Kim CS, Ihm SH, Choi MG, Yoo HJ, Lee SJ (2016) Synergistic cytotoxicity of BIIB021 with triptolide through suppression of PI3K/Akt/mTOR and NF-κB signal pathways in thyroid carcinoma cells. Biomed Pharmacother 83:22–32
Kim SL, Kim SH, Park YR, Liu YC, Kim EM, Jeong H-J, Kim YN, Seo SY, Kim IH, Lee SO, Lee ST, Kim SW (2017) Combined parthenolide and balsalazide have enhanced antitumor efficacy through blockade of NF-kB activation. Mol Cancer Res 15:141–151
Kinch MS, Haynesworth A, Kinch SL, Hoyer D (2014) An overview of FDA-approved new molecular entities: 1827–2013. Drug Discov Today 19:1033–1039
King AE, Southam KA, Dittmann J, Vickers JC (2013) Excitotoxin-induced caspase-3 activation and microtubule disintegration in axons is inhibited by taxol. Acta Neuropathol Commun 1:59. https://doi.org/10.1186/2051-5960-1-59
Klinger NV, Mittal S (2016) Therapeutic potential of curcumin for the treatment of brain tumors. Oxi Med Cell Long 2016:9324085. https://doi.org/10.1155/2016/9324085
Klippstein R, Bansal SS, Al-Jamal KT (2016) Doxorubicin enhances curcumin's cytotoxicity in human prostate cancer cells *in vitro* by enhancing its cellular uptake. Int J Pharma 514:169–175
Kocaadam B, Sanlier N (2017) Curcumin, an active component of turmeric (*Curcuma longa*), and its effects on health. Crit Rev Food Sci Nutr 57:2889–2895
Koduru S, Grierson DS, van de Venter M, Afolayan AJ (2007) Anticancer activity of steroid alkaloids isolated from *Solanum aculeastrum*. Pharm Biol 45:613–618
Komakech R, Kang Y, Lee J-H, Omujal F (2017) A review of the potential of phytochemicals from *Prunus africana* (Hook f.) kalkman stem bark for chemoprevention and chemotherapy of prostate cancer. Evid-Based Compl Altern Med 2017:3014019. https://doi.org/10.1155/2017/3014019
Koňariková K, Ježovičová M, Keresteš J, Gbelcová H, Ďuračková Z, Žitňanová IS (2015) Anticancer effect of black tea extract in human cancer cell lines. Springerplus 4:127. https://doi.org/10.1186/s40064-015-0871-4
Kondo M, MacKinnon SL, Craft CC, Matchett MD, Hurta RA, Neto CC (2011) Ursolic acid and its esters: occurrence in cranberries and other *Vaccinium* fruit and effects on matrix metalloproteinase activity in DU145 prostate tumor cells. J Sci Food Agric 91:789–796
Koprowska K, Hartman ML, Sztiller-Sikorska M, Czyz ME (2013) Parthenolide enhances dacarbazine activity against melanoma cells. Anti-Cancer Drugs 24:835–845
Kumar G, Mittal S, Sak K, Tuli HS (2016) Molecular mechanisms underlying chemopreventive potential of curcumin: current challenges and future perspectives. Life Sci 148:313–328
Lam SS, Ho ES, He BL, Wong WW, Cher CY, Ng NK, Man CH, Gill H, Cheung AM, Ip HW, So CC, Tamburini J, So CW, Ho DN, Au CH, Chan TL, Ma ES, Liang R, Kwong YL, Leung AY (2016) Homoharringtonine (omacetaxine mepesuccinate) as an adjunct for FLT3-ITD acute myeloid leukemia. Sci Transl Med 8:359ra129
Lanasa MC, Andritsos L, Brown JR, Gabrilove J, Caligaris-Cappio F, Ghia P, Larson RA, Kipps TJ, Leblond V, Milligan DW, Janssens A, Johnson AJ, Heerema NA, Bühler A, Stilgenbauer S, Devin J, Hallek M, Byrd JC, Grever MR (2015) Final results of EFC6663: a multicenter, international, phase 2 study of alvocidib for patients with fludarabine-refractory chronic lymphocytic leukemia. Leuk Res 39:495–500

Lee SY, Kim HH, Park SU (2015) Recent studies on betulinic acid and its biological and pharmacological activity. EXCLI J 14:199–203
Lewinska A, Adamczyk-Grochala J, Ewa Kwasniewicz E, Deregowska A, Wnuk M (2017) Ursolic acid-mediated changes in glycolytic pathway promote cytotoxic autophagy and apoptosis in phenotypically different breast cancer cells. Apopstosis 22:800–815
Li XJ, Jiang ZZ, Zhang LY (2014) Triptolide: progress on research in pharmacodynamics and toxicology. J Ethnopharmacol 155:67–79
Li F, Zhang J, Arfuso F, Chinnathambi A, Zayed ME, Alharbi SA, Kumar AP, Ahn KS, Sethi G (2015) NF-κB in cancer therapy. Arch Toxicol 89:711–731
Li X, Mao Y, Li K, Shi T, Yao H, Yao J, Wang S (2016) Pharmacokinetics and tissue distribution study in mice of triptolide-loaded lipid emulsion and accumulation effect on pancreas. Drug Deliv 23:1344–1354
Liang X, Grue-Sørensen G, Petersen AK, Högberg T (2012) Semisynthesis of ingenol 3-angelate (PEP005): efficient stereoconservative angeloylation of alcohols. Synlett 23:2647–2652
Liang X, Grue-Sørensen G, Mansson K, Vedsø P, Soor A, Stahlhut M, Bertelsen M, Engell KM, Högberg T (2013) Syntheses, biological evaluation and SAR of ingenol mebutate analogues for treatment of actinic keratosis and non-melanoma skin cancer. Bioorg Med Chem Lett 23:5624–5629
Lien JC, Lin MW, Chang SJ, Lai KC, Huang AC, Yu FS, Chung JG (2017) Tetrandrine induces programmed cell death in human oral cancer CAL 27 cells through the reactive oxygen species production and caspase-dependent pathways and associated with beclin-1-induced cell autophagy. Environ Toxicol 32:329–343
Lin YT, Huang AC, Kuo CL, Yang JS, Lan YH, Yu CC, Huang WW, Chung JG (2013) Induction of cell cycle arrest and apoptosis in human osteosarcoma U-2 OS cells by *Solanum lyratum* extracts. Nutr Cancer 65:469–479
Liu YH, Ou Y (2013) Progress in the research of structure and pharmacological activity of parthenolide. Chin J Pharm Biotechnol 20:586–589
Liu J, Wang Z (2015) Increased oxidative stress as a selective anticancer therapy. Oxidative Med Cell Longev 2015:294303. https://doi.org/10.1155/2015/294303
Liu T, Liu X, Li W (2016a) Tetrandrine, a Chinese plant-derived alkaloid, is a potential candidate for cancer chemotherapy. Oncotarget 7:40800–40815
Liu W, Zhai Y, Heng X, Che FY, Chen W, Sun D, Zhai G (2016b) Oral bioavailability of curcumin: problems and advancements. J Drug Target 24:694–702
Liu WC, Gonga T, Zhu P (2016c) Advances in exploring alternative Taxol sources. RSC Adv 6:48800–48809
Liu R, Ji P, Liu B, Qiao H, Wang X, Zhou L, Deng T, Ba Y (2017) Apigenin enhances the cisplatin cytotoxic effect through p53-modulated apoptosis. Oncol Lett 13:1024–1030
Long J, Ding YH, Wang PP, Zhang Q, Chen Y (2013) Protection-group-free semisyntheses of parthenolide and its cyclopropyl analogue. J Org Chem 78:10512–10518
Long J, Zhang SF, Wang PP, Zhang XM, Yang ZJ, Zhang Q, Chen Y (2014) Total syntheses of parthenolide and its analogues with macrocyclic stereocontrol. J Med Chem 57:7098–7112
Long J, Ding YH, Wang PP, Zhang Q, Chen Y (2016) Total syntheses and structure–activity relationship study of parthenolide analogues. Tetrahedron Lett 57:874–877
Lü S, Wang J (2014) Homoharringtonine and omacetaxine for myeloid hematological malignancies. J Hematol Oncol 7:2. https://doi.org/10.1186/1756-8722-7-2
Lu Y, Li F, Xu T, Sun J (2017) Tetrandrine prevents multidrug resistance in the osteosarcoma cell line, U-2OS, by preventing pgp overexpression through the inhibition of NF-κB signaling. Int J Mol Med 39:993–1000
Luo R, Fang D, Chu P, Wu H, Zhang Z, Tang Z (2016) Multiple molecular targets in breast cancer therapy by betulinic acid. Biomed Pharmacother 84:1321–1330
Ma D, Lu B, Feng C, Wang C, Wang Y, Luo T, Feng J, Jia H, Chi G, Luo Y, Ge P (2016) Deoxypodophyllotoxin triggers parthanatos in glioma cells via induction of excessive ROS. Cancer Lett 371:194–204

Ma JW, Zhang Y, Ye JC, Li R, Wen YL, Huang JX, Zhong XY (2017) Tetrandrine exerts a radiosensitization effect on human glioma through inhibiting proliferation by attenuating ERK phosphorylation. Biomol Ther 25:186–193

Mahoney S, Arfuso F, Millward M, Dharmarajan A (2014) The effects of phenoxodiol on the cell cycle of prostate cancer cell lines. Cancer Cell Int 14:110. https://doi.org/10.1186/s12935-014-0110-z

Maiyo F, Moodley R, Singha M (2016) Phytochemistry, cytotoxicity and apoptosis studies of β-sitosterol-3-*O*-glucoside and β-amyrin from *Prunus africana*. Afr J Tradit Compl Altern Med 13:105–112

Mao JT, Xue B, Smoake J, Lu QY, Park H, Henning SM, Burns W, Bernabei A, Elashoff D, Serio KJ, Massie L (2016) MicroRNA-19a/b mediates grape seed procyanidin extract-induced antineoplastic effects against lung cancer. J Nutr Biochem 34:118–125

Matias D, Bessa C, Simões MF, Reis CP, Saraiva L, Rijo P (2017) Natural products as lead protein kinase c modulators for cancer therapy Atta-ur-Rahman FRS Studies in natural products chemistry, bioactive natural products. Elsevier, Amsterdam, 45–79

Meng C, Zhu H, Song H, Wang Z, Huang G, Li D, Ma Z, Ma J, Qin Q, Sun X, Ma J (2014) Targets and molecular mechanisms of triptolide in cancer therapy. Chin J Cancer Res 26:622–626

Meng Z, Lv Q, Lu J, Yao H, Lv X, Jiang F, Lu A, Zhang G (2016) Prodrug strategies for paclitaxel. Int J Mol Sci 17:E796. https://doi.org/10.3390/ijms17050796

Millsop JW, Sivamani RK, Fazel N (2013) Botanical agents for the treatment of nonmelanoma skin cancer. Dermatol Res Pract 2013:837152. https://doi.org/10.1155/2013/837152

Mishra BB, Tiwari VK (2011) Natural products: an evolving role in future drug discovery. Eur J Med Chem 46:4769–4807

Mock CD, Jordan BC, Selvam C (2015) Recent advances of curcumin and its analogues in breast cancer prevention and treatment. RSC Adv 5:75575–75588

Modi S, Kir D, Giri B, Majumder K, Arora N, Dudeja V, Banerjee S, Saluja AK (2016) Minnelide overcomes oxaliplatin resistance by downregulating DNA repair pathway in pancreatic cancer. J Gastrointest Surg 20:13–24

Mohandas KM, Desai DC (1999) Epidemiology of digestive tract cancers in India. V. Large and small bowel. Indian J Gastroenterol 18:118–121

Moore A, Pinkerton R (2009) Vincristine: can its therapeutic index be enhanced? Pediatr Blood Cancer 53:1180–1187

Mora E, Smith EML, Donohoe C, Hertz D (2016) Vincristine-induced peripheral neuropathy in pediatric cancer patients. Am J Cancer Res 6:2416–2430

Moselhy J, Srinivasan S, Ankem MK, Damodaran C (2015) Natural products that target cancer stem cells. Anticancer Res 35:5773–5788

Mujumdar N, Mackenzie TN, Dudeja V, Chugh R, Antonoff MB, Borja-Cacho D, Sangwan V, Dawra R, Vickers SM, Saluja AK (2010) Triptolide induces cell death in pancreatic cancer cells by apoptotic and autophagic pathways. Gastroenterology 139:598–608

Mujumdar N, Banerjee S, Chen Z, Sangwan V, Chugh R, Dudeja V, Yamamoto M, Vickers SM, Saluja AK (2014) Triptolide activates unfolded protein response leading to chronic ER stress in pancreatic cancer cells. Am J Physiol Gastrointest Liver Physiol 306:1011–1020

Muthaura CN, Keriko JM, Mutai C, Yenesew A, Gathirwa JW, Irungu BN, Nyangacha R, Mungai GM, Derese S (2015) Antiplasmodial potential of traditional phytotherapy of some remedies used in treatment of malaria in Meru-Tharaka Nithi County of Kenya. J Ethnopharmacol 175:315–323

Muto N, Tomokuni T, Haramoto M, Tatemoto H, Nakanishi T, Inatomi Y, Murata H, Inada A (2008) Isolation of apoptosis- and differentiation-inducing substances toward human promyelocytic leukemia HL-60 cells from leaves of *Juniperus taxifolia*. Biosci Biotechnol Biochem 72:477–484

Nakshatri H, Appaiah HN, Anjanappa M, Gilley D, Tanaka H, Badve S, Crooks PA, Mathews W, Sweeney C, Bhat-Nakshatri P (2015) NF-κB-dependent and independent epigenetic modulation using the novel anticancer agent DMAPT. Cell Death Dis 6:e1608. https://doi.org/10.1038/cddis.2014.569

Nazha A, Kantarjian H, Cortes J, Quintás-Cardama A (2013) Omacetaxine mepesuccinate (synribo)-newly launched in chronic myeloid leukemia. Exp Opin Pharmacother 14:1977–1986
Nie XP, Yao F, Yue XD, Li GS, Dai SJ (2014) New eudesmane-type sesquiterpenoid from *Solanum lyratum* with cytotoxic activity. Nat Prod Res 28:641–645
Niedzwiecki A, Roomi MW, Kalinovsky T, Rath M (2016) Anticancer efficacy of polyphenols and their combinations. Forum Nutr 8:E552. https://doi.org/10.3390/nu8090552
Njoya EM, Munvera AM, Mkounga P, Nkengfack AE, McGaw LJ (2017) Phytochemical analysis with free radical scavenging, nitric oxide inhibition and antiproliferative activity of *Sarcocephalus pobeguinii* extracts. BMC Compl Altern Med 17:199. https://doi.org/10.1186/s12906-017-1712-5
Nomura A, McGinn O, Dudeja V, Sangwan V, Saluja AK, Banerjee S (2015) Minnelide effectively eliminates $CD133^+$ side population in pancreatic cancer. Mol Cancer 14:200. https://doi.org/10.1186/s12943-015-0470-6
Nwodo JN, Ibezim A, Simoben CV, Ntie-Kang F (2016) Exploring cancer therapeutics with natural products from african medicinal plants, Part II: Alkaloids, terpenoids and flavonoids. Anti Cancer Agents Med Chem 16:108–127
Ochwangi DO, Kimwele CN, Oduma JA, Gathumbi PK, Mbaria JM, Kiama SG (2014) Medicinal plants used in treatment and management of cancer in Kakamega County, Kenya. J Ethnopharmacol 151:1040–1055
Ogbourne SM, Suhrbier A, Jones B, Cozzi SJ, Boyle GM, Morris M, McAlpine D, Johns J, Scott TM, Sutherland KP, Gardner JM, Le TT, Lenarczyk A, Aylward JH, Parsons PG (2004) Antitumor activity of 3-ingenyl angelate: plasma membrane and mitochondrial disruption and necrotic cell death. Cancer Res 64:2833–2839
Oing C, Seidel C, von Amsberg G, Oechsle K, Bokemeyer C (2016) Pharmacotherapeutic treatment of germ cell tumors: standard of care and recent developments. Exp Opin Pharmacother 17:545–560
Omosa LK, Midiwo JO, Masila VM, Gisacho BM, Munayi R, Francisca-Kamakama, Chemutai KP, Elhaboob G, Saeed ME, Hamdoun S, Kuete V, Efferth T (2016) Cytotoxicity of 91 Kenyan indigenous medicinal plants towards human CCRF-CEM leukemia cells. J Ethnopharmacol 179:177–196
Orhan IE, Tosun F, Gülpınar AR, Kartal M, Duran A, Mihoglugil F, Akalgan D (2015) LC-MS quantification of parthenolide and cholinesterase inhibitory potential of selected *Tanacetum* L. (Emend. Briq.) taxa. Phytochem Lett 11:347–352
Orofino MR, Grootjans S, Biavatti MW, Vandenabeele P, D'Herde K (2012) Sesquiterpene lactones as drugs with multiple targets in cancer treatment: focus on parthenolide. Anti-Cancer Drugs 23:883–896
Ortensi B, Setti M, Osti D, Pelicci G (2013) Cancer stem cell contribution to glioblastoma invasiveness. Stem Cell Res Ther 4:18. https://doi.org/10.1186/scrt166
Osterman CJD, Wall NR (2015) Curcumin and pancreatic cancer: a research and clinical update. J Nat Sci 1:e124
Ostrom QT, Gittleman H, Stetson L, Virk SM, Barnholtz-Sloan JS (2015) Epidemiology of gliomas. Cancer Treat Res 163:1–14
Pareek A, Suthar M, Rathore GS, Bansal V (2011) Feverfew *Tanacetum parthenium* L.: a systematic review. Pharmacogn Rev 5:103–110
Patel JJ, Acharya SR, Acharya NS (2014) *Clerodendrum serratum* (L.) Moon. – a review on traditional uses, phytochemistry and pharmacological activities. J Ethnopharmacol 154:268–285
Pattnaik B, Lakshmi JK, Kavitha R, Jagadeesh B, Bhattacharjee D, Jain N, Mallavadhani UV (2017) Synthesis, structural studies, and cytotoxic evaluation of novel ursolic acid hybrids with capabilities to arrest breast cancer cells in mitosis. J Asian Nat Prod Res 19:260–271
Pavan AR, Silva GD, Jornada DH, Chiba DE, Fernandes GF, Man Chin C, Dos Santos JL (2016) Unraveling the anticancer effect of curcumin and resveratrol. Forum Nutr 8:E628. https://doi.org/10.3390/nu8110628

Pawar VK, Panchal SB, Singh Y, Meher JG, Sharma K, Singh P, Bora HK, Singh A, Datta D, Chourasia MK (2014) Immunotherapeutic vitamin e nanoemulsion synergies the antiproliferative activity of paclitaxel in breast cancer cells via modulating Th1 and Th2 immune response. J Control Release 196:295–306
Peese K (2010) New agents for the treatment of leukemia: discovery of DMAPT (LC-1). Drug Discov Today 15:322. https://doi.org/10.1016/j.drudis.2009.08.008
Peng W, Qin R, Li X, Zhou H (2013) Botany, phytochemistry, pharmacology, and potential application of *Polygonum cuspidatum* Sieb.et Zucc.: a review. J Ethnopharmacol 148:729–745
Perrone D, Ardito F, Giannatempo G, Dioguardi M, Troiano G, Lo Russo L, De Lillo A, Laino L, Lo Muzio L (2015) Biological and therapeutic activities and anticancer properties of curcumin. Exp Ther Med 10:1615–1623
Pichette A, Liu H, Roy C, Tanguay S, Simard F, Lavoie S (2004) Selective oxidation of botulin for the preparation of betulinic acid, an antitumoral compound. Synth Commun 34:3925–3937
Pimentel-Gutiérrez HJ, Bobadilla-Morales L, Barba-Barba CC, Ortega-De-La-Torre C, Sánchez-Zubieta FA, Corona-Rivera JR, González-Quezada BA, Armendáriz-Borunda JS, Silva-Cruz R, Corona-Rivera A (2016) Curcumin potentiates the effect of chemotherapy against acute lymphoblastic leukemia cells via downregulation of NF-κB. Oncol Lett 12:4117–4124
Pisha E, Chai H, Lee IS, Chagwedera TE, Farnsworth NR, Cordell GA, Beecher CW, Fong HH, Kinghorn AD, Brown DH (1995) Discovery of betulinic acid a selective inhibitor of human-melanoma that functions by induction of apoptosis. Nat Med 10:1046–1051
Potze L, Di Franco S, Grandela C, Pras-Raves ML, Picavet DI, van Veen HA, van Lenthe H, Mullauer FB, van der Wel NN, Luyf A, van Kampen AH, Kemp S, Everts V, Kessler JH, Vaz FM, Medema JP (2016a) Betulinic acid induces a novel cell death pathway that depends on cardiolipin modification. Oncogene 35:427–437
Potze L, Di Franco S, Kessler JH, Stassi G, Medema JP (2016b) Betulinic acid kills colon cancer stem cells. Curr Stem Cell Res Ther 11:427–433
Puneeth HR, Ananda H, Kumar KSS, Rangappa KS, Sharada AC (2016) Synthesis and antiproliferative studies of curcumin pyrazole derivatives. Med Chem Res 25:1842–1851
Qiu W, Zhang AL, Tian Y (2017) Tetrandrine triggers an alternative autophagy in DU145 cells. Oncol Lett 13:3734–3738
Rahimi HR, Nedaeinia R, Shamloo AS, Nikdoust S, Oskuee RK (2016) Novel delivery system for natural products: nano-curcumin formulations. Avicenna J Phytomed 6:383–398
Ramasamy TS, Ayob AZ, Myint HHL, Thiagarajah S, Amini F (2015) Targeting colorectal cancer stem cells using curcumin and curcumin analogues: insights into the mechanism of the therapeutic efficacy. Cancer Cell Int 15:96. https://doi.org/10.1186/s12935-015-0241-x
Ramsay JR, Suhrbier A, Aylward JH, Ogbourne S, Cozzi SJ, Poulsen MG, Baumann KC, Welburn P, Redlich GL, Parsons PG (2011) The sap from *Euphorbia peplus* is effective against human nonmelanoma skin cancers. Br J Dermatol 164:633–636
Redondo-Blanco S, Fernández J, Gutiérrez-del-Río I, Villar CJ, Lombó F (2017) New insights toward colorectal cancer chemotherapy using natural bioactive compounds. Front Pharmacol 8:109. https://doi.org/10.3389/fphar.2017.00109
Réthy B, Csupor-Löffler B, Zupkó I, Hajdú Z, Máthé I, Hohmann J, Rédei T, Falkay G (2007) Antiproliferative activity of Hungarian Asteraceae species against human cancer cell lines. Part I. Phytother Res 21:1200–1208
Rivard C, Geller M, Schnettler E, Saluja M, Vogel RI, Saluja A, Ramakrishnan S (2014) Inhibition of epithelial ovarian cancer by Minnelide, a water-soluble pro-drug. Gynecol Oncol 135:318–324
Rizk AM, Hammouda FM, El-Missiry MM, Radwan HM, Evans FJ (1985) Biologically active diterpene esters from *Euphorbia peplus*. Phytochemistry 24:1605–1606
Rosen RH, Gupta AK, Tyring SK (2012) Dual mechanism of action of ingenol mebutate gel for topical treatment of actinic keratoses: rapid lesion necrosis followed by lesion-specific immune response. J Am Acad Dermatol 66:486–493
Roshan SK, Spano AD, McKinney AM, Nascene DR (2017) Potentially reversible acute cerebellar toxicity associated with minnelide. Neuroradiology 59:419–421

Rosshandler Y, Shen AQ, Cortes J, Khoury HJ (2016) Omacetaxine mepesuccinate for chronic myeloid leukemia. Expert Rev Hematol 9:419–424

Rousalova I, Banerjee S, Sangwan V, Evenson K, McCauley JA, Kratzke R, Vickers SM, Saluja A, D'Cunha J (2013) Minnelide: a novel therapeutic that promotes apoptosis in non-small cell lung carcinoma *in vivo*. PLoS One 8:e77411. https://doi.org/10.1371/journal.pone.0077411

Sales JM, Resurreccion AVA (2014) Resveratrol in peanuts. Crit Rev Food Sci Nutr 54:734–770

Sarek J, Kvasnica M, Vlk M, Urban M, Dzubak P, Hajduch M (2011) The potential of triterpenoids in the treatment of melanoma. In: Murph M (ed) Research on melanoma-a glimpse into current directions and future trends. InTech, Rijeka, pp 125–158

Schwertheim S, Wein F, Lennartz K, Worm K, Schmid KW, Sheu-Grabellus SY (2017) Curcumin induces G2/M arrest, apoptosis, NF-κB inhibition, and expression of differentiation genes in thyroid carcinoma cells. J Cancer Res Clin Oncol 143:1143–1154

Seca AML, Pinto DCGA, Silva AMS (2015) The current status of bioactive metabolites from the genus *Juniperus*. In: Gupta VK (ed) Bioactive phytochemicals: perspectives for modern medicine, vol 3. Daya Publishing House, New Delhi, pp 367–410

Seca AML, Silva AMS, Pinto DCGA (2017) Parthenolide and parthenolide-like sesquiterpene lactones as multiple targets drugs: current knowledge and new developments. In: Atta-Ur-Rahman FRS (ed) Studies in natural products chemistry (bioactive natural products). Elsevier, Amsterdam, pp 337–372

Sen S, Hassane DC, Corbett C, Becker MW, Jordan CT, Guzman ML (2013) Novel mTOR inhibitory activity of ciclopirox enhances parthenolide antileukemia activity. Exp Hematol 41:799–807

Seo HS, Jo JK, Ku JM, Choi HS, Choi YK, Woo JK, Kim HI, Kang SY, Lee KM, Nam KW, Park N, Jang BH, Shin YC, Ko SG (2015) Induction of caspase dependent extrinsic apoptosis through inhibition of signal transducer and activator of transcription 3 (STAT3) signalling in HER2-overexpressing BT-474breast cancer cells. Biosci Rep 35:e00276. https://doi.org/10.1042/BSR20150165

Serova M, Ghoul A, Benhadji KA, Faivre S, Le Tourneau C, Cvitkovic E, Lokiec F, Lord J, Ogbourne SM, Calvo F, Raymond E (2008) Effects of protein kinase C modulation by PEP005, a novel ingenol angelate, on mitogen-activated protein kinase and phosphatidylinositol 3-kinase signaling in cancer cells. Mol Cancer Ther 7:915–922

Shah NN, Cole DE, Lester-McCully CM, Wayne AS, Warren KE, Widemann BC (2016a) Plasma and cerebrospinal fluid pharmacokinetics of vincristine and vincristine sulfate liposomes injection (VSLI, marqibo®) after intravenous administration in non-human primates. Investig New Drugs 34:61–65

Shah NN, Merchant MS, Cole DE, Jayaprakash N, Bernstein D, Delbrook C, Richards K, Widemann BC, Wayne AS (2016b) Vincristine sulfate liposomes injection (VSLI, Marqibo®): results from a phase I study in children, adolescents, and young adults with refractory solid tumors or leukemias. Pediatr Blood Cancer 63:997–1005

Shankar E, Zhang A, Franco D, Gupta S (2017) Betulinic acid-mediated apoptosis in human prostate cancer cells involves p53 and nuclear factor-kappa b (NF-κB) pathways. Molecules 22:E264. https://doi.org/10.3390/molecules22020264

Shanmugam MK, Dai X, Kumar AP, Tan BK, Sethi G, Bishayee A (2013) Ursolic acid in cancer prevention and treatment: molecular targets, pharmacokinetics and clinical studies. Biochem Pharmacol 85:1579–1587

Shanmugam MK, Rane G, Kanchi MM, Arfuso F, Chinnathambi A, Zayed ME, Alharbi SA, Tan BK, Kumar AP, Sethi G (2015) The multifaceted role of curcumin in cancer prevention and treatment. Molecules 20:2728–2769

Shen AQ, Munteanu M, Khoury HJ (2014) Updated product label allows home administration of omacetaxine mepesuccinate. Oncologist 19:e14. https://doi.org/10.1634/theoncologist.2014-0230

Singh S, Sharma B, Kanwar SS, Kumar A (2016) Lead phytochemicals for anticancer drug development. Front Plant Sci 7:1667. https://doi.org/10.3389/fpls.2016.01667

Sinha R, Anderson DE, McDonald SS, Greenwald P (2003) Cancer risk and diet in India. J Postgrad Med 49:222–228
Sinha D, Sarkar N, Biswas J, Bishayee A (2016) Resveratrol for breast cancer prevention and therapy: preclinical evidence and molecular mechanisms. Semin Cancer Biol 40–41:209–232
Sivakumar P, Mengani S, Vijayabaskaran M, Kumar RS, Perumal P, Jayakar B (2010) Antitumor and antioxidant activities of *Triumfetta rhomboidea* against Ehrlich ascites carcinoma bearing Swiss Albino mice. Res J Pharm Biol Chem Sci 1:486–494
Soladoye MO, Amusa NA, Raji-Esan SO, Chukwuma EC, Taiwo AA (2010) Ethnobotanical survey of anticancer plants in Ogun State, Nigeria. Ann Biol Res 1:261–273
Soliman HH (2017) *nab*-paclitaxel as a potential partner with checkpoint inhibitors in solid tumors. Onco Target Ther 10:101–112
Soliman MM, Abdo Nassan M, Ismail TA (2014) Immunohistochemical and molecular study on the protective effect of curcumin against hepatic toxicity induced by paracetamol in Wistar rats. BMC Complement Altern Med 14:457. https://doi.org/10.1186/1472-6882-14-457
Song JM, Qian X, Upadhyayya P, Hong KH, Kassie F (2014) Dimethylaminoparthenolide, a water soluble parthenolide, suppresses lung tumorigenesis through down-regulating the STAT3 signaling pathway. Curr Cancer Drug Tar 14:59–69
Song XL, Liu S, Jiang Y, Gu LY, Xiao Y, Wang X, Cheng L, Li XT (2017a) Targeting vincristine plus tetrandrine liposomes modified with DSPE-PEG2000-transferrin in treatment of brain glioma. Eur J Pharm Sci 96:129–140
Song JM, Molla K, Anandharaj A, Cornax I, O'Sullivan MG, Kirtane AR, Panyam J, Kassie F (2017b) Triptolide suppresses the *in vitro* and *in vivo* growth of lung cancer cells by targeting hyaluronan-CD44/RHAMM signalling. Oncotarget 8:26927–26940
Sordillo PP, Helson L (2015) Curcumin and cancer stem cells: curcumin has asymmetrical effects on cancer and normal stem cells. Anticancer Res 35:599–614
Sri Ramya PV, Angapelly S, Guntuku L, Singh Digwal C, Nagendra Babu B, Naidu VG, Kamal A (2017) Synthesis and biological evaluation of curcumin inspired indole analogues as tubulin polymerization inhibitors. Eur J Med Chem 127:100–114
Strimpakos AS, Sharma RA (2008) Curcumin: preventive and therapeutic properties in laboratory studies and clinical trials. Antioxid Redox Signal 10:511–546
Sung B, Chung HY, Kim ND (2016) Role of apigenin in cancer prevention via the induction of apoptosis and autophagy. J Cancer Prev 21:216–226
Tang D, Chen K, Huang L, Li J (2017) Pharmacokinetic properties and drug interactions of apigenin, a natural flavone. Exp Opin Drug Metab Toxicol 13:323–330
Tariq A, Sadia S, Pan K, Ullah I, Mussarat S, Sun F, Abiodun OO, Batbaatar A, Li Z, Song D, Xiong Q, Ullah R, Khan S, Basnet BB, Kumar B, Islam R, Adnan M (2017) A systematic review on ethnomedicines of anticancer plants. Phytother Res 31:202–264
Teoh PL, Cheng AYF, Liau M, Lem FF, Kaling GP, Chua FN, Cheong BE (2017) Chemical composition and cytotoxic properties of *Clinacanthus nutans* root extracts. Pharm Biol 55:394–401
Torquato HFV, Goettert MI, Justo GZ, Paredes-Gamero EJ (2017) Anticancer phytometabolites targeting cancer stem cells. Curr Genomics 18:156–174
Tsai HY, Ho CT, Chen YK (2017) Biological actions and molecular effects of resveratrol, pterostilbene, and 3′-hydroxypterostilbene. J Food Drug Anal 25:134–147
Turek M, Krzyczmonik M, Balczewski P (2016) New hopes in cancer battle-a review of new molecules and treatment strategies. Med Chem 12:700–719
Tzogani K, Nagercoil N, Hemmings RJ, Samir B, Gardette J, Demolis P, Salmonson T, Pignatti F (2014) The European medicines agency approval of ingenol mebutate (Picato) for the cutaneous treatment of non-hyperkeratotic, non-hypertrophic actinic keratosis in adults: summary of the scientific assessment of the committee for medicinal products for human use (CHMP). Eur J Dermatol 24:457–463
Uwagawa T, Yanaga K (2015) Effect of NF-κB inhibition on chemoresistance in biliary-pancreatic cancer. Surg Today 45:1481–1488

Varoni EM, Lo Faro AF, Sharifi-Rad J, Iriti M (2016) Anticancer molecular mechanisms of resveratrol. Front Nutr 3:8. https://doi.org/10.3389/fnut.2016.00008

Vazquez-Santillan K, Melendez-Zajgla J, Jimenez-Hernandez L, Martínez-Ruiz G, Maldonado V (2015) NF-κB signaling in cancer stem cells: a promising therapeutic target. Cell Oncol 38:327–339

Velde ME, Kaspers GL, Abbink FCH, Wilhelm AJ, Ket JCF, Berg MH (2017) Vincristine-induced peripheral neuropathy in children with cancer: a systematic review. Crit Rev Oncol Hematol 114:114–130

Vergote I, Leamon CP (2015) Vintafolide: a novel targeted therapy for the treatment of folate receptor expressing tumors. Ther Adv Med Oncol 7:206–218

Vipasha S, Hardeep K, Tarun K, Tullika M (2016) Traditional Indian herb *Catharanthus roseus* used as cancer treatment: a review. Int J Pharmacogn Phytochem Res 8:1926–1928

Viral D, Shivanand P, Jivani NP (2011) Anticancer evaluation of *Adiantum venustum* Don. J Young Pharma 3:48–54

Waechter F, Silva GS, Willig J, de Oliveira CB, Vieira BD, Trivella DB, Zimmer AR, Buffon A, Pilger DA, Gnoatto S (2017) Design, synthesis, and biological evaluation of betulinic acid derivatives as new antitumor agents for leukemia. Anticancer Agents Med Chem (Online). https://doi.org/10.2174/1871520617666170412143638

Wang YJ, Pan MH, Cheng AL, Lin LI, Ho YS, Hsieh CY, Lin JK (1997) Stability of curcumin in buffer solutions and characterization of its degradation products. J Pharm Biomed Anal 15:1867–1876

Wang YR, Xu Y, Jiang ZZ, Guerram M, Wang B, Zhu X, Zhang LY (2015a) Deoxypodophyllotoxin induces G2/M cell cycle arrest and apoptosis in SGC-7901 cells and inhibits tumor growth *in vivo*. Molecules 20:1661–1675

Wang Y, Wang B, Guerram M, Sun L, Shi W, Tian C, Zhu X, Jiang Z, Zhang L (2015b) Deoxypodophyllotoxin suppresses tumor vasculature in HUVECs by promoting cytoskeleton remodeling through LKB1-AMPK dependent Rho A activation. Oncotarget 6:29497–29512

Wang S, Wang W, Lv H, Zhou J, Shen L, Yin L, Ding Y (2016a) Preparation and evaluation of long-circulating deoxypodophyllotoxin-loaded liposomes using poly(ethylene glycol)-distearoylphosphatidylethanolamine. J Pharm Innov 11:134–142

Wang H, Ma D, Wang C, Zhao S, Liu C (2016b) Triptolide inhibits invasion and tumorigenesis of hepatocellular carcinoma MHCC-97H cells through NF-κB signaling. Med Sci Monit 22:1827–1836

Wang X, Song Y, Su Y, Tian Q, Li B, Quan J, Deng Y (2016c) Are PEGylated liposomes better than conventional liposomes. A special case for vincristine. Drug Deliv 23:1092–1100

Wang M, Zhao T, Liu Y, Wang Q, Xing S, Li L, Wang L, Liu L, Gao D (2017a) Ursolic acid liposomes with chitosan modification: promising antitumor drug delivery and efficacy. Mater Sci Eng C 71:1231–1240

Wang S, Wang H, Li T, Li C, Zhou Y, Zhong X (2017b) The selection and stability analysis of stable and high Taxol-producing cell lines from *Taxus cuspidate*. J For Res. https://doi.org/10.1007/s11676-017-0406-7

Watkins R, Wu L, Zhang C, Davis RM, Xu B (2015) Natural product-based nanomedicine: recent advances and issues. Int J Nanomedicine 10:6055–6074

Wen G, Qu XX, Wang D, Chen XX, Tian XC, Gao F, Zhou XL (2016) Recent advances in design, synthesis and bioactivity of paclitaxel-mimics. Fitoterapia 110:26–37

WHO (2017) Who factsheets. http://www.who.int/mediacentre/factsheets/fs297/en/. Assessed on 18 Oct 2017

Wozniak M, Szulawska-Mroczek A, Hartman ML, Nejc D, Czyz M (2013) Parthenolide complements the cell death-inducing activity of doxorubicin in melanoma cells. Anticancer Res 33:3205–3212

Woźniak Ł, Skąpska S, Marszałek K (2015) Ursolic acid-A pentacyclic triterpenoid with a wide spectrum of pharmacological activities. Molecules 20:20614–20641

Wu M, Jiang Z, Duan H, Sun L, Zhang S, Chen M, Wang Y, Gao Q, Song Y, Zhu X, Zhang L (2013) Deoxypodophyllotoxin triggers necroptosis in human non-small cell lung cancer NCI-H460 cells. Biomed Pharmacother 67:701–706

Wyrebska A, Gach K, Janecka A (2014) Combined effect of parthenolide and various anticancer drugs or anticancer candidate substances on malignant cells *in vitro* and *in vivo*. Mini Rev Med Chem 14:222–228

Xi C, Peng S, Wu Z, Zhou Q, Zhou J (2017) Toxicity of triptolide and the molecular mechanisms involved. Biomed Pharmacother 90:531–541

Xiao Z, Morris-Natschke SL, Lee KH (2016) Strategies for the optimization of natural leads to anticancer drugs or drug candidates. Med Res Rev 36:32–91

Xie CQ, Zhou P, Zuo J, Li X, Chen Y, Chen JW (2016) Triptolide exerts pro-apoptotic and cell cycle arrest activity on drug-resistant human lung cancer A549/Taxol cells via modulation of MAPK and PI3K/Akt signaling pathways. Oncol Lett 12:3586–3590

Xin Y, Yin F, Qi S, Shen L, Xu Y, Luo L, Lan L, Yin Z (2013) Parthenolide reverses doxorubicin resistance in human lung carcinoma A549 cells by attenuating NF-B activation and HSP70 up-regulation. Toxicol Lett 221:73–82

Xu H, Hou Z, Zhang H, Kong H, Li X, Wang H, Xie W (2014) An efficient Trojan delivery of tetrandrine by poly(Nvinylpyrrolidone)-block-poly(epsilon-caprolactone) (PVPb-PCL) nanoparticles shows enhanced apoptotic induction of lung cancer cells and inhibition of its migration and invasion. Int J Nanomedicine 9:231–242

Xu Y, Li J, Li QJ, Feng YL, Pan F (2017) Betulinic acid promotes TRAIL function on liver cancer progression inhibition through p53/Caspase-3 signaling activation. Biomed Pharmacother 88:349–358

Yallapu MM, Nagesh PKB, Jaggi M, Chauhan SC (2015) Therapeutic applications of curcumin nanoformulations. AAPS J 1:1341–1356

Yang J, Liang Q, Wang M, Jeffries C, Smithson D, Tu Y, Boulos N, Jacob MR, Shelat AA, Wu Y, Ravu RR, Gilbertson R, Avery MA, Khan IA, Walker LA, Guy RK, Li XC (2014a) UPLC-MS-ELSD-PDA as a powerful dereplication tool to facilitate compound identification from small-molecule natural product libraries. J Nat Prod 77:902–909

Yang Y, Chen Y, Zhong ZY, Zhang J, Li F, Jia LL, Liu L, Zhu X, Liu XD (2014b) Validated LC–MS/MS assay for quantitative determination of deoxypodophyllotoxin in rat plasma and its application in pharmacokinetic study. J Pharm Biomed Anal 88:410–415

Yang ZJ, Ge WZ, Li QY, Lu Y, Gong JM, Kuang BJ, Xi X, Wu H, Zhang Q, Chen Y (2015) Syntheses and biological evaluation of costunolide, parthenolide, and their fluorinated analogues. J Med Chem 58:7007–7020

Yang CY, Lin CK, Lin GJ, Hsieh CC, Huang SH, Ma KH, Shieh YS, Sytwu HK, Chen YW (2017a) Triptolide represses oral cancer cell proliferation, invasion, migration, and angiogenesis in co-inoculation with U937 cells. Clin Oral Investig 21:419–427

Yang C, Ma X, Wang Z, Zeng X, Hu Z, Ye Z, Shen G (2017b) Curcumin induces apoptosis and protective autophagy in castration-resistant prostate cancer cells through iron chelation. Drug Des Devel Ther 11:431–439

Yao H, Liu J, Xu S, Zhu Z, Xu J (2017) The structural modification of natural products for novel drug discovery. Exp Opin Drug Discov 12:121–140

Ye LY, Hu S, Xu HE, Xu RR, Kong H, Zeng XN, Xie WP, Wang H (2017) The effect of tetrandrine combined with cisplatin on proliferation and apoptosis of A549/DDP cells and A549 cells. Cancer Cell Int 17:40. https://doi.org/10.1186/s12935-017-0410-1

Yi JM, Huan XJ, Song SS, Zhou H, Wang YQ, Miao ZH (2016) Triptolide induces cell killing in multidrug-resistant tumor cells via CDK7/RPB1 rather than XPB or p44. Mol Cancer Ther 15:1495–1503

Yip-Schneider MT, Wu H, Stantz K, Agaram N, Crooks PA, Schmidt CM (2013) Dimethylaminoparthenolide and gemcitabine: a survival study using a genetically engineered mouse model of pancreatic cancer. BMC Cancer 13:194. https://doi.org/10.1186/1471-2407-13-194

Yogeeswari P, Sriram D (2005) Betulinic acid and its derivatives: a review on their biological properties. Curr Med Chem 12:657–666

Yoshihara H, Yoshimoto Y, Hosoya Y, Hasegawa D, Kawano T, Sakoda A, Okita H, Manabe A (2017) Infantile fibrosarcoma treated with postoperative vincristine and dactinomycin. Pediatr Int 59:371–374

Yu HJ, Ma L, Jiang J, Sun SQ (2016) Protective effect of curcumin on neural myelin sheaths by attenuating interactions between the endoplasmic reticulum and mitochondria after compressed spinal cord. J Spine 5:4. https://doi.org/10.4172/2165-7939.1000322

Yuan S, Wang L, Chen X, Fan B, Yuan Q, Zhang H, Yang D, Wang S (2016) Triptolide inhibits the migration and invasion of human prostate cancer cells via Caveolin-1/CD147/MMPs pathway. Biomed Pharmacother 84:1776–1782

Zhang Q, Lu Y, Ding Y, Zhai J, Ji Q, Ma W, Yang M, Fan H, Long J, Tong Z, Shi Y, Jia Y, Han B, Zhang W, Qiu C, Ma X, Li Q, Shi Q, Zhang H, Li D, Zhang J, Lin J, Li LY, Gao Y, Chen Y (2012) Guaianolide sesquiterpene lactones, a source to discover agents that selectively inhibit acute myelogenous leukemia stem and progenitor cells. J Med Chem 55:8757–8769

Zhang DM, Xu HG, Wang L, Li YJ, Sun PH, Wu XM, Wang GJ, Chen WM, Ye WC (2015) Betulinic acid and its derivatives as potential antitumor agents. Med Res Rev 35:1127–1155

Zhang TJ, Han S, Zhang W, Zhao FJ, Li J (2016a) Triptolide promoting colon cancer CT26 cell death through induction of autophagy. Acta Anat Sin 47:774–778

Zhang W, Kang M, Zhang T, Li B, Liao X, Wang R (2016b) Triptolide combined with radiotherapy for the treatment of nasopharyngeal carcinoma via NF-κB-related mechanism. Int J Mol Sci 17:2139. https://doi.org/10.3390/ijms17122139

Zhang S, Wang S, Zhan J (2016c) Engineered biosynthesis of medicinally important plant natural products in microorganisms. Curr Top Med Chem 16:1740–1754

Zhang Y, Wen YL, Ma JW, Ye JC, Wang X, Huang JX, Meng CY, Xu XZ, Wang SX, Zhong XY (2017a) Tetrandrine inhibits glioma stem-like cells by repressing β-catenin expression. Int J Oncol 50:101–110

Zhang PR, Zhang JH, Yu M, Zhang X (2017b) Triptolide reverses MCF-7/ADR cell resistance by down-regulating P-glycoprotein expression. Int J Clin Exp Med 10:1513–1521

Zhao Y, Yang Y, Chen Q, Kasimu R, Akber Aisa H (2016a) Isolation of deoxypodophyllotoxin and podophyllotoxin from *Juniperus sabina* by high speed counter current chromatography. Afinidad 73:237–240

Zhao L, Wu P, Zhang PG, Xie DZ, Gao D, Zhou NJ (2016b) Effect of triptolide on human colorectal cancer HCT116 cell proliferation, autophagy and apoptosis. Chin Pharmacol Bull 32:1399–1404

Zhao G, Han X, Cheng W, Ni J, Zhang Y, Lin J, Song Z (2017) Apigenin inhibits proliferation and invasion, and induces apoptosis and cell cycle arrest in human melanoma cells. Oncol Rep 37:2277–2285

Zhou ZL, Yang YX, Ding J, Li YC, Miao ZH (2012) Triptolide: structural modifications, structure–activity relationships, bioactivities, clinical development and mechanisms. Nat Prod Rep 29:457–475

Zhou Z, Tang M, Liu Y, Zhang Z, Lu R, Lu J (2017) Apigenin inhibits cell proliferation, migration, and invasion by targeting Akt in the A549 human lung cancer cell line. Anti-Cancer Drugs 28:446–456

Zhu W, Hu J, Wang X, Tian J, Komatsu S (2015) Organ-specific analysis of mahonia using gel-free/label-free proteomic technique. J Proteome Res 14:2669–2685

Zong Y, Wu J, Shen K (2017) Nanoparticle albumin-bound paclitaxel as neoadjuvant chemotherapy of breast cancer: a systematic review and meta-analysis. Oncotarget 8:17360–17372

Zulkipli IN, Rajabalaya R, Idris A, Sulaiman NA, David SR (2017) *Clinacanthus nutans*: a review on ethnomedicinal uses, chemical constituents and pharmacological properties. Pharm Biol 55:1093–1113

Chapter 5
Designing of Natural Anticancerous Drugs and Their Delivery System

Manisha Singh, Rashi Rajput, Ramneek Kaur, Sachin Kumar, and Rachana

Abstract Chemoprevention of cancer with conventional therapeutic approach has shown severe side effects along with high possibility of recurrence of the same. This results in decreased therapeutic efficacy of the chemotherapeutic agents used currently. Therefore, to reduce their ill effects, an alternative gateway for cancer treatment and prevention needs to be explored. The natural, plant-based anticancer compounds are proving to be a substantial target in replacing the common approaches which employs chemical drugs. The natural compounds derived from various medicinal plants like vinca alkaloids, cyanidin, curcumin, fisetin, rosmarinic acid, etc. have already been documented with a remarkable antineoplastic property. The aim of the chapter is to focus on the various phytocompounds and their mechanism of action to treat the tumorigenic growth accompanied with the drug delivery system. Furthermore, the limitations and barriers associated with the formulation of phytocompounds and their possible enhanced therapeutic efficiency by nanoparticle-based drug delivery mechanics are also discussed.

Keywords Chemoprevention · Nanoparticles · Medicinal plants · Phytocompounds · Therapeutics

5.1 Introduction

Natural products have been used for over 40 years to treat cancer. The important sources of the compounds are plants and microbes from marine and terrestrial environments. Plant alkaloids and podophyllotoxins have the antitumor property (Verdine 1996), while the other major contribution on natural compounds is shown by microbes. There is always an increasing demand for the discovery of potential drugs to treat cancer, because of the development of resistance to chemotherapeutic drugs. Additionally, undesirable side effects and high toxicity are also associated

M. Singh (✉) · R. Rajput · R. Kaur · S. Kumar · Rachana
Department of Biotechnology, Jaypee Institute of Information Technology, Noida, Uttar Pradesh, India

M. S. Akhtar, M. K. Swamy (eds.), *Anticancer Plants: Clinical Trials and Nanotechnology*, https://doi.org/10.1007/978-981-10-8216-0_5

with the drugs, which increase the surge for the development of new anticancer drugs with high therapeutic index or lesser side effects (Holland 1997; Manju et al. 2017). Natural products have brought about a great success in the society. The use of microbial and plant secondary metabolites has helped in increasing the life span in the twentieth century (Bandyopadhyay et al. 2017). They have reduced suffering and pain and helped in revolutionizing medicines by assisting the transplantation of genes or organs. The chemical diversity is based on the geographical and biological diversity. Thus, the researchers are exploring the entire globe for natural product-based therapeutics because of their easy access to terrestrial life and pharmaceutically active compounds (Henkel et al. 1999; Gouvea and Kassicieh 2005; Skinnider et al. 2017). There are approximately one million natural products known till now. There are about 500,000–600,000 plants known to produce therapeutically active products. The chemical structures of 160,000 natural products have been known, and the number is increasing by 10,000 per year (Knight et al. 2003). Out of these, 100,000 are from plants. So, the need of the hour is to identify and isolate plant products with anticancer property with high therapeutic index and fewer side effects (Fenical and Jensen 1993; Desai et al. 2008).

Cancer is a group of diseases involving the growth of unwanted cells and having the ability to spread to different parts of the body. There were approximately 1.5 million new cases in the USA in 2008, and the number is increasing exponentially. About US$ 60 million worth of antitumor compounds have been traded in the year 2007 alone. Natural products are the most significant anticancer agents (Roessner and Scott 1996). The natural products account for three quarters of antitumor agents used in cancer treatment (Cragg et al. 1997). The current treatment of cancer includes radiotherapy, chemotherapy, and surgery (Harvey 2008). They have proven to be useful in different types of cancer such as colorectal, breast, ovarian, pancreatic, and lung cancer (McCaskill and Croteau 1997). But, the effectiveness of chemotherapy is restrained by the toxic effects produced on the nontargeted cells. Hence, alternative and complementary therapies like herbal therapies are used increasingly (Raskin et al. 2002). Such intrusions have not been studied or tested rigorously. The use of plants for the treatment of cancer is not new; indeed, they are one of the most important sources of bioactive compounds for treatment of many ailments. Industrialized societies rely on plants for development of bioactive molecules from rural herbal plants (Cragg and Newman 2005). Plant extracts either cause the target cells to undergo cell death or target dividing cells by impeding mitosis. The aim of the chapter is to focus on the various phytocompounds and their mechanism of action to treat the tumorigenic growth accompanied with the drug delivery system. Furthermore, the limitations and barriers associated with the formulation of phytocompounds and their possible enhanced therapeutic efficiency by nanoparticle-based drug delivery mechanics are also discussed.

5.2 Limitations of Synthetic Chemotherapeutic Drugs

Various approaches have been generated for treating different types of cancers, out of which some main approaches are surgical excision, irradiation, and chemotherapy which are used effectively depending on the type of tumors, their location, and their stage of development. Chemotherapy is one of the efficient therapeutic approaches applicable for treatment of various localized and metastasized tumors which can be used alone or with a combination of other therapies (Jemal et al. 2008, 2010). Different series of the chemically synthesized drugs are used to treat tumors by delivering them in the form of a beam to kill cancerous cells in the patient's body. However, being a potential therapeutic approach, there are some limitations to it like lack of aqueous selectivity and solubility along with development of multidrug resistance (Li 2002). Many chemotherapeutics derived either from natural or synthetic source have hydrophobic characteristics and need solvents to prepare the dosage form which leads to severe toxicity. Lack of selectivity or targeting of anticancer drugs toward tumor cells also contributes in causing significant damage to normal proliferating cells (Kwon 2003). Also, another limitation like multidrug resistance (MDR) caused due to efflux pumps like P-glycoprotein (Pgp) in the membrane of the cells, diverting various anticancer drugs out of the cells and restricting cell damage, causes negative impact (Callaghan et al. 2014). Secondly, nontargeted delivery of these anticancer drugs in the human body also affects the availability of dosage within the tumorous cells and results in suboptimal treatment caused by excessive toxicity (Agostinis et al. 2011). Therefore, an efficient natural product-based drug delivery system with targeted approach can help in culminating all the barriers and limitations (Das and Khuda-Bukhsh 2016).

5.3 Need for Natural Product-Based Drug for Cancer Therapy

The established synthetic chemotherapeutic drugs sometimes show the toxicity that creates a significant barrier in the cancer treatment. Many approaches have been proposed for treating cancers using plant-derived or natural products, and a large number of natural products have been explored for their potential against cancerous cells. It was largely observed that natural products show encouraging characteristics in terms of their easy absorption and metabolism in the human body with a lesser toxicity in comparison with synthetic drugs. Plants are the major source of bioactive compounds with numerous health benefits and anticancer activities. Natural compounds obtained from plant resources are well-utilized since ancient history for treating different kinds of human diseases in India, China, Egypt, and Greece, and several chemo-drugs have been developed from the traditionally utilized plants

(Arumugam et al. 2016; Swamy et al. 2016, 2017; Swamy and Sinniah 2016; Mohanty et al. 2017). Several classes of phytocompounds such as phenolics, alkaloids, flavonoids, etc., isolated from various medicinal plant species, have shown a noteworthy antiproliferative efficacy and can be used as a source to develop novel anticancer agents against different cancers (Mohanty et al. 2014, 2017; Swamy and Sinniah 2015). Approximately, 50–60% of patients diagnosed with cancer in the United States are utilizing plant-derived agents or nutrients (alternative medicines) for their treatment outstating the traditional therapeutics like chemotherapy (Gutheil et al. 2012). These plant-derived bioactive compounds include genistein (soybean), curcumin (turmeric), sulforaphane (broccoli), tea polyphenols (green tea), isothiocyanates (cruciferous vegetables), resveratrol (grapes), diallyl sulfide (garlic), apigenin (parsley), and the list continues.

5.4 Drug Discovery from Natural Products for Cancer Therapy

From 1981 to 2002, the United States Food and Drug Administration (US FDA) had approved over 62% natural source-based cancer drugs (Newman et al. 2003), and thereafter, again till 2006, the approvals and inclusions of phytocompounds reach to 47% (Newman and Cragg 2007), and the reassessment in 2010 exhibited the rise of 50%, showing the existence of efficacious phytocompounds equally potent to treat cancer (Newman and Cragg 2012). These natural products exhibit various approaches to combat the tumor growth like angiogenesis inhibition, blocking signaling pathway, induce apoptosis, etc.

Apigenin, a flavonoid found in parsley, has been reported to exhibit potential cytotoxic properties, comparable to doxorubicin to counter breast cancer and colon cancer (Hoensch and Oertel 2011; El-Alfy et al. 2011). Also, it has been extensively discussed by Chung et al. (2007) and later by Turktekin et al. (2011) that apigenin showed cell death in human colon cancer cell lines in male Sprague-Dawley rats, by decreasing azoxymethane (AOM)-induced aberrant crypt foci (ACF) synthesis which increases apoptosis and contributes in preventing cancer of the colon (Chung et al. 2007; Leonardi et al. 2010; Turktekin et al. 2011). Apigenin also alters leptin receptor pathway and causes cellular death in lung adenocarcinoma cell line (Bruno et al. 2011). The other most explored phytocompound is curcumin, showing an immense non-proliferative effect on colon, breast, brain, and lung cancer in various studies (Bachmeier et al. 2010; Senft et al. 2010). Curcumin has a potential of inducing apoptosis in cancer cells without leading to toxicity to the healthier cells and can affect many cancerous compounds or pathways like inhibiting NF-κB which cause inflammatory diseases inducing cancer, dispersion of raptor from mTOR, preventing mTOR complex I, etc. (Aggarwal and Shishodia 2004; Bachmeier et al. 2007; Beevers et al. 2009). Ravindran et al. (2009) stated that curcumin can regulate the proliferation of cancerous cells by controlling various cell signaling pathways such as cell growth pathway (cyclin D1, c-myc), cell survival pathway (Bcl-x, Bcl-2, XIAP, cFLIP, c-IAP1), caspase activation pathway (caspase-3, caspase-8, and caspase-9), death receptor pathway (DR4 and 5), protein kinase pathway (AMPK, JNK, and Akt), tumor suppressor pathway (p53, p21), and mitochondrial pathways (Ravindran et al. 2009).

Saffron, a food coloring component extracted from dry stigma of *Saffron crocus* flower, is marked as an efficient compound for its novel anticancer properties against hepatocellular and pancreatic carcinoma (Abdullaev and Espinosa-Aguirre 2004; Amin et al. 2011), skin carcinoma (Das et al. 2010), and colorectal and breast cancer (Chryssanthi et al. 2011). Crocetin (an active compound of saffron) disrupts the cancerous cell growth by preventing synthesis of nucleic acid, decreasing the free radical formation process, causing cell death and disturbing signal-mediated pathway of growth factors (Aung et al. 2007). In a study by Nam et al. (2010), crocetin showed its potential in inhibiting LPS-induced nitric oxide release, reducing the synthesis of IL-1β, TNF-α, and ROS and causing the suppression of NF-κB pathway and LPS effect on hippocampal cell death (Chryssanthi et al. 2011) (Table 5.1).

Cyanidin is a pigment extracted from grapes or red berries which contain antioxidant and radical scavenging properties that reduces the risk of cancer by limiting the proliferation of cells and COX-2 and iNOS gene expression in cells with colonic cancer (Bathaie and Mousavi 2010). In the study conducted by Lim et al. 2011, C3G disabled benzo[a]pyrene-7,8-diol-9, 10 epoxide-induced NF-κB and AP-1 activation and MKK4, MEK, MAPK, and AKt phosphorylation, preventing Fyn kinase signaling pathway activation, which makes it suitable for cancer treatment (Nam et al. 2010). NF-C3G also resists the activation of ethanol-induced ErbB2/cSrc/FAK pathway in the metastasis of breast cancer (Kim et al. 2008). Cyanidin significantly inhibits UVB-induced COX-2 expression and PGE2 secretion in epithelial skin cells by repressing NF-κB and AP-1 which are controlled by MAPK. It also targets MKK-4, MEK1, and Raf-1 to suppress UVB-induced COX-2 expression (Lim et al. 2011).

Brassica vegetables like cauliflower, broccoli, collard greens, etc. contain indole-3-carbinol (I3C) which converts into a digestive form called diindolylmethane (DIM) when exposed to the acidic condition of the stomach (Xu et al. 2010). Both derivatives showed anti-carcinogenic potential in different cancer types. I3C showed effective signs in preventing lung adenocarcinoma induced by tobacco smoke by modulating receptor tyrosine kinase/PI3K/Akt signaling pathway partially (Kim et al. 2010). I3C and DIM also showed remarkable anticancerous effect in different hormone-responsive cancers such as ovarian, breast, and prostate cancers (Kim and Milner 2005). Gingerol, a potential compound derived from different spices of ginger, demonstrated effective anticancer properties in treatment of tumors present in pancreas, breast, ovarian, and colon (Jeong et al. 2009; Acharya et al. 2010; Qian et al. 2011). It has antioxidant, antitumor, and anti-inflammation characteristics and reduces the expression of iNOS and TNF-alpha by inhibiting phosphorylation of IκBα and NF-κB nuclear translocation (Park et al. 2006a, b; Rhode et al. 2007; Oyagbemi et al. 2010).

Table 5.1 Bioactive compounds derived from natural resources eliciting anticancer activities

Plant	Botanical name	Bioactive compound	Type of cancer targeting	References
Saffron crocus	*Crocus sativus*	Saffron	Lung cancer, pancreatic cancer, skin carcinoma, colorectal cancer, and breast cancer	Abdullaev and Espinosa-Aguirre (2004) and Amin et al. (2011)
Cruciferous vegetables	*Brassica* spp.	Indole-3-carbinol (I3C), diindolylmethane (DIM)	Lung, breast, prostate, and ovarian cancers	Kim and Milner (2005) and Xu et al. (2010)
Tea	*Camellia sinensis*	Kaempferol	Pancreatic cancer and lung cancer	Cui et al. (2008) and Gacche et al. (2011)
Broccoli	*Brassica oleracea*			
Grapefruit	*Citrus paradisi*			
Red grapes skin	*Vitis vinifera*	Resveratrol	Skin and gastro-intestinal tract tumor, colon cancer	Kim et al. (2008)
Peanuts	*Arachis hypogaea*			
Ginger	*Zingiber officinale*	Gingerol	Colon cancer, breast and ovarian cancer, and pancreatic cancer	Jeong et al. (2009), Acharya et al. (2010), and Qian et al. (2011)
Catclaw acacia	*Acacia greggii*	Fisetin	Colon cancer	Acharya et al. (2010)
Guajillo	*Acacia berlandieri*			
Strawberries	*Fragaria ananassa*			
Apple	*Malus*			
Turmeric	*Curcuma longa*	Curcumin	Colon cancer, breast cancer, lung metastases, and brain tumor	Bachmeier et al. (2010) and Senft et al. (2010)
Grapes	*Vitis vinifera*	Cyanidin-3-glucoside (C3G)	Brest cancer, colon cancer	Bathaie and Mousavi (2010) and Lim et al. (2011)
		Cyanidin-3-O-glucoside, cyanidin-3-O-rutinoside		
Green tea	*Camellia sinensis*	Epigallocatechin gallate (EGCG)	Brain, prostate, cervical, and bladder cancers	Das et al. (2010)
Rosemary	*Rosmarinus officinalis*	Rosmarinic acid	Colon cancer, breast cancer, leukemia	Xu et al. (2010)

(continued)

Table 5.1 (continued)

Plant	Botanical name	Bioactive compound	Type of cancer targeting	References
Parsley	*Petroselinum crispum*	Apigenin	Breast cancer, colon cancer, lung cancer	El-Alfy et al. (2011) and Hoensch and Oertel (2011)
Celery	*Apium graveolens*			
Chamomile	*Matricaria chamomilla*			
Egyptian plant				
Moringa peregrine	*Moringa peregrina*			

A natural flavonol extracted from tea, broccoli, grapefruit, etc. known as kaempferol was explored for its anticancerous, antiangiogenic, and radical scavenging effect (Park et al. 2006a; Gacche et al. 2011; Calderon-Montano et al. 2011). Kaempferol was also inspected for treating lung and pancreatic cancer (Nöthlings et al. 2007; Cui et al. 2008). The studies revealed that kaempferol plays a role of aryl hydrocarbon receptor (AhR) antagonist to restrict the upregulation of ABCG2 which alters the resistance of ABCG2-mediated multi-drug and effectively treat esophageal cancer (To et al. 2012). Another study also disclosed that kaempferol can cause cell death of ovarian cancerous cells by activating the p53 in intrinsic pathway (Luo et al. 2011).

5.5 Designing Plant-Based Drugs to Encounter Cancerous Growth

Plants are used as a source of medicine and for health benefits since the ancient history. Approximately, 80–85% of the population depends on conventional therapy for their routine health-related issues. The traditional medicines involve the use of plant extracts or bioactive components (Elujoba et al. 2005). Much development has been done for treatment and control of cancer, but the scope of improvement still remains. The main limitation of synthetic drugs is associated with their negative impacts as mentioned above. Therefore, natural therapies involving plants or drugs derived from plants are used to treat cancer. The investigation of anticancer agents derived from plants commenced in the 1950s wherein vinca alkaloid (vincristine and vinblastin) was discovered and developed and cytotoxic podophyllotoxin was isolated (Cragg and Newman 2005). The following are some plant-derived antitumor agents which are in clinical use.

5.5.1 Vinca Alkaloids

Vinca alkaloids are a class of organic compounds and are made of hydrogen, carbon, oxygen, and nitrogen. These are a part of therapeutics which are obtained from the Madagascar periwinkle plant (Moudi et al. 2013). The alkaloids which are naturally extracted from *Catharanthus roseus* have cytotoxic and hypoglycemic effects (Tiong et al. 2015). They are used to treat diabetes and high blood pressure and can be used as disinfectants. They are also used to treat cancer. The vinca alkaloids in clinical use are vinblastine (VBN), vincristine (VCR), vindesine (VDS), and vinorelbine (VRL). VRL, VBL (vinblastine), and VCR have been approved for their use in USA. The mechanism of action of cytotoxicity of vinca alkaloid is the interaction with tubulin and, thus, disruption of the function of microtubules, particularly those microtubules which comprise the mitotic spindle apparatus, leading to metaphase arrest (Moudi et al. 2013). The alkaloids can have other effects rather than the disruption of microtubule. Many other biochemical activities that do not include the effect on microtubule occur only after treating the cells with therapeutically irrelevant dose of alkaloid. Both the vinca alkaloids and anti-microtubule agents have their influence on malignant and nonmalignant cells in non-mitotic cell functions, because microtubules are involved in many non-mitotic cell cycles.

The alkaloids connect to the binding sites of tubulin, and the connection between them occurs rapidly and it can reverse too. The evidence demonstrates the existence of two vinca alkaloid binding sites per one mole of tubulin dimer (Correia and Lobert 2001). This leads to interruption of microtubule congregation. One of the most crucial aspects of low drug concentration is shortening at the end of microtubule leading to production of kinetic cap thereby suppressing its function and decrease in the rate of bone growth (Stanton et al. 2011). The disturbing effect of vinca alkaloid especially at the end of spindles leads to metaphase arrest and occurs at drug concentrations lower than those that decrease the microtubule mass. It has been observed that microtubule-disrupting agents and vinca alkaloids have the capability to inhibit proliferation of malignant cells *in vitro*. For example, VBL within a range between 0.1 and 1.0 pmol/L leads to the blockage of chemotaxis, malignant angiogenesis, and the spreading of fibronectin which are all the important steps of angiogenesis. Also, other lymphoid tumors and normal fibroblasts were unaffected at the minute concentrations. VBL in combination with Ab against VEGF (vascular endothelial growth factor) leads to increased antitumor response; this is also observed even in tumors which are resistant to the cytotoxic effect of the drug at low concentrations of VBL. The cell proliferation is inhibited by vinca alkaloids by binding to microtubules, which leads to mitotic blockage and apoptosis. The destabilization of microtubules is caused by binding of VCR and related compounds to tubulin and, thus, blocking the polymerization.

Vinca alkaloids are included in chemotherapy regimens for medicinal therapy. VBL is used to treat Hodgkin's and non-Hodgkin's lymphomas and testicular carcinoma. VBL is same to VRL. They have antitumor activity against bone and breast cancer and osteosarcoma. VRL helps to decrease the stability of lipid bilayer membranes. VRL has been approved in the USA for managing and treating lung cancer. VCR is approved for curing Wilms' tumor, leukemia, neuroblastoma, Hodgkin's disease, rhabdomyosarcoma, and several other lymphomas (Rowinsky and Donehower 1995). VCR is used for to treat various nonmalignant hematologic disorders, for example, hemolytic uremic syndrome, thrombotic thrombocytopenia purpura, and refractory autoimmune thrombocytopenia (Rowinsky and Donehower 1995). VDS has similar effects as VBL. Antineoplastic activity of VDS is mentioned in pediatric solid tumors, malignant melanoma, blast crisis of chronic myeloid leukemia, acute lymphocytic leukemia, metastatic breast, and colorectal, renal, and esophageal tumors (Joel 1996). Vinflunine was developed recently adding two fluor molecules by the superacidic chemistry and is the first fluorinated microtubule inhibitor that belongs to the family of vinca alkaloids. It is used to treat second-line TCCU (transitional cell carcinoma of the urothelium), carcinoma of the breast, and lung carcinoma (Bennouna et al. 2008).

5.5.2 *Podophyllotoxin Derivatives*

The two species of Podophyllaceae family, *Podophyllum peltatum* and *P. emodi*, are reported to be used in the treatment of warts and skin cancer (Kelly and Hartwell 1954). *P. peltatum* is a perennial and native herb found in North America. *P. emodi* grows over the range of the Himalayas and contains white flowers growing below the leaves and has paprika-like fruits. And the parts are highly noxious. Thus, the utilization of podophyllin formulations provides an example of how an extract with ethnopharmacological properties despite of causing toxicity can be utilized for drug development. In 1980, Podwissotzki isolated podophyllotoxin from the crude mixture of podophyllin and established the structural formula. But, the correct structural formula was elucidated in the 1950s after the discovery and advancement in spectroscopic techniques. Lignan (closely related podotoxin) was also isolated during the time and was introduced in clinical trials, but due to unacceptable toxicity and lack of efficacy, it was dropped. In the 1960s, etoposide and teniposide were developed at Sandoz Laboratory in Switzerland which can be used for treating testicular and bronchial cancers (Gordaliza et al. 2004) (Table 5.2).

5.5.2.1 Pharmacological Activity

Podophyllotoxin derivatives bind reversibly to solid or soluble tubulin and inhibit the binding of colchicine to tubulin by competitively binding to it. Tubulin regulates the assembly of microtubules.

Table 5.2 Phytochemicals used for cancer therapy

Phytochemicals	Source	Drug delivery system	Medicinal properties	References
Podophyllotoxin	*Podophyllum peltatum*	Lipid nanoparticles	Brain tumors, small cell lung cancer, testicular cancer, lymphoblastic cancer	Kelly et al. (1954)
Taxol	*Taxus brevifolia*	Polymeric nanoparticles	Breast cancer, ovarian cancer, and lung cancer	Rowinsky and Donehower (1995) and Jennewein and Croteau (2001)
Vinblastine	*Vinca rosea*	Vinblastine-loaded folic acid-functionalized meso-porous silica nanoparticles	Hodgkin's lymphoma, non-Hodgkin's lymphoma	Correia and Lobert (2001) and Moudi et al. (2013)
Etoposide	*Podophyllum peltatum*	Polymeric nanoparticles	Ewing's sarcoma, testicular cancer, glioblastoma multiforme, Kaposi's sarcoma, lung cancer	Gordaliza et al. (2004)
Teniposide	*Podophyllum peltatum*	Albumin nanoparticles	Hodgkin's lymphoma, acute lymphocytic leukemia, brain tumors	
Homoharringtonine	*Cephalotaxus*	Lipid nanoparticles	Chronic myeloid leukemia	Cragg and Newman (2005)
Elliptinium	*Bleekeria vitiensis*	Polymeric nanoparticles	Breast cancer	Kaur et al. (2011)
Camptothecin	*Camptotheca acuminata*	Liposomal nanoparticles	Lung cancer, ovarian cancer, and cervical cancer	Nirmala et al. (2011)
Docetaxel	*Taxus baccata*	Nanoliposomes, activated nanoparticles	Breast cancer, stomach, head, neck, non-small cell lung cancer	Lawania and Mishra (2013)
Topotecan	*Camptotheca acuminata*	Liposomal nanoparticles	Ovarian, small cell lung cancer	Qin et al. (2017)
Irinotecan	*Camptotheca acuminata*	Silica nanoparticles	Colorectal cancer	

Microtubule Inhibition

Tubulin (a 6S dimeric protein of 110,000 M_1) is the building block of microtubules. Tubulin consists of two nonidentical chains designated as α and β of approximately 50,000 M_2. The tubulin dimers polymerize to form a prototype, which further organizes to produce microtubules. Microtubules undergo changes rapidly; the equilibrium between free tubulin and microtubules leads to constant assembly/disassembly at both the ends of microtubule (Margolis and Wilson 1981; Bohlin and Rosen 1996). Microtubule inhibitor causes arrestation of malignant cells, mitotic spindle, and acceleration/inhibition of release of hormones and enzymes. They affect the plasma membrane directly. Also, it is observed that the microtubules inhibit the release of chemotactic substances from phagocytic leukocytes in dot-like attractant assay (Xu et al. 2005). As low as 5 μM concentration of podophyllotoxin causes complete inhibition of tubulin polymerization.

Neutrophil Chemotaxis

It was observed that neutrophil chemotaxis was inhibited by podophyllotoxin (by 15%); inhibition was maximum at podophyllotoxin concentration of 0.01 μg/ml. A different concentration of CPH 82 (podophyllotoxin glycosides) was chosen by Rantapaa-Dahlqvist et al. (1994) to mimic the clinical conditions when they studied the cell cycle *in vitro*. Out of nine cell lines, eight presented cell accumulation in the G_2 phase of the cell cycle. It was observed that the progression through G phase slowed down in various cell lines. Further, the blockage was seen in G_1 and G_2 phase in all the three cell lines. For some cell lines, the development of M cells appeared considerably hindered. A dose-dependent and different pattern of cell cycle arrest was seen in podophyllotoxin CPH 82 and taxol (Dahlqvist et al. 1994).

Cell-Mediated Lympholysis

Podophyllotoxins inhibit the division of mouse spleen cells against stimulator cells and induce cytolytic activity. A sudden fall in cell-mediated lympholysis was caused by podophyllotoxins at the concentration of 1.5–3.0 ng/ml. The effect of drug was effective when added within 72 h. Large doses of podophyllotoxin had an impact on *in vivo* and *in vitro* immune responses. Because of many adverse events which were caused by contaminants in crude extracts, therefore, it is available in highly pure form. CPH 82 is a therapeutically active compound which is made up of two similar benzylidenated podophyllotoxin glucosides (a derivative of podophyllotoxin) and has been formulated to increase the absorption of the intestine. Various anti-inflammatory screening tests are done on CPH 82 (Bohlin and Rosen 1996).

5.5.3 *Taxanes*

It is also effective against Kaposi's sarcoma. Paclitaxel was first isolated from the bark of the *Taxus brevifolia* Nutt. (Taxaceae) (Nirmala et al. 2011). Several species of *Taxus* (*T. baccata* L. and *T. canadensis* Marshall.) and different parts of *T. brevifolia* are used by the tribes of America for the treatment of cancer, and this has been documented (Prakash et al. 2013). In Ayurveda, the leaves of *T. baccata* are used for treating cancer. The biosynthesis of paclitaxel takes place in the leaf region of different *Taxus* species, wherein baccatins are converted to paclitaxel and analogues of paclitaxel. For example, docetaxel (Taxotere) is an important natural renewable source of this class of drug (Jennewein and Croteau 2001). It has shown efficacy against rheumatoid arthritis, multiple sclerosis, and psoriasis. Docetaxel is used for treating breast cancer. Approximately 12 taxane analogues are in preclinical and clinical trials. Out of 2069 clinical trials of cancer by NCI (National Cancer Institute), 248 have been recorded as drugs derived from taxanes, which includes 134 with paclitaxel (Taxol), 105 with docetaxel (Taxotere), and 10 with miscellaneous taxanes (which includes single agent or combination with other anticancer agents). Twenty-three taxanes are in preclinical development (Lawania and Mishra 2013). Taxanes bind to β-tubulin in the microtubules and promote α,β-tubulin assembly. The binding mimics with a nucleotide "GTP" with some significant variations. The main cause of cytotoxic effect of paclitaxel is mitotic arrest because of the inhibition of microtubule dynamics. The inductions of apoptosis by biochemical events which may bind the paclitaxel to microtubules are still not known. The apoptosis might occur following a typical exit from mitosis to G_1 multinucleated state or directly after mitotic arrest. However, evidences indicate that arrestation of G_2/M phase of the cell cycle is not the only mechanism for apoptosis stimulated by paclitaxel. At high paclitaxel concentrations, formation of multipolar or monopolar spindles corresponded to a drop in fraction of HeLa cell line, whereas, at low concentration, the inhibition of spindle fibers was correlated to altered bipolar spindles, thereby proposing that spindle dynamics are also compromised (Abal et al. 2003).

5.5.4 *Camptothecin Derivatives*

Camptothecin (CPT) is the class of anticancer drugs which was isolated from the bark and stem of Chinese ornamental tree *Camptotheca acuminata* Decne (Nyssaceae) (Nirmala et al. 2011). This tree is used for treating cancer in traditional Chinese medicine. CPT is a quinolone alkaloid which is toxic to cells and inhibits topoisomerase 1 (DNA enzyme). CPT possesses an excellent anticancer property, but the solubility is low and produces an adverse drug reaction in the body. Because of these drawbacks of CPT, medicinal and synthetic chemists have developed different derivatives of CPT, i.e., irinotecan and topotecan. It consists of a planar pentacyclic ring which includes pyrrolo [3,4-β]-quinoline moiety (rings A, B, and C), pyridine (ring D), and a chiral center (position 20) within α-hydroxy-lactone ring (S configuration, E-ring) (Qin et al. 2017). It is observed that, because of its planar structure, it inhibits topoisomerase.

CPT binds to DNA and topo 1 (via hydrogen bond) thus producing a ternary and a stable complex. It averts religation of DNA and consequently begins damage of nucleic acid resulting in cell death. E-ring of CPT interacts with three varied locations of the enzyme. The hydroxyl group at the location 20 forms hydrogen bonding to a side chain present on aspartic acid (Asp533) of topo 1. It is worth to be noted that the configuration of chiral carbon is (S) because of the fact that (R) is inactive. Further, lactone ring interacts with amino groups on arginine 364 (Arg364) by forming a hydrogen bond. The pyridine ring interacts with +1 cytosine present on the noncleaved strand and helps to stabilize DNA-topo 1 by making a hydrogen bond. The hydrogen bonding is between the amino group in the ring D of +1 cytosine and carbonyl group in position 17. The cytotoxicity of the complex is due to the alteration of single-stranded breaks to double-stranded breaks in the course of S phase (synthesis phase) (Liu et al. 2000).

5.5.4.1 Structure-Activity Relationship (SAR)

Some modifications can be done to improve the activity of CPT. It has been noted that the substitution at position 7, 9, 10, and 11 improved the physical properties (metabolic stability and potency) and activity of CPT. On the other hand, replacement at position 12 and 14 results in inactive derivative. Addition of an alkyl group increases its efficacy. Substitution of alkyl group (chloromethyl and ethyl) at position 7 exhibited increase in cytotoxicity. It is because of the reason that these groups react with DNA in the presence of topo 1 resulting in increase in tumorigenic activity (Liu et al. 2000). Also, increasing the number of carbon atoms at position 7 results in enhanced hydrophobicity and, ultimately, higher stability and potency in plasma. Examples of analogues of CPT that have been modified at position 7 are karenitecins and silatecans (Liu et al. 2000). They are inhibitors of topo 1 and have alkylsilyl group at position 7 thereby making them more stable and lipophilic. Silyl camptothecins or silatecans lead to reduction in drug-HAS interactions resulting in plasma stability, and also, they are able to cause transient opening in the blood-brain barrier (BBB). The most active silatecan is DB-67 (a 10-hydroxy derivative) (Baker Jr. et al. 2014). BNP1350 belongs to the class of karenitecins and is capable of overcoming drug resistance and is toxic to cells. Lipophilic moieties such as oxyiminomethyl or iminomethyl can make CPT lipohillic. The activity of CPT can be considerably improved by adding a functional group "hydroxyl" at position 11 or 10 and the groups which are able to withdraw electrons (bromo, chloro, and nitro) at position 10 and 9. But the problem with them is that they are insoluble in aqueous solutions, thus causing difficulty in administrating them (Van Hattum et al. 2000).

5.5.5 Homoharringtonine

Homoharringtonine (omacetaxine mepesuccinate or HTT) is a semisynthetic preparation which is isolated from the Chinese plant, *Cephalotaxus harringtonia* var. drupacea (*Cephalotaxus*) which has a potential anticancerous activity (Cragg and Newman 2006). HTT (a cephalotaxine ester) inhibits the synthesis of protein and is a well-known agent to treat hematological cancer. It is synthesized from the extract of *Cephalotaxus* species of plant (mainly from leaves). Elliptinium was extracted from species of various genera of the Apocynaceae family which includes *Bleekeria vitiensis* (Fijian plant) with anticancer property (Kaur et al. 2011). For treating chronic myeloid leukemia (CML), HTT received orphan drug designation from EMEA in October 2005. Further, in March 2006, HTT obtained orphan drug status from the Food and Drug Administration (FDA) for curing CML. CML was approved as fast track designation by the FDA in November 2006. HTT was marketed under the brand name Synribo (TM) in October 2012 and has been accepted by the FDA for patients who are resistant or intolerant to tyrosine kinase inhibitors. The purified HTT is known to show efficiency against various cancers. Elliptinium is a therapeutically active compound which is sold in France and is used for treating breast cancer. The pharmacodynamics of omacetaxine is not clearly understood. HTT is involved in inhibition of protein synthesis by binding to 80S ribosomes in eukaryotic cells and interfering with the chain elongation which leads to its antineoplastic activity. This molecule also causes initiation of apoptosis and differentiation in cancerous cells (Chan et al. 2004).

5.6 Types of Drug Delivery Routes for Distribution of Anticancer Drugs

At present, the cancer therapy, especially in regard to drug delivery, has undergone an enormous change from the conventional approach. This evolution is based on the need to increase the therapeutic index of the bioactive compounds. Even though cancerous cells are more susceptible to chemotherapy drugs as compared to normal cells, the drugs can cause injury to normal cells and are therefore nonselective. Some other issues associated with conventional approach are low therapeutic index, low solubility, nonspecific distribution, lack of targeting ability, and systemic toxicity. Efforts are being made to kill the cancerous cells and restrict the death of normal cells. To attain this goal, there is a need for development of a carrier system for existing and new drugs and improved therapeutic targets in correspondence to molecular changes in cancerous cells. Nanotechnology, an emerging field, unravels the various possibilities associated with it. Nanoparticles (NPs) are 1–100 nm in size. The size of the particle is made in this range to bypass the fenestrations of epithelium of cancer cells causing the high concentration of NPs at the site of tumor. The nanoparticles are used for targeted drug delivery. Further, NPs have the ability of crossing various biological barriers (blood-brain barrier, cerebrospinal fluid, etc.). They have a large surface area, due to which a large amount of drugs can be loaded onto them, and have modifiable electronic, optical, biological, and magnetic properties. After the binding of nanoparticles to the receptor of the cell, the NPs undergo phagocytosis or receptor-mediated endocytosis and release the drug in the cytoplasm. NPs were developed in the late 1960s. The first nanoparticle conjugated with drug lipid vesicle was developed during the 1970s. Sandimmune® was the first micellar drug to be approved by the FDA for systemic administration of NPs in humans in 1983 (Siddiqui et al. 2015). Further, in the 1990s, the first polymeric nanoparticle-drug conjugate Adagen® was clinically approved for use in humans (Allen and Cullis 2004).

5.7 Mechanism of Targeting of Nanoparticles

Nanoparticles target tumor cells in two ways: active and passive targeting. Active targeting occurs by specific interaction between the target cell and the drug/drug carrier by receptor ligand interaction, antigen (Ag), and antibody (Ab) recognition for localization of drug intracellularly. On the other hand, passive targeting of drugs refers to localization of drugs in the area surrounded by tumor cells with leaky vasculature. The phenomenon is known as enhanced permeation and retention (EPR). It enables targeted delivery of nanoparticles to the tumor site. It is based on the size of NPs. The surface of NPs can be modified with various ligands that enables it to interact with specific receptors present on the tumor cells, thereby, imparting active targeting (Danhier et al. 2010). Nanoparticles for delivery of drugs include various designs in terms of shape, size, and materials. Each particle differs in terms of release kinetics, capacity of drug loading, stability of drugs, drug and particle stability, and the ability of targeted delivery. So, the following are examples of nanoparticles used for targeted delivery:

5.7.1 Polymeric Nanoparticles

The polymeric nanoparticles are generally produced from organic polymers of either natural or synthetic origin. They are biocompatible and biodegradable and are accepted for drug delivery of nanomaterials (Soppimath et al. 2001). Their surface can be modified by chemical transformations; they have an outstanding pharmacokinetic control and are appropriate for encapsulation and delivery of a wide variety of drugs. The polymers used in the formulation of polymeric nanoparticles are poly(butyl)cyanoacrylate, polylactic acid, gelatins, poly(lactic-co-glycolic acid) copolymer, poly(alkylcyanoacrylate), polyglycolic acid, chitosan, and poly(methyl methacrylate) (Kumari et al. 2010). Additionally, the coatings of polymeric-based nanoparticles can be functionalized onto different types of NPs so as to vary and enhance their biodistribution properties. PLGA and PLA have been approved by the FDA for human use. Various classes of drugs such as low-molecular-weight compounds can be encapsulated in the polymeric nanoparticles. Furthermore, these polymeric NPs have been used to treat breast cancer, thereby reducing the proliferation of cancer cells. The matrix which is made up of polymers prevents the degradation of drugs and also provides a system for controlled release of the drug. The level and extent of drug release are based on the drug-to-polymer ratio, the composition and molecular weight of the drug, and the polymer used (Alexis et al. 2008). The process of internalization and localization of NPs intracellularly is governed by the surface properties of the NPs. The advantages of polymeric nanoparticles are as follows: biodegradable polymers are used for the preparation of NPs, there is sustained and controlled release of the drug at the targeted site, the degradation of particles can be controlled by the constituents of the matrix, the encapsulation efficiency is high, bioactive compounds can be integrated into the delivery vehicle without any chemical reaction, and the vehicle can be used for different routes of administration including nasal, oral, parenteral, intraocular, etc.

In recent decades, various drug-linked polymeric nanoformulations have been fabricated for the treatment of cancer. PLA/PLGA and PCL are the most common polymers used as a carrier system. They are nontoxic, biodegradable, and biocompatible and have been used in medical sciences for more than 20 years. MTZ (mitoxantrone) is an antineoplastic agent which is shown to have activity against malignant glioma cell line but is unable to cross the BBB and causes adverse systemic side effects. Doxorubicin (DOX)-loaded PLGA microspheres (Yoo and Park 2001), cisplatin-loaded PLGA-mPEG nanoparticles (Gryparis et al. 2007), TX-loaded poly(lactic)-tocopheryl PEG succinate NPs (PLA-TPGS NPs), and poly(ethylene oxide)-modified polycaprolactone nanoparticles (PEO-PCL NPs) are some examples of polymeric nanoparticles used for cancer therapy. Yemisci prepared PLGA microspheres loaded with MTZ and studied on rat models. It was observed that MTZ-loaded microspheres efficiently delivered the bioactive compound to the site of the tumor, thereby preventing the proliferation of glioma cells without producing any side effect (Yemisci et al. 2006). Another polymeric nanocarrier, chitosan, is shown to be toxic against numerous tumorigenic cell lines both *in vivo* and *in vitro*. Chitosan possesses many remarkable properties like nontoxicity, biodegradability, and biocompatibility. Qi et al. formulated chitosan nanoparticles (CNPs), and the effect of CNPs was detected on BEL7402 (human hepatocellular carcinoma cell line). The results revealed that CNP acts as a promising candidate for treating hepatocellular carcinoma.

5.7.2 Gold Nanoparticles

Gold nanoparticles (AuNPs) enjoy a long history dating back to the Roman times, where they were used for staining glasses for decorative purposes. They have interesting physicochemical properties which make them ideal candidates for countless biomedical applications ranging from biodiagnostics, cellular imaging, immunostaining, drug/DNA delivery, and biosensing (Khlebtsov and Dykman 2011; Zeng et al. 2011; Rudramurthy et al. 2016). Hans Heinrich Helcher published a comprehensive treatise on AuNP in 1718. It was indicated that the utilization of boiled starch in drinkable gold preparation prominently increased the stability. Faraday did pioneering work in terms of the synthesis of stable aqueous dispersions of AuNP. The preparation of colloidal AuNP was first described by him in 1857 (Link and El-Sayed 1999). He basically defined the formation of varied colored solutions that were prepared by bonding $AuCl_3$with sodium citrate. Unknown to him, the reaction produced AuNPs ranging from a size of 12–60 nm. In the 1950s, it was discovered that AuNPs could be attached to protein biologics without affecting the activity leading to their usage in histopathology and immunodiagnostics. It is not surprising that there are countless methods of gold nanoparticle synthesis of controllable size, monodispersity, and shape in the literature, which include synthesis both in the form of aqueous and organic solutions. The modification of gold nanoparticle surface has also garnered a lot of interest lately. The strong affinity of gold for thiol and amine groups has enabled the fabrication of gold nanoparticles with biomolecules like amino acids and proteins. AuNPs, when used in conjunction with specific surface moieties, can be taken up by cells. Their unique properties make them potential candidates for transporting and unloading pharmaceuticals, essentially because they are inert and nontoxic and can be easily synthesized.

AuNP can be functionalized with a bioactive compound and a targeting ligand that particularly identifies a specific receptor (active targeting) (Ghosh et al. 2008). For example, transferrin can be functionalized onto AuNP and can be used for targeting tumor cells that overexpress the transferrin receptor (Wang et al. 2013). They have been employed for the delivery of antitumor agents like tumor necrosis factor (TNF), through the enhanced permeation and retention (EPR) effect which is characteristic of tumor cells. On exposure to laser photoradiation, AuNPs with specific aspect ratios or compositions have the potential to produce local heat that can destroy the diseased tissues in case of solid tumors. AuNP was covalently bound to gemcitabine and cetuximab (active agents to treat pancreatic cancer). EGFR is overexpressed in 60% of pancreatic cancers, and the nanocomplex is under investigation under phase II clinical trial. The AuNP-gemcitabine-cetuximab combination was superior to any of the agents used alone. Thirty percent inhibition of tumor was observed when nonconjugated agents were used in combination, whereas >80% of inhibition was noticed when a low dose of complex gemcitabine was used (2 mg/kg). However, Jiang et al. (2008) reported that citrate-coated AuNPs were synthesized with numerous trastuzumab Ab to allow cross linking and targeting with human epidermal growth factor receptor (HER-2) in breast cancerous cells (SK-BR-3).

5.7.3 Liposomes

Artificial phospholipid vesicles ranging from 500 to 1000 nm in size can be used not only for the loading of a number of water-soluble drugs but also for water-insoluble drugs. The pH-sensitive liposomes are now being utilized to achieve the release of active compounds. Liposomes need to be derived from pH-sensitive compounds in order to become pH-sensitive. They get endocytosed in the intact form by fusing with the membrane because of the acidic pH of the endosome and then discharge their components in the cytoplasm. New findings are focused upon the construction of liposomes with lipid compositions that render them pH-sensitive and usage of various pH-sensitive polymers. The class of pH-sensitive liposomes that have been extensively studied offer a number of advantages like the prevention of damaging action of external media on the drugs because of being enclosed within the liposomes, biocompatibility, biological inertness, low toxicity, less antigenic reactions, selective accumulation in the target organ or tissue, increase in efficiency of the bioactive compound, and decrease in liposome loss in the reticuloendothelial system (RES).

Currently, there are many products which have been marketed, and some are undergoing preclinical and clinical trials for treating cancer. Doxil, a PEGylated liposome, was the foremost liposomal formulation accepted by the FDA for treating Kaposi's sarcoma in AIDS patients. It encapsulates doxorubicin, an anticancerous drug, and is commercialized by Johnson & Johnson. It is approved for treating ovarian and breast cancer. DaunoXome is another liposomal formulation of daunorubicin which is accepted by the FDA for treating Kaposi's sarcoma related to AIDS. CPX-351 is another liposomal formulation developed by Celator Pharmaceuticals which is made with a combination of daunorubicin and cytarabine (Lasic 1996). It presented favorable results in phase III clinical trial on subjects suffering from acute myeloid leukemia (AML). Lipoplatin is another formulation of cisplatin which is developed by Regulon Inc. and is in phase III clinical trial and is used for treating lung cancer.

5.7.4 Iron Oxide Nanoparticles

Magnetic nanoparticles comprise of a core of magnetic iron oxide (usually magnetite-[Fe_3O_4] or maghemite [γ-Fe_2O_3]) and a shell of polymers such as dextran, silica, or PVA or metals like gold onto which attachment of functional groups can be done using various cross-linkers (Sun et al. 2008). Recently cobalt/gold nanoparticles with the same core-shell strategy have been synthesized with the advantage of having a higher magnetic moment than magnetite or maghemite (Peng et al. 2008; Rahman et al. 2011; Espinosa et al. 2016; Kanagesan et al. 2016a, b).

Methotrexate (MTX) has both therapeutic effect in cancerous cells that overexpress folate receptor on their surface and targeting role. Kohler and colleagues synthesized iron oxide nanoparticles conjugated with MTX (MTX-IO NP) by amidation between amine group on the NP and carboxylic acid on MTX. The results presented that cancer cells internalized a high level of MTX-IO NP than the control cells (negative) (Kohler et al. 2006). Multifunctional polymeric micelle was developed by Nasongkla and colleagues and loaded with superparamagnetic iron oxide nanoparticles (SPIO NP) in the micelle at 6.7 w/w %. Doxorubicin was loaded inside the micelle at 2.7 w/w % which would release in a pH-dependent manner. The advantage of a multifunctional nanoparticle is the encapsulation of SPIO and DOXO NP in the micelle core that avoids the direct contact of hydrophobic SPIO surfaces and absorption of blood proteins, thereby reducing nonspecific uptake by the reticuloendothelial system (RES). Additionally, cRGD ligand (Arg-Glyc-Asp) was linked onto the surface of micelle by thiol-maleimide linkage so that it can target $\alpha_v\beta_3$integrins which are expressed on tumor epithelial cells. The targeted cells internalize the complex, and high concentration of DOXO is released in the cell nuclei (Nasongkla et al. 2006).

5.7.5 *Dendrimers*

Dendrimers are macromolecular compounds with an inner core with a series of branches and the shape of which can be altered as and when required. Convergent or divergent growth polymerization is used to fabricate these from AB-type monomers. The number of functional groups doubles or triples with every layer. The surface groups can be functionalized with the drug molecules, or they can be loaded in the interior of the core. Various types of polymers can be used like polyamidoamine (PAMAM), polyethyleneimine, polypropyleneimine, poly(L-glutamic acid) (PLGA), and polyethylene glycol (PEG). They have a nanometric size, their surface can be modified easily, and they can be easily prepared and cause amplification of response due the easy functionalization of their surface with different functional groups (Caminade and Turrin 2014). Moreover, a therapeutic nanodevice like PAMAM dendrimer is used as a drug delivery carrier system. Fractional acetylation of G_5 (generation 5) PAMAM dendrimer was done so as to neutralize a portion of primary amino functional groups, improve the solubility of PAMAM during the conjugation reaction of FITC (fluorescein isothiocyanate), and stop nonspecific targeting interactions. The remaining non-acetylated amino group was used for conjugating FITC, and folic acid targets overexpressed folate receptors on cancer cells and MTX (methotrexate, a chemotherapeutic drug). The nanodevice was utilized for targeted drug delivery to treat carcinoma (Baker 2009; Madaan et al. 2014).

5.7.6 *Polymeric Micelle*

These are constructed by the spontaneous compilation of block copolymers that have an amphiphilic character in hydrophilic media, and these micelles have a diameter of some nanometers. These consist of a structure called core-shell with the inner core which assists as a storehouse of hydrophobic drugs which is bounded by an outer shell of hydrophilic carriers. Their advantages include controlled drug release, easy preparation, and effective modification without affecting the drug (Nishiyama and Kataoka 2006). Kabanov et al. (1998) synthesized micelle-based carrier system by forming micelles which consisted of Pluronic P85 and murine polyclonal Ab which are targeted against α_2-glycoprotein for delivery of haloperidol (a neuroleptic drug). The polyclonal Ab was attached to the micelle using analog butylpoly(25)(oxypropylene)poly(20)(oxyethylene) ether of 2-hydroxyacetaldehyde (BPEA). For cancer therapy, monoclonal Ab can be conjugated to the micelle. EGFR, a transmembrane glycoprotein, is overexpressed on the cells of solid tumors, and it was targeted with mAb C225. The external domain of EGFR binds to Ab and is appropriate for micelles to target various cancer cells. Folate can also be conjugated to the micelle because the folate receptor is overexpressed in different cancers including breast, ovary, kidney, lung, and myeloid cell (Oerlemans et al. 2010; Alexander-Bryant et al. 2013).

5.7.7 *Graphene Oxide Nanosheets*

Graphene oxide is a sheet made by oxidation of graphite. The carbon layers in graphite are interspersed with oxygen molecules, and then reduced, to separate each of the carbon layers or a few layers of graphene. So, the oxidation and reduction processes lead to formation of graphene oxide. This leads to increase in interplanar spacing between the carbon layers. And then, the completely oxidized product is dispersed in a base solution to make graphene oxide. Graphene oxide is considered as a starting material to make various processable materials (superconductor, composite for antibacterial activity, supercapacitors, carrier system to deliver drugs, base material for future application of water purification). Graphene oxide, a carbon nanomaterial (CNT), has an enormous potential in medical (Depan et al. 2011; Premkumar and Geckeler 2012) and material (Lian et al. 2011; Shahil and Balandin 2012) science because of its unique physicochemical, mechanical, and optical property. The popularity of graphene oxide as a drug carrier (Liu et al. 2011), imaging probe (Markovic et al. 2011), and biomedical device is increasing nowadays. The properties that can be exploited for loading of poorly soluble drugs are π–π stacking, high surface area, and hydrophobic or electrostatic interactions of graphene without compromising efficiency and potency. Graphene and GO have many unique and advantageous properties and have turned out to be potential candidates for targeted, local, and systemic drug delivery systems (Liu et al. 2013).

In an experiment by Chen et al. (2011), GO was functionalized with PEI (cationic polymer), and then, it was utilized for drug delivery. PEI is a non-viral gene vector and has an affinity for phosphates present in DNA or RNA, and this polymer can be modified chemically. The PEI modification resulted in enhanced cell selectivity, transfection efficacy, and decrease in cell toxicity. But, the limitation of biocompatibility still persists in its biomedical application (Li et al. 2013). GO functionalized with PEI (PEI-GO) was used for the delivery of gene, and it was noticed that cytotoxicity of PEI-GO was less than PEI alone. PEI-GO facilitated in achieving high transfection of drug gene (Chen et al. 2011). In a recent study, graphene foams are used as 3D porous scaffolds in neural stem cell cultures (NSCs). The porous scaffolds have an ability of stimulating NSCs electrically and help in proliferation and differentiation of neural stem cells. Graphene oxide can be utilized as a nanocarrier for delivery of anticancer drugs to a target site. For example, doxorubicin, an anticancer drug, is adsorbed on the surface of GO sheet by hydrogen bonding and π–π stacking, thus suggesting that graphene oxide can be used as a nanocarrier. There is an urgent need to assess the toxicity of graphene oxide, since the risk evaluation of graphene oxide is lacking. Recently, it was found that graphene causes low toxicity in macrophages (Li et al. 2012), stem cells (Akhavan et al. 2012), and lung cells (Vallabani et al. 2011). Toxicity in adenocarcinoma cells was observed, using A549 cell line (Chang et al. 2011). Toxicity in fibrosarcoma cells was also observed using L929 cells. *Ginkgo biloba*-graphene oxide nanocomposite (Gb-rGO) was synthesized and tested for the biocompatibility of GO, and Gb-rGO was evaluated on human breast cancer cells by various assays like apoptosis, cell viability, and alkaline phosphatase assay. The results exhibited that the synthesized Gb-rGO is cytocompatible to the cells. Further, the nanocomplex can be used in different biomedical applications like drug delivery, imaging, biosensing, and tissue engineering (Gurunathan et al. 2014).

5.8 Conclusions and Future Prospects

Synthetically derived anticancer drugs used for the conventional approaches such as chemotherapy have many limitations because they exhibit a toxic effect on the noncancerous cells in the human body which causes health-related problems. Thus, there is a need for other therapeutic approaches that can treat cancer effectively with minimal side effects. The anticancer agents derived from natural products are the desired approach for nontoxic treatment of cancerous cells. The secondary derivatives of plant extracts like flavonoids, polyphenols, etc. were broadly explored for their potential characteristics to treat cancer. Some of them showed valuable anticancer properties like antioxidant activity, induction of apoptosis, cytotoxicity to cancer cells, anti-inflammatory properties, target specificity, etc. Naturally derived drugs were generated and progressed to clinical examination for further studies about their secondary metabolites, stability, and long-run procedures. The natural products derived from plants have the ability to decrease cancer. These may show anticancer activity in various animal models of skin cancer, leukemia, and sarcomas. The uses of nanoparticles have increased exponentially over the past few years for cancer therapy and drug delivery. In the future, more researches are desired for the development of sophisticated and novel applications that may result in regression of tumor, early detection, and effective response to chemotherapy. Moreover, it also has new perspectives for efficient and conventional formulations of drugs because of instability or poor bioavailability.

References

Abal M, Andreu J, Barasoain I (2003) Taxanes: microtubule and centrosome targets, and cell cycle dependent mechanisms of action. Curr Cancer Drug Targ 3:193–203

Abdullaev F, Espinosa-Aguirre J (2004) Biomedical properties of saffron and its potential use in cancer therapy and chemoprevention trials. Cancer Detect Prev 28:426–432

Acharya A, Das I, Singh S, Saha T (2010) Chemopreventive properties of indole-3-carbinol, diindolylmethane and other constituents of cardamom against carcinogenesis. Recent Pat Food Nutr Agric 2:166–177

Aggarwal BB, Shishodia S (2004) Suppression of the nuclear factor-kappa B activation pathway by spice-derived phytochemicals: reasoning for seasoning. Ann N Y Acad Sci 1030:434–441

Agostinis P, Berg K, Cengel KA, Foster TH, Girotti AW, Gollnick SO, Hahn SM, Hamblin MR, Juzeniene A, Kessel D, Korbelik M (2011) Photodynamic therapy of cancer: an update. CA Cancer J Clin 61:250–281

Akhavan O, Ghaderi E, Akhavan A (2012) Size-dependent genotoxicity of graphene nanoplatelets in human stem cells. Biomaterials 33:8017–8025

Alexander-Bryant AA, Vanden Berg-Foels WS, Wen X (2013) Bioengineering strategies for designing targeted cancer therapies. Adv Cancer Res 118:1–59

Alexis F, Pridgen E, Molnar LK, Farokhzad OC (2008) Factors affecting the clearance and biodistribution of polymeric nanoparticles. Mol Pharm 5:505–515

Allen TM, Cullis PR (2004) Drug delivery systems: entering the mainstream. Science 303:1818–1822

Amin A, Hamza AA, Bajbouj K, Ashraf SS, Daoud S (2011) Saffron: a potential candidate for a novel anticancer drug against hepatocellular carcinoma. Hepatology 54:857–867

Arumugam G, Swamy MK, Sinniah UR (2016) *Plectranthus amboinicus* (Lour.) Spreng: botanical, phytochemical, pharmacological and nutritional significance. Molecules 21:369

Aung H, Wang C, Ni M, Fishbein A, Mehendale SR, Xie JT, Shoyama AY, Yuan CS (2007) Crocin from *Crocus sativus* possesses significant anti-proliferation effects on human colorectal cancer cells. Exp Oncol 29:175–180

Bachmeier BE, Mohrenz IV, Mirisola V, Schleicher E, Romeo F, Höhneke C, Jochum M, Nerlich AG, Pfeffer U (2007) Curcumin downregulates the inflammatory cytokines CXCL1 and-2 in breast cancer cells via NFκB. Carcinogenesis 29:779–789

Bachmeier BE, Mirisola V, Romeo F, Generoso L, Esposito A, Dell'Eva R, Blengio F, Killian PH, Albini A, Pfeffer U (2010) Reference profile correlation reveals estrogen-like transcriptional activity of curcumin. Cell Physiol Biochem 26:471–482

Baker JR Jr (2009) Dendrimer-based nanoparticles for cancer therapy. Hematol Am Soc Hematol Educ Program 2009:708–719

Baker Jr JR, Huang BM, Thomas TP (2014) Pro-drug complexes and related methods of use. US Patent No. 8853151

Bandyopadhyay AA, Khetan A, Malmberg LH, Zhou W, Hu WS (2017) Advancement in bioprocess technology: parallels between microbial natural products and cell culture biologics. J Indian Microbiol Biotechnol 44:785–797
Bathaie SZ, Mousavi SZ (2010) New applications and mechanisms of action of saffron and its important ingredients. Crit Rev Food Sci Nutr 50:761–786
Beevers CS, Chen L, Liu L, Luo Y, Webster NJ, Huang S (2009) Curcumin disrupts the mammalian target of rapamycin-raptor complex. Cancer Res 69:1000–1008
Bennouna J, Delord JP, Campone M, Nguyen L (2008) Vinflunine: a new microtubule inhibitor agent. Clin Cancer Res 14:1625–1632
Bohlin L, Rosen B (1996) Podophyllotoxin derivatives: drug discovery and development. Drug Discov Today 1:343–351
Bruno A, Siena L, Gerbino S, Ferraro M, Chanez P, Giammanco M, Gjomarkaj M, Pace E (2011) Apigenin affects leptin/leptin receptor pathway and induces cell apoptosis in lung adenocarcinoma cell line. Eur J Cancer 47:2042–2051
Calderon-Montano JM, Burgos-Morón E, Pérez-Guerrero C, López-Lázaro M (2011) A review on the dietary flavonoid kaempferol. Mini Rev Med Chem 11:298–344
Callaghan R, Luk F, Bebawy M (2014) Inhibition of the multidrug resistance P-glycoprotein: time for a change of strategy. Drug Metab Dispos 42:623–631
Caminade AM, Turrin CO (2014) Dendrimers for drug delivery. J Mater Chem B 2:4055–4066
Chan J, Khan SN, Harvey I, Merrick W, Pelletier J (2004) Eukaryotic protein synthesis inhibitors identified by comparison of cytotoxicity profiles. RNA 10:528–543
Chang Y, Yang ST, Liu JH, Dong E, Wang Y, Cao A, Liu Y, Wang H (2011) *In vitro* toxicity evaluation of graphene oxide on A549 cells. Toxicol Lett 200:201–210
Chen B, Liu M, Zhang L, Huang J, Yao J, Zhang Z (2011) Polyethylenimine-functionalized graphene oxide as an efficient gene delivery vector. J Mater Chem 21:7736–7741
Chryssanthi DG, Dedes PG, Karamanos NK, Cordopatis P, Lamari FN (2011) Crocetin inhibits invasiveness of MDA-MB-231 breast cancer cells via downregulation of matrix metalloproteinases. Planta Med 77:146–151
Chung C, Jiang Y, Cheng D, Birt DF (2007) Impact of adenomatous polyposis coli (APC) tumor suppressor gene in human colon cancer cell lines on cell cycle arrest by apigenin. Mol Carcinog 46:773–782
Correia JJ, Lobert S (2001) Physiochemical aspects of tubulin-interacting antimitotic drugs. Curr Pharm Des 7:1213–1228
Cragg GM, Newman DJ (2005) Plants as a source of anti-cancer agents. J Ethnopharmacol 100:72–79
Cragg G, Newman D (2006) Natural products as sources of antitumor agents. In: Ethnopharmacology. EOLSS Publishers, Oxford
Cragg GM, Newman DJ, Snader KM (1997) Natural products in drug discovery and development. J Nat Prod 60:52–60
Cui Y, Morgenstern H, Greenland S, Tashkin DP, Mao JT, Cai L, Cozen W, Mack TM, Lu QY, Zhang ZF (2008) Dietary flavonoid intake and lung cancer – a population based case control study. Cancer 112:2241–2248
Dahlqvist SR, Landberg G, Roos G, Norberg B (1994) Cell cycle effects of the anti-rheumatic agent CPH82. Rheumatology 33:327–331
Danhier F, Feron O, Préat V (2010) To exploit the tumor microenvironment: passive and active tumor targeting of nanocarriers for anti-cancer drug delivery. J Control Release 148:135–146
Das S, Khuda-Bukhsh AR (2016) PLGA-loaded nanomedicines in melanoma treatment: future prospect for efficient drug delivery. Indian J Med Res 144:181–193
Das I, Das S, Saha T (2010) Saffron suppresses oxidative stress in DMBA-induced skin carcinoma: a histopathological study. Acta Histochem 112:317–327
Depan D, Shah J, Misra R (2011) Controlled release of drug from folate-decorated and graphene mediated drug delivery system: synthesis, loading efficiency, and drug release response. Mater Sci Eng C 31:1305–1312

Desai AG, Qazi GN, Ganju RK, El-Tamer M, Singh J, Saxena AK, Bedi YS, Taneja SC, Bhat HK (2008) Medicinal plants and cancer chemoprevention. Curr Drug Metab 9:581–591

El-Alfy TS, Ezzat SM, Hegazy AK, Amer AM, Kamel GM (2011) Isolation of biologically active constituents from *Moringa peregrina* (Forssk.) Fiori. (family: Moringaceae) growing in Egypt. Pharmacogn Mag 7:109–115

Elujoba AA, Odeleye O, Ogunyemi C (2005) Traditional medicine development for medical and dental primary health care delivery system in Africa. Afr J Tradit Complement Alternat Med 2:46–61

Espinosa A, Di Corato R, Kolosnjaj-Tabi J, Flaud P, Pellegrino T, Wilhelm C (2016) Duality of iron oxide nanoparticles in cancer therapy: amplification of heating efficiency by magnetic hyperthermia and photothermal bimodal treatment. ACS Nano 10:2436–2446

Fenical W, Jensen PR (1993) Marine microorganisms: a new biomedical resource. In: Attaway DH, Zaborsky OR (eds) Pharmaceutical and bioactive natural products. Springer, Boston, pp 419–457

Gacche RN, Shegokar HD, Gond DS, Yang Z, Jadhav AD (2011) Evaluation of selected flavonoids as antiangiogenic, anticancer, and radical scavenging agents: an experimental and in silico analysis. Cell Biochem Biophys 61:651–663

Ghosh P, Han G, De M, Kim CK, Rotello VM (2008) Gold nanoparticles in delivery applications. Adv Drug Deliv Rev 60:1307–1315

Gordaliza M, Garcia P, Del Corral JM, Castro MA, Gómez-Zurita MA (2004) Podophyllotoxin: distribution, sources, applications and new cytotoxic derivatives. Toxicon 44:441–459

Gouvea R, Kassicieh S (2005) Using resources in R & D policy planning: Brazil, the Amazon and biotechnology. Technol Forecast Soc Chang 72:535–547

Gryparis EC, Hatziapostolou M, Papadimitriou E, Avgoustakis K (2007) Anticancer activity of cisplatin-loaded PLGA-mPEG nanoparticles on LNCaP prostate cancer cells. Eur J Pharm Biopharm 67:1–8

Gurunathan S, Han JW, Park JH, Eppakayala V, Kim JH (2014) *Ginkgo biloba*: a natural reducing agent for the synthesis of cytocompatible graphene. Int J Nanomedicine 9:363

Gutheil W, Reed G, Ray A, Anant S, Dhar A (2012) Crocetin: an agent derived from saffron for prevention and therapy for cancer. Curr Pharm Biotechnol 13:173–179

Harvey AL (2008) Natural products in drug discovery. Drug Discov Today 13:894–901

Henkel T, Brunne RM, Müller H, Reichel F (1999) Statistical investigation into the structural complementarity of natural products and synthetic compounds. Angew Chem Int Ed Eng 38:643–647

Hoensch HP, Oertel R (2011) Emerging role of bioflavonoids in gastroenterology: especially their effects on intestinal neoplasia. World J Gastrointest Oncol 3:71–75

Holland HD (1997) Evidence for life on Earth more than 3850 million years ago. Science 275:38–39

Jemal A, Siegel R, Ward E, Hao Y, Xu J, Murray T, Thun MJ (2008) Cancer statistics 2008. CA: Cancer J Clin 58:71–96

Jemal A, Siegel R, Xu J, Ward E (2010) Cancer statistics 2010. CA: Cancer J Clin 60:277–300

Jennewein S, Croteau R (2001) Taxol: biosynthesis, molecular genetics, and biotechnological applications. Appl Microbiol Biotechnol 57:13–19

Jeong CH, Bode AM, Pugliese A, Cho YY, Kim HG, Shim JH, Jeon YJ, Li H, Jiang H, Dong Z (2009) [6]-Gingerol suppresses colon cancer growth by targeting leukotriene A_4 hydrolase. Cancer Res 69:5584–5591

Jiang W, Kim BY, Rutka JT, Chan WC (2008) Nanoparticle-mediated cellular response is size-dependent. Nat Nanotechnol 3:145–150

Joel S (1996) The comparative clinical pharmacology of vincristine and vindesine: does vindesine offer any advantage in clinical use. Cancer Treat Rev 21:513–525

Kanagesan S, Hashim M, Aziz ABS, Ismail I, Tamilselvan S, Alitheen N, Swamy MK, Purna Chandra Rao B (2016a) Evaluation of antioxidant and cytotoxicity activities of copper ferrite ($CuFe_2O_4$) and zinc ferrite ($ZnFe_2O_4$) nanoparticles synthesized by sol-gel self-combustion method. Appl Sci 6:184

Kanagesan S, Hashim M, Aziz SAB, Ismail I, Tamilselvan S, Alitheen NB, Swamy MK, Rao BPC (2016b) Synthesis, characterization and *in vitro* evaluation of manganese ferrite ($MnFe_2O_4$) nanoparticles for their biocompatibility with murine breast cancer cells (4T1). Molecules 21:312. https://doi.org/10.3390/molecules21030312

Kaur R, Kapoor K, Kaur H (2011) Plants as a source of anticancer agents. J Nat Prod Plant Resour 1:119–124

Kelly MG, Hartwell JL (1954) The biological effects and the chemical composition of podophyllin. Rev J Natl Cancer Inst 14:967–1010

Khlebtsov N, Dykman L (2011) Biodistribution and toxicity of engineered gold nanoparticles: a review of *in vitro* and *in vivo* studies. Chem Soc Rev 40:1647–1671

Kim YS, Milner J (2005) Targets for indole-3-carbinol in cancer prevention. J Nutr Biochem 16:65–73

Kim JM, Kim JS, Yoo H, Choung MG, Sung MK (2008) Effects of black soybean [*Glycine max* (L.) Merr.] seed coats and its anthocyanidins on colonic inflammation and cell proliferation *in vitro* and *in vivo*. J Agric Food Chem 56:8427–8433

Kim JE, Kwon JY, Seo SK, Son JE, Jung SK, Min SY, Hwang MK, Heo YS, Lee KW, Lee HJ (2010) Cyanidin suppresses ultraviolet B-induced COX-2 expression in epidermal cells by targeting MKK4, MEK1, and Raf-1. Biochem Pharmacol 79:1473–1482

Knight V, Sanglier JJ, DiTullio D, Braccili S, Bonner P, Waters J, Hughes D, Zhang L (2003) Diversifying microbial natural products for drug discovery. Appl Microbiol Biotechnol 62:446–458

Kohler N, Sun C, Fichtenholtz A, Gunn J, Fang C, Zhang M (2006) Methotrexate-immobilized poly (ethylene glycol) magnetic nanoparticles for MR imaging and drug delivery. Small 2:785–792

Kumari A, Yadav SK, Yadav SC (2010) Biodegradable polymeric nanoparticles based drug delivery systems. Colloids Surf B Biointerfaces 75:1–18

Kwon GS (2003) Polymeric micelles for delivery of poorly water-soluble compounds. Crit Rev Ther Drug Carrier Syst 20:357–403

Lasic DD (1996) Doxorubicin in sterically stabilized liposomes. Nature 380:561–562

Lawania R, Mishra A (2013) Anticancer potential of plants and natural products: a review. J Pharmacol Biomed Anal 1:104–115

Leonardi T, Vanamala J, Taddeo SS, Davidson LA, Murphy ME, Patil BS, Wang N, Carroll RJ, Chapkin RS, Lupton JR, Turner ND (2010) Apigenin and naringenin suppress colon carcinogenesis through the aberrant crypt stage in azoxymethane-treated rats. Exp Biol Med 235:710–717

Li C (2002) Poly (L-glutamic acid)-anticancer drug conjugates. Adv Drug Deliv Rev 54:695–713

Li Y, Liu Y, Fu Y, Wei T, Le Guyader L, Gao G, Liu RS, Chang YZ, Chen C (2012) The triggering of apoptosis in macrophages by pristine graphene through the MAPK and TGF-beta signaling pathways. Biomaterials 33:402–411

Li N, Zhang Q, Gao S, Song Q, Huang R, Wang L, Liu L, Dai J, Tang M, Cheng G (2013) Three-dimensional graphene foam as a biocompatible and conductive scaffold for neural stem cells. Sci Rep 3:1604. https://doi.org/10.1038/srep01604

Lian P, Zhu X, Liang S, Li Z, Yang W, Wang H (2011) High reversible capacity of SnO_2/graphene nanocomposite as an anode material for lithium-ion batteries. Electrochim Acta 56:4532–4539

Lim TG, Kwon JY, Kim J, Song NR, Lee KM, Heo YS, Lee HJ, Lee KW (2011) Cyanidin-3-glucoside suppresses B[a]PDE-induced cyclooxygenase-2 expression by directly inhibiting Fyn kinase activity. Biochem Pharmacol 82:167–174

Link S, El-Sayed MA (1999) Size and temperature dependence of the plasmon absorption of colloidal gold nanoparticles. J Phys Chem B 103:4212–4217

Liu LF, Desai SD, Li TK, Mao Y, Sun ME, SIM SP (2000) Mechanism of action of camptothecin. Ann N Y Acad Sci 922:1–10

Liu Z, Robinson JT, Tabakman SM, Yang K, Dai H (2011) Carbon materials for drug delivery and cancer therapy. Mater Today 14:316–323

Liu J, Cui L, Losic D (2013) Graphene and graphene oxide as new nanocarriers for drug delivery applications. Acta Biomater 9:9243–9257

Luo H, Rankin GO, Li Z, DePriest L, Chen YC (2011) Kaempferol induces apoptosis in ovarian cancer cells through activating p53 in the intrinsic pathway. Food Chem 128:513–519

Madaan K, Kumar S, Poonia N, Lather V, Pandita D (2014) Dendrimers in drug delivery and targeting: drug-dendrimer interactions and toxicity issues. J Pharm Bioallied Sci 6:139–150

Manju K, Jat R, Anju G (2017) A review on medicinal plants used as a source of anticancer agents. Int J Drug Res Technol 2:177–183

Margolis RL, Wilson L (1981) Microtubule treadmills-possible molecular machinery. Nature 293:705–711

Markovic ZM, Harhaji-Trajkovic LM, Todorovic-Markovic BM, Kepić DP, Arsikin KM, Jovanović SP, Pantovic AC, Dramićanin MD, Trajkovic VS (2011) *In vitro* comparison of the photothermal anticancer activity of graphene nanoparticles and carbon nanotubes. Biomaterials 32:1121–1129

McCaskill D, Croteau R (1997) Prospects for the bioengineering of isoprenoid biosynthesis. In: Berger RG (ed) Biotechnology of aroma compounds. Springer, Berlin, pp 107–146

Mohanty SK, Mallappa KS, Godavarthi A, Subbanarasiman B, Maniyam A (2014) Evaluation of antioxidant, *in vitro* cytotoxicity of micropropagated and naturally grown plants of *Leptadenia reticulata* (Retz.) Wight & Arn.-an endangered medicinal plant. Asian Pac J Trop Med 7:S267–S271

Mohanty SK, Swamy MK, Sinniah UR, Anuradha M (2017) *Leptadenia reticulata* (Retz.) Wight & Arn.(Jivanti): botanical, agronomical, phytochemical, pharmacological, and biotechnological aspects. Molecules 22:1019. https://doi.org/10.3390/molecules22061019

Moudi M, Go R, Yien CYS, Nazre M (2013) Vinca alkaloids. Int J Prev Med 4:1231–1235

Nam KN, Park YM, Jung HJ, Lee JY, Min BD, Park SU, Jung WS, Cho KH, Park JH, Kang I, Hong JW, Lee EH (2010) Anti-inflammatory effects of crocin and crocetin in rat brain microglial cells. Eur J Pharmacol 648:110–116

Nasongkla N, Bey E, Ren J, Ai H, Khemtong C, Guthi JS, Chin SF, Sherry AD, Boothman DA, Gao J (2006) Multifunctional polymeric micelles as cancer-targeted, MRI-ultrasensitive drug delivery systems. Nano Lett 6:2427–2430

Newman DJ, Cragg GM (2007) Natural products as sources of new drugs over the last 25 years. J Nat Prod 70:461–477

Newman DJ, Cragg GM (2012) Natural products as sources of new drugs over the 30 years from 1981 to 2010. J Nat Prod 75:311–335

Newman DJ, Cragg GM, Snader KM (2003) Natural products as sources of new drugs over the period 1981-2002. J Nat Prod 66:1022–1037

Nirmala MJ, Samundeeswari A, Sankar PD (2011) Natural plant resources in anticancer therapy: a review. Res Plant Biol 1:1–14

Nishiyama N, Kataoka K (2006) Current state, achievements, and future prospects of polymeric micelles as nanocarriers for drug and gene delivery. Pharmacol Ther 112:630–648

Nöthlings U, Murphy SP, Wilkens LR, Henderson BE, Kolonel LN (2007) Flavonols and pancreatic cancer risk: the multiethnic cohort study. Am J Epidemiol 166:924–931

Oerlemans C, Bult W, Bos M, Storm G, Nijsen JF, Hennink WE (2010) Polymeric micelles in anticancer therapy: targeting, imaging and triggered release. Pharm Res 27:2569–2589

Oyagbemi AA, Saba AB, Azeez OI (2010) Molecular targets of [6]-gingerol: its potential roles in cancer chemoprevention. Biofactors 36:169–178

Park JS, Rho HS, Kim DH, Chang IS (2006a) Enzymatic preparation of kaempferol from green tea seed and its antioxidant activity. J Agric Food Chem 54:2951–2956

Park YJ, Wen J, Bang S, Park SW, Song SY (2006b) [6]-Gingerol induces cell cycle arrest and cell death of mutant p53-expressing pancreatic cancer cells. Yonsei Med J 47:688–697

Peng XH, Qian X, Mao H, Wang AY (2008) Targeted magnetic iron oxide nanoparticles for tumor imaging and therapy. Int J Nanomedicine 3:311–321

Prakash O, Kumar A, Kumar P (2013) Anticancer potential of plants and natural products: a review. Am J Pharma Sci 1:104–115

Premkumar T, Geckeler KE (2012) Graphene-DNA hybrid materials: assembly, applications, and prospects. Prog Polym Sci 37:515–529

Qian X, Melkamu T, Upadhyaya P, Kassie F (2011) Indole-3-carbinol inhibited tobacco smoke carcinogen-induced lung adenocarcinoma in A/J mice when administered during the post-initiation or progression phase of lung tumorigenesis. Cancer Lett 311:57–65

Qin SY, Zhang AQ, Cheng SX, Rong L, Zhang XZ (2017) Drug self-delivery systems for cancer therapy. Biomaterials 112:234–247

Rahman MM, Khan SB, Jamal A, Faisal M, Aisiri AM (2011) Iron oxide nanoparticles, nanomaterials. In: Rahman M (ed) Nanomaterials. InTech. http://www.intechopen.com/books/nanomaterials/iron-oxide-nanoparticles. Assessed 18 Sept 2017

Raskin I, Ribnicky DM, Komarnytsky S, Ilic N, Poulev A, Borisjuk N, Brinker A, Moreno DA, Ripoll C, Yakoby N, O'Neal JM (2002) Plants and human health in the twenty-first century. Trends Biotechnol 20:522–531

Ravindran J, Prasad S, Aggarwal BB (2009) Curcumin and cancer cells: how many ways can curry kill tumor cells selectively. AAPS J 11:495–510

Rhode J, Fogoros S, Zick S, Wahl H, Griffith KA, Huang J, Liu JR (2007) Ginger inhibits cell growth and modulates angiogenic factors in ovarian cancer cells. BMC Complement Alternat Med 7:44–51

Roessner CA, Scott AI (1996) Achieving natural product synthesis and diversity via catalytic networking ex vivo. Chem Biol 3:325–330

Rowinsky EK, Donehower RC (1995) Paclitaxel (taxol). New Engl J Med 332:1004–1014

Rudramurthy GR, Swamy MK, Sinniah UR, Ghasemzadeh A (2016) Nanoparticles: alternatives against drug-resistant pathogenic microbes. Molecules 21:836. https://doi.org/10.3390/molecules21070836

Senft C, Polacin M, Priester M, Seifert V, Kögel D, Weissenberger J (2010) The nontoxic natural compound curcumin exerts anti-proliferative, anti-migratory, and anti-invasive properties against malignant gliomas. BMC Cancer 10:491. https://doi.org/10.1186/1471-2407-10-491

Shahil KM, Balandin AA (2012) Thermal properties of graphene and multilayer graphene: applications in thermal interface materials. Solid State Commun 152:1331–1340

Siddiqui IA, Sanna V, Ahmad N, Sechi M, Mukhtar H (2015) Resveratrol nanoformulation for cancer prevention and therapy. Ann N Y Acad Sci 1348:20–31

Skinnider MA, Dejong CA, Franczak BC, McNicholas PD, Magarvey NA (2017) Comparative analysis of chemical similarity methods for modular natural products with a hypothetical structure enumeration algorithm. J Cheminform 9:46. https://doi.org/10.1186/s13321-017-0234-y

Soppimath KS, Aminabhavi TM, Kulkarni AR, Rudzinski WE (2001) Biodegradable polymeric nanoparticles as drug delivery devices. J Control Release 70:1–20

Stanton RA, Gernert KM, Nettles JH, Aneja R (2011) Drugs that target dynamic microtubules: a new molecular perspective. Med Res Rev 31:443–481

Sun C, Lee JS, Zhang M (2008) Magnetic nanoparticles in MR imaging and drug delivery. Adv Drug Deliv Rev 60:1252–1265

Swamy MK, Sinniah UR (2015) A comprehensive review on the phytochemical constituents and pharmacological activities of *Pogostemon cablin* Benth.: an aromatic medicinal plant of industrial importance. Molecules 20:8521–8547

Swamy MK, Sinniah UR (2016) Patchouli (*Pogostemon cablin* Benth.): botany, agrotechnology and biotechnological aspects. Ind Crop Prod 87:161–176

Swamy MK, Sinniah UR, Akhtar MS (2016) Antimicrobial properties of plant essential oils against human pathogens and their mode of action: an updated review. Evidence-Based Complement Alternat Med 22:1019. https://doi.org/10.3390/molecules22061019

Swamy MK, Arumugam G, Kaur R, Ghasemzadeh A, Yusoff MM, Sinniah UR (2017) GC-MS based metabolite profiling, antioxidant and antimicrobial properties of different solvent extracts of Malaysian *Plectranthus amboinicus* leaves. Evidence-Based Complement Alternat Med 2017:1517683. https://doi.org/10.1155/2017/1517683

Tiong SH, Looi CY, Arya A, Ong WF, Hazni H, Mustafa MR, Awang K (2015) Vindogentianine, a hypoglycemic alkaloid from *Catharanthus roseus* (L.) G. Don (Apocynaceae). Fitoterapia 102:182–188

To KK, Yu L, Liu S, Fu J, Cho CH (2012) Constitutive AhR activation leads to concomitant ABCG2-mediated multidrug resistance in cisplatin-resistant esophageal carcinoma cells. Mol Carcinog 51:449–464

Turktekin M, Konac E, Onen HI, Alp E, Yilmaz A, Menevse S (2011) Evaluation of the effects of the flavonoid apigenin on apoptotic pathway gene expression on the colon cancer cell line (HT29). J Med Food 14:1107–1117

Vallabani N, Mittal S, Shukla RK, Pandey AK, Dhakate SR, Pasricha R, Dhawan A (2011) Toxicity of graphene in normal human lung cells (BEAS-2B). J Biomed Nanotechnol 7:106–107

Van Hattum AH, Pinedo HM, Schlüper HM, Hausheer FH, Boven E (2000) New highly lipophilic camptothecin BNP1350 is an effective drug in experimental human cancer. Int J Cancer 88:260–266

Verdine GL (1996) The combinatorial chemistry of nature. Nature 384:11–13

Wang Y, Chen JT, Yan XP (2013) Fabrication of transferrin functionalized gold nanoclusters/graphene oxide nanocomposite for turn-on near-infrared fluorescent bioimaging of cancer cells and small animals. Anal Chem 85:2529–2535

Xu J, Wang F, Van Keymeulen A, Rentel M, Bourne HR (2005) Neutrophil microtubules suppress polarity and enhance directional migration. Proc Natl Acad Sci U S A 102:6884–6889

Xu M, Bower KA, Wang S, Frank JA, Chen G, Ding M, Wang S, Shi X, Ke Z, Luo J (2010) Cyanidin-3-glucoside inhibits ethanol-induced invasion of breast cancer cells overexpressing ErbB2. Mol Cancer 9:285

Yemisci M, Bozdag S, Çetin M, Söylemezoglu F, Çapan Y, Dalkara T, Vural I (2006) Treatment of malignant gliomas with mitoxantrone-loaded poly (lactide-co-glycolide) microspheres. Neurosurgery 59:1296–1303

Yoo HS, Park TG (2001) Biodegradable polymeric micelles composed of doxorubicin conjugated PLGA-PEG block copolymer. J Control Release 70:63–70

Zeng S, Yong KT, Roy I, Dinh XQ, Yu X, Luan F (2011) A review on functionalized gold nanoparticles for biosensing applications. Plasmonics 6:491–506

Chapter 6
Applications of Nano-based Novel Drug Delivery Systems in Herbal Medicine-Mediated Cancer Therapy

Saumya Srivastava and Anjana Pandey

Abstract Nanotechnology is a fast-growing field with numerous applications in the field of medical science. One such application comprises nanoparticle biosynthesis from plant extracts and their compounds with their potential applications in cancer therapy. These plant-based nanoparticles have been observed to be effective against various types of cancerous cells both *in vitro* and animal models. Another application of nanotechnology is the herbal therapeutics comprising the use of novel nano-based drug delivery systems in the treatment of cancer. These novel drug delivery systems aid in increasing the therapeutic value and bioavailability of the herbal medicine. The application of nanoherbal formulations as novel drug delivery systems (NDDS) is more valuable as compared to other therapies. These novel drug delivery systems include phytosomes, liposomes, microsphere, nanocapsules, ethosomes, transfersomes, nanoemulsions, and polymeric nanoparticles. The effectiveness of these different plant-based nanodrug delivery systems has been studied against various cancer types. These alternative drug delivery systems help in increasing the efficiency of a drug delivery and safeguard the drug from metabolic processes alongside its sustained delivery, proper distribution, and protection from physical and chemical deterioration. In addition, they reduce the possible side effects of the drugs. In spite of the advancements, cancer endures to be a predominant and fatal disease. This has led to the increased use of nano-based anticancer drugs and their delivery systems, also known as nanotherapies against tumors due to their ability of site-specific targeting and multifunctionality. In this chapter, recent advancements in application of plant-based nanomaterials in cancer therapy and impending strategies are discussed.

Keywords Drug delivery systems · Ethosomes · Nanocapsules · Nanoemulsions · Nanotechnology

S. Srivastava · A. Pandey (✉)
Department of Biotechnology, Motilal Nehru National Institute of Technology, Allahabad, Uttar Pradesh, India
e-mail: anjanap@mnnit.ac.in

M. S. Akhtar, M. K. Swamy (eds.), *Anticancer Plants: Clinical Trials and Nanotechnology*, https://doi.org/10.1007/978-981-10-8216-0_6

6.1 Introduction

Cancer is characterized by a condition when there is an uncontrolled growth of cells, having the capacity to invade and spread to distant body parts. The World Health Organization (WHO) has reported that among the main determinants of morbidity and mortality throughout the world, cancer is a major cause and it is projected to increase by 70% in the next two decades. The worldwide deaths estimated due to cancer till 2012 are 8.2 million with lung cancer topping the list, followed by liver, stomach, colorectal, and breast cancers (WHO 2015). Natural products from plant resources are being used in the treatment of different disorders in Egypt, China, India, and Greece from ancient times, and numerous drugs have been developed from medicinal plants (Arumugam et al. 2016; Swamy et al. 2016, 2017; Swamy and Sinniah 2016; Mohanty et al. 2017). The first documented proofs on the uses of medicinal plants date back to 2600 BC. Researchers throughout the world are focusing on use of herbal medicines to modulate immune system against cancer. Thorough understanding of the synergistic interactions of different constituents of herbs possessing anticancer activity and novel formulations can be useful to kill the cancerous cells without any other cellular toxicity. The very first phytochemicals that are being used clinically include vinca alkaloids, vinblastine, and vincristine from the Madagascar periwinkle, *Catharanthus roseus* (Apocynaceae). The isolation and identification of paclitaxel from the bark of Pacific yew, *Taxus brevifolia* Nutt. (Taxaceae), also represent the successful application of a natural product in the field of drug discovery. Epipodophyllotoxin, an isomer of podophyllotoxin, has been isolated and is reported to have antitumor activity. Etoposide and teniposide are another two semisynthetic derivatives of epipodophyllotoxin which are being utilized in the treatment of lymphomas and bronchial and testicular cancers. *Bleekeri avitensis* A.C. Sm., a medicinal plant of Fijian origin, is being used in France for treatment of breast cancer (Cragg and Newman 2005). Similarly *Allium sativum* consists of more than 100 biologically active secondary metabolites including alliinase, alliin, S-allyl cysteine (SAC), allicin, diallyl trisulfide (DATS), diallyl disulfide (DADS), and methyl allyl trisulfide. *Aloe vera* contains aloe emodin, a compound reported to activate the macrophages against cancer. Silvestrol isolated from the fruits of *Aglaia silvestris* (M. Roemer) Merrill. (Meliaceae) is reported to exhibit cytotoxicity against lung and breast cancer cell lines (Cragg and Newman 2005). Acetogenins isolated from *Annona* species possess significant anticancer activity against leukemia and sarcoma. *Arctium lappa* consists of various potent anticancer agents and is being used in the treatment of malignant melanoma, lymphoma, and cancers of the breast, pancreas, esophagus, ovary, bile duct bladder, and bone (Shabani 2016).

Different alkaloids and flavonoids isolated from numerous medicinal plants exhibit cytotoxic efficacy against different types of cancerous cells both *in vitro* and *in vivo* (Mohanty et al. 2014, 2017; Swamy and Sinniah 2015). One of the modes of action of these compounds includes inhibiting cancerous cell growth. Nanotechnology has helped in the development of new dosage forms in medicine. Till now, researchers have majorly concentrated on the use of metallic nanoparticles due to their rapid actions. Metallic nanoparticles possess unique physical and chemical properties due to their size leading to diverse biomedical applications. Silver, gold, iron oxide, zinc oxide, aluminum oxide, and copper oxide are some of the major metallic nanoparticles synthesized (Rudramurthy et al. 2016; Kanagesan et al. 2016a, b).

AgNPs synthesized using *Premna serratifolia* leaves exhibited noteworthy anticancer activity in carbon tetrachloride (CCl_4)-induced liver cancer in a study performed on Swiss albino (BALB/c) mice. In another study by Sre et al. (2015), the cytotoxic activity of biologically synthesized AgNPs using *Erythrina indica* on MCF-7 (breast cancer) cells and HepG2 (hepatocellular carcinoma) cells was reported. The *in vivo* and *in vitro* cytotoxic effects of AuNPs have also been stated in numerous studies, exhibiting the anticancer properties of AuNPs by the induction of oxidative stress. AuNPs synthesized using *Gymnema sylvestre* leaf extracts have also been investigated for their anticancer activity against hepatocellular carcinoma (HepG2) cell lines at concentrations of 250 μg/ml (Rao et al. 2016). Iron oxide nanoparticles have been observed for their antitumor activity both directly and indirectly via nontoxic wavelength radiation, which gets readily absorbed by toxic stimuli of reactive oxygen species production. The synthesis of iron oxide (Fe_3O_4) nanoparticles from seaweed (*Sargassum muticum*) has also been stated recently. Titanium oxide is another inorganic nanoparticle which can be surface-modified for inhibition of tumor cell growth. The advancements in nanotechnological approaches have the capability to deliver a solution to limitations faced by other dosage forms. It can be surface-engineered to inhibit tumor growth. The innovative nanotechnological approaches may provide a solution to limitations faced by many of phytochemical, physicochemical, and pharmacokinetic properties (Rao et al. 2016). This chapter discusses on the role of nano-based formulations and their applications in the field of herbal medicine and applications of nanotechnology in designing novel drug delivery systems containing bioactive compounds from the plant origin.

6.2 Hallmarks of Cancer

Cancer reconceptualization has come due to research advancements that have been made in the past decades at the molecular, cellular, and biochemical levels. Some of the specific traits which are acquired by the cancer cells result in the complexity of treating this disease. These features include self-reliance in growth signals, anti-growth signal insensitivity, unlimited replication capacity, persistent angiogenesis, avoidance to apoptosis, invasion, and metastatic potential (Fig. 6.1). These characters are shared by almost all types of cancer (Flavahan et al. 2017). Several growth

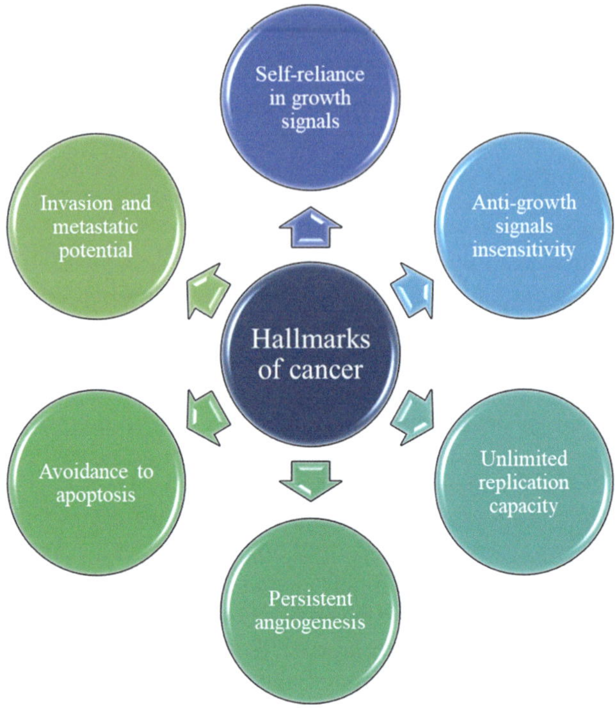

Fig. 6.1 Diagrammatic representation of various cancer hallmarks

and anti-growth signals regulate the cell cycle carefully in normal cells. Cancer cells disrupt these regulatory mechanisms and impart their own proliferative signals. Growth factors are supplied to the cancer cells via either growth factor ligands produced by themselves or by stimulation of the normal tumor-associated stromal cells which supply growth factors necessary for the cancer cells in return (Bhowmick et al. 2004). The cancer cells have the property to evade the growth suppressors in order to have prolonged growth-promoting factors. Various tumor suppressor genes have been investigated and certified, out of which two proteins, namely, RB (retinoblastoma-associated) and TP53 (tumor protein p53), play crucial roles in cell senescence and apoptosis. The cellular signals from varied extracellular ligands incorporated by RB protein decides the fate of a cell, i.e., whether a cell should proceed in the cell cycle or not (Burkhart and Sage 2008; Hanahan and Weinberg 2011; Flavahan et al. 2017). TP53 gene can stop the cell cycle in the presence of abnormal signals until the reappearance of normal conditions or otherwise in conditions of a permanent damage to DNA and apoptosis induction (Evan and Littlewood 1998; Xia et al. 2017).

Cancer cells have the ability to escape the natural cell barrier to cancer growth, i.e., apoptosis. Different members of Bcl-2 family molecules regulate apoptosis.

Cancer cells have different mechanisms to avoid apoptosis including damage to TP53 function, enhancement in anti-apoptotic factor expression, and decreased expression of pro-apoptotic factors (Adams and Cory 2007; Dzutsev et al. 2017). Due to senescence and crisis phase, normal cells undergo a limited number of cell cycle, but these obstacles can be bypassed by cancer cells resulting in the immortalization and infinite potential of replication. At the end of the chromosomes, telomeres are present that prevent the end-to-end fusion of normal cell DNA which can frighten the cell viability. The length of telomeric DNA is also regulated, but in the cancer cells, telomerase (a DNA polymerase enzyme) responsible for addition of the telomere at the DNA ends and prevents the telomere erosion (Dagg et al. 2017).

Cancer cells are capable of developing new blood vessels by activating an "angiogenic switch." Pro- and anti-angiogenic members regulate this process (Folkman and Shing 1992). Angiogenesis can be inhibited by TSP-1 (thrombospondin-1), endostatin, and angiostatin; these are natural barriers. Tumor progression in mice has been seen after deleting the genes encoding these inhibitors (Nyberg et al. 2005). Tumor cells invade and metastasize by establishing the communication between tumor cells, tumor microenvironment, and stromal cells (Ost et al. 2015). Cancer cells also require the nutrients for growth. Recent discoveries have investigated that the cancer cells also have the capability to reprogram the cell's metabolism in order to get the fuel (van der Heiden and De Berardinis 2017). Potential of cancer cells to evade the immune system has also been reported (Flavahan et al. 2017). These features provide the basic stage for most of the cancer research advancement and are helpful in revealing the cancer concepts.

The worldwide load of cancer augmented to approximately 14 million new cases per year in 2012, which is further expected to increase at a rate of 22 million annually in the next two decades. It is also projected that cancer deaths are going to increase from 8.2 to 13 million annually. Worldwide, most common types of cancers diagnosed in 2012 were lung (1.8 million cases, 13.0% of the total), breast (1.7 million, 11.9%), and large bowel (1.4 million, 9.7%). The cancers most commonly responsible for deaths include lung (1.6 million, 19.4% of the total), liver (0.8 million, 9.1%), and stomach (0.7 million, 8.8%) (GLOBOCAN 2012). In this regard, phytomedicines have been playing as important source of pharmacologically active compounds since ancient times, and their usage has increased owing to their therapeutic activity and lower effects as compared to other drugs. Nanotechnology can be employed to improve targeted delivery of drugs, delivery of poorly water-soluble herbal drugs, co-delivery of two or more drugs, release of large herbal molecules, and observation of sites of drug delivery by incorporating herbal drugs with imaging modalities (Liong et al. 2008; Lambert 2010; Gunasekaran et al. 2014). Researchers in the fields of herbal medicine and nanomedicine have observed that therapeutic nanoparticles (NPs) are more effective drug delivery system as compared to traditional drug delivery systems.

6.3 Role of Plant-Based Remedies and Herbal Medicine in Cancer Treatment

Herbal products, specifically medicinal plants, are being utilized in the treatment of various diseases since thousands of years. Terrestrial plants are being employed as medicines in India, China, Egypt, and Greece from thousands of years, and a large number of modern chemo drugs against cancers have been discovered and synthesized from them. The first compounds/agents taken into clinical use for different types of cancer treatment including lymphomas, leukemia, lung and breast cancers, advanced testicular cancer, etc. include vinca alkaloids (vinblastine and vincristine) isolated from *C. roseus* which ascertained a new era for utilization of plant materials as anticancer drugs. The revelation of the compound paclitaxel isolated from barks of the Pacific yew, *T. brevifolia* (Taxaceae), also narrates a success story in natural product drug discovery (Cragg and Newman 2005; Stathis et al. 2017). *Taxus baccata* is also reported for its utilization in the Indian system Ayurvedic medicine for treatment of cancer. Paclitaxel is expressively active against advanced breast and small and non-small cell lung cancer. Irinotecan and topotecan are two semisynthetic derivatives synthesized from camptothecin, which are being used in the treatment of ovarian and small cell lung cancer and colorectal cancer, respectively (Bhokare et al. 2016; Stathis et al. 2017). Epipodophyllotoxin, an isomer of podophyllotoxin, is isolated as an antitumor compound from the roots of *Podophyllum peltatum* and *P. emodi* (Berberidaceae). Etoposide and teniposide, two semisynthetic derivatives of epipodophyllotoxin, are being utilized in the treatment of bronchial and testicular cancers and lymphomas (Shabani 2016). Homoharringtonine, isolated from the Chinese tree *Cephalotaxus harringtonia* (Cephalotaxaceae), is also a plant-derived agent take into clinical use (Itokaw and Wang 2005). Combretastatins, isolated from the bark of *Combretum caffrum* (Combretaceae), is active against leukemia and colon and lung cancers (Ohsumi et al. 1998; Sherbet 2017).

6.4 Nanotechnology and Nanomedicine

Technology at the nanoscale is playing a crucial role to unfold solutions for many biological problems that were rigid a few years ago. In recent times, silicon chip technology has helped to manipulate the atoms and molecules resulting in revolution in the field of life sciences. Nanotechnology and nanomedicine are regarded as two complementary disciplines, targeting for betterment of life. Applications of nanotechnology in medicine consist of both materials and devices, which can be designed to interact with living systems at molecular scale with high specificity. This eventually results in their potential applications in targeted cellular- and tissue-specific uses and thus helps in acquiring higher therapeutic efficacy (Gunasekaran et al. 2014; Gortzi et al. 2008). Hence, in the existing situation, nanotechnology is beneficial in solving problems in different fields of medical science (Bhati-Kushwaha and Malik 2017). Different types of nanomaterials used in therapeutic applications are explained below.

6.4.1 Polymer Nanoparticles

Polymer nanoparticles are colloidal and solid particles having their size in the range of 10–1000 nm. These are also called as nanocapsules and nanospheres. Polymeric nanoparticles are synthesized by using preformed polymers or by polymerization of monomer units. For this purpose, different methods like salting out, solvent evaporation, supercritical fluid evaporation, rapid expansion of supercritical solution, and dialysis are being employed (Ma et al. 2017). The selection of preparation method is done on the basis of numerous factors including the area of application, type of polymeric system, size obligation, etc. The polymeric nanoparticles, prepared by these techniques, have been observed for their efficient therapeutic activities.

Lectins are known to bind carbohydrate moieties. This lectin-carbohydrate interaction is highly specific and can be utilized in development of NPs directed to certain lectins (De Mejia and Prisecaru 2005; Sharon 2007). Various lectins have been observed to exhibit anticancer activities *in vitro*, *in vivo*, and also in human case studies and are being used as therapeutic agents, for inhibition of tumor growth (Parveen and Sahoo 2008). Likewise, Tsutsui et al. (2007) developed a drug delivery system capable of targeting brain tumors using bionanocapsules (BNCs). In their study, they substituted the pre-S1 peptide with the antibody affinity motif of protein A and prepared hybrid BNCs conjugated with antihuman EGFR antibody recognizing EGFRVIII. The results exhibited that the hybrid BNCs were efficiently delivered to glioma cells and not to the normal glial cells. PLG [Poly(lactide-co-glycolide)] and its monomers have been found to have potential in wide applications for encapsulation and delivery of various anticancer drugs. In another study, Zhang and Feng (2006) analyzed the drug encapsulation efficiency, *in vitro* drug release, cellular uptake, and cytotoxicity of TX-loaded poly(lactide)-tocopheryl PEG succinate NPs. The drug encapsulation efficiency and *in vitro* drug release profile of this drug were analyzed using HPLC and the cancer cell lines HT-29 and Caco-2 (Zhang and Feng 2006; Xiong et al. 2017).

6.4.2 Metallic Nanoparticles

Metallic nanoparticles are nanosized metals with size ranging from 1 to 100 nm. There are numerous liquid phase approaches for synthesis of metallic nanoparticles, including chemical reduction, reverse micelle and sol gel, etc. Novel metallic nanoparticles having spherical shape and size were synthesized by different chemical reduction methods (Schwarz et al. 2004; Wang et al. 2008). MNPs have a wide range of applications owing to distinctive features including large surface area and process a large number of low coordination sites and electronic structure between molecular and metallic states. MNPs are being used in drug delivery, magnetic separation of labeled cells and other biological entities, and also as agents for contrast enhancement in magnetic resonance imaging.

Gold-silver (Au-Ag) nanorods conjugated with molecular aptamers were shown to need up to six times lower laser power irradiation to induce cell death compared to Au nanoshells or Au nanorods. These aptamer Scg8-AuAg nanorod conjugates exhibited exceptional hyperthermia efficiency and selectivity toward CEM cells, larger than the affinity of aptamer probes alone. Bimetallic Au-Ag nanostructures having dendrite morphology have been examined to destroy A549 lung cancer cells. The photothermal activity of these dendrites needed lower NP concentrations and laser power for efficient damage to the cancer cells. AuNPs have been reported for their application as vehicles for the delivery of different anticancer agents, including paclitaxel. The hydrophobic drug administration necessitates molecular encapsulation, and the efficiency of nanosized particles in evading the reticuloendothelial system is well studied (Conde et al. 2012; Hnawate and Deore 2017).

6.4.3 Magnetic Nanoparticles

Magnetic nanoparticles have been produced with numerous different phases and compositions including both pure metals and metal alloys (Schwarz et al. 2004). A number of methods have been reported including sonochemistry, coprecipitation, solvothermal, combustion synthesis, colloidal method, hydrothermal method, thermal decomposition, and microemulsion methods (Iwaki et al. 2003). The key applications of magnetic NPs are in the field of bioseparation processes, in which conjugation of the target biomolecules with magnetic nanoparticles, functionalized with specific receptors, produces complexes and hence can be easily separated using applied magnetic field, thus providing a suitable and time-saving method for bioseparation (Kharisov et al. 2014).

6.4.4 Nanomedicine in Therapeutics

During the past decades, much consideration has been paid to the enhancement of new novel drug delivery systems (NDDS) for herbal medicines. Plant-based unique drug delivery system has advantages over traditional drug delivery systems for delivery of herbal drug at optimal rate and targeted drug delivery, thus minimizing toxicity and enhancing bioavailability of the drug. In these drug delivery systems, the dissemination control of drug is attained by loading the drug in a carrier (Biju et al. 2006). Herbal compounds have become more popular with time due to their extensive applications in the treatment of a variety of diseases with lower toxicity and enhanced therapeutic effects (Atmakuri and Dathi 2010). But herbal medicines possess some short comings including, instability at acidic pH and metabolism occurring in lever resulting in lower levels of drug in blood. These constrains can be overcome by utilization of nano based novel drug delivery systems (Uhumwangho and Okor 2005). Amalgamation of novel drug delivery technology with herbal compounds reduces the drug deterioration rates, metabolism, and also side effects due to drug accumulation. Numerous NDDS including microspheres, liposomes, and phytosomes are being studied for the delivery of herbal drugs. Combination of herbal compounds in these delivery systems also assists in the upsurge in solubility, improved stability, reduced toxicity, greater biological activity, enhanced distribution, sustained drug delivery, and also guarding from degradation in metabolic processes. One such example is the use of liposomes in delivery of anticancer compounds by increasing the drug concentration in tumor-affected area and lowering the drug exposure in normal cells and tissues, hence checking toxicity effects (Yadav et al. 2011).

Nanotechnology has a wide range of applications in the development of new drug delivery systems which can overcome the drawbacks of conventional drug delivery systems including immediate requirement for drugs with improved therapeutic efficacy against CSCs (cancer stem cells). Nanoparticle (NP)-based carriers (nanocarriers) including micelles, liposomes, polymeric nanoparticles, dendrimers, etc. are being preferred as drug delivery systems. Though various anticancer agents have been identified, most of these have limitations when rendered to clinical studies such as off-target effects, low pharmacokinetics, a hydrophobic nature, inconsistent stability, and lower distribution of the drug. Nanotechnology can help to overcome these limitations. In the past years, researchers have concentrated on developing a single nanoformulation capable to carry dual drugs (one specific against CSCs and a second against bulk tumor cells) for their delivery to the target site through active or passive targeting (Singh et al. 2017).

6.5 Novel Drug Delivery Systems for Herbal Drugs

Before entering in the blood stream, the phytoconstituents of the herbal drugs face highly acidic pH of the stomach, and some of them also get metabolized in the liver, due to which the optimum quantity of the biologically active compounds does not reach the blood leading to lower or negligible therapeutic effects of the drug. Use of nanocarriers with herbal drugs carry optimal quantity of the drug to the action site, detouring barriers including acidic pH and metabolism in the liver, and upsurge the sustained drug circulation in the blood (Kuntal et al. 2005). Some of the types of novel herbal drug delivery systems are discussed below.

6.5.1 Phytosome

Phytosomes are phospholipid-based NDDS, which have been observed to be favorable for herbal drug delivery. Mixing the phytoconstituents at specific molar ratios with phosphatidylcholine leads to synthesis of phytosomes. It is the phytolipid-based drug delivery system which bridges the conventional and novel delivery systems. The word phytosome narrates to phyto (plant), whereas some represents cell-like. Phytosomes are unconventional types of herbal preparations with better absorption characteristics, as compared to conventional herbal compounds. Phytosomes also exhibit better therapeutic and pharmacokinetic profiles as compared to conventional extracts obtained from medicinal plants. Phytosomes can be synthesized by mixing the polyphenolic phytoconstituents at specific ratios with phosphatidylcholine. The phytosomal analysis has been focused on *Silybum marianum*, which consists of liver-protective flavonoids. The fruit of the *S. marianum* plant (family, Asteraceae) consists of flavonoids exhibiting hepatoprotective activities (Atmakuri and Dathi 2010). The phytosome plays a role in protecting phytoconstituents from degradation in digestive systems and gut microorganisms by synthesizing little compartment capable of being transferred to the lipid-friendly surroundings of the enterocyte cell membranes and then entering the blood stream (Sharma 2014).

Phytosomes can be identified as complex between a natural product and natural phospholipids. This complex can be prepared by reactions of stoichiometric quantities of substrate and phospholipids in a suitable solvent. It has been observed that the phospholipid substrate interaction can be attributed to the formation of hydrogen bonds between the polar head of phospholipids and the polar functional groups present on substrate using spectroscopic techniques. In aqueous environments, phytosomes acquire micellar liposomal-like structures. In phytosomes the active compound is attached to the polar head of phospholipids, thus becoming an integral part of the membrane (Kumar et al. 2017).

6.5.2 Liposomes

Liposomes are regarded as concentric and bilayered compartments where the aqueous phase is surrounded by a lipid bilayer membrane majorly consisting natural or/and synthetic phospholipids. The liposomes are sphere-shaped particles which encapsulate the drugs in the interior. Liposomes are synthesized using phospholipids, having hydrophilic polar head and hydrophobic tail (Ju Qun and Guo 2007). The polar end comprises of phosphoric atom bound to a hydrophilic molecule. Liposomes have the capability to encapsulate both hydrophilic and lipophilic compounds. Liposome exhibits properties which make them capable to augment the bioavailability, bio-distribution, ingredient solubility, altered pharmacokinetics, and both *in vitro* and *in vivo* stability. Liposome-based NDDS can increase the therapeutic efficacy of herbal compounds (Saraf 2010).

Paulis et al. (2012) observed that liposomes having size of 100 nm were capable to move out of the vasculature (extravasation) slowly and were slower to go to the tissue and release the active constituents over a period of time resulting in controlled and sustained release of the drug and also helped in maintaining the higher drug concentrations over a longer period of time being suitable in treatment of myocardial infarction and also to enhance the cardioprotection in a rat model of ischemia/reperfusion (IR) injury. The lipophilic interior of the liposome helps in the inclusion of drugs having lipophilic nature. Liposomes also help in improving the bioavailability of the drug as compared to conventional drug delivery systems (Geldenhuys et al. 2017).

6.5.3 *Nanoparticles*

Nanoparticles are nanosized structures comprising of semisynthetic or polymers. Recently, nanoparticles for herbal compounds have engrossed much consideration. Nanoparticles represent colloidal systems having particles of size distribution from 10 to 1000 nm. It is regarded as an efficient system as the compound is encapsulated in the nanoparticle and gets delivered at site of action easily. The nanospheres are spherical particulates with solid core having dimensions in nanosize. Nanospheres consist of herbal compounds implanted in the matrix or engrossed onto the surface, whereas nanocapsules exhibit vesicular system, where the herbal drug is encompassed inside (Vyas and Khar 2002). The nanoparticulate system is advantageous due to its increased solubility as compared to traditional drug delivery systems. Microencapsulation of herbal drugs in nanoparticles is an efficient way employed for the protection of drug or food ingredients from degradation, premature interaction, and volatile losses. The advantages of the nanoparticle include improvement in absorption of the herbal compounds, reduction in dosage formulation, and increase in solubility (Prabhu et al. 2010).

The nanoparticles can be classified into two categories, viz., hard and soft, dependent on their physicochemical characteristics and the method by which the drug is loaded into the nanoparticles. Hard NPs can be classified as materials on which the drug can be loaded only on the surface through passive adsorption or covalent attachment. Examples of hard nanoparticles include quantum dots, silver (Ag) NPs, gold (Au) NPs, and metal oxide NPs. Soft nanoparticles are the materials in which the drug can be loaded into the central core. The examples of soft NPs include liposomes, micelles, and polymeric NPs such as poly(D,L lactic-co-glycolic acid) (PLGA) (Sangtani et al. 2017).

6.5.4 *Emulsions*

Emulsions are biphasic systems where one phase is dispersed into the other as minute droplets with diameter ranging between 0.1 and 100 μm. In emulsion-based drug delivery system, the drug is distributed properly due to lymph affinity.

Microemulsions can be defined as solutions consisting of nanosized droplets of an insoluble liquid disseminated in an aqueous buffer where the droplets are covered with a surfactant for reducing the surface tension among both the phases. Emulsions result in targeted sustained release and also enhance the permeability of drugs into the mucous and skin (Cui et al. 2009).

Microemulsions have vast range of applications both in drug targeting and controlled release of the drug (Garg and Goyal 2012). The major problem faced in the delivery of oil-soluble drugs through oral route is its low aqueous solubility. Microemulsions exhibit imitable capability to evade problems related to solubility (Goyal et al. 2013). Microemulsions can encapsulate the drugs having variable solubility due to the presence of polar, nonpolar, and interfacial domains present in them. Microemulsions safeguard the amalgamated drugs from both oxidative and enzymatic degradation. Microemulsion formulations consisting of cyclosporin A, ritonavir, and saquinavir are available commercially (Fricker et al. 2010). Microemulsions help in reducing drug dosage and also enhance the bioavailability (Goyal et al. 2014). Betamethasone dipropionate exhibits antiproliferative, immunomodulatory, and anti-inflammatory activity (Kaur et al. 2016a, b). The addition of corticosteroids and keratolytic agents in microemulsions results in an improved and sustained corticosteroid delivery rate, resulting in superior antipsoriatic activity. Use of hydrogels as immunotherapeutic agents has abundant potential for refining the efficacy of vaccines and immunotherapeutics in disease treatment (Kaur et al. 2016a, b).

6.5.5 *Microsphere*

Microspheres consist of fine particles, having diameters ranging between few micrometers. Microspheres are also regarded as microparticles. Microspheres can be prepared using numerous synthetic and natural materials. Microspheres can be either biodegradable or non-biodegradable. Examples of biodegradable microspheres include altered starch microspheres, albumin microspheres, gelatin microspheres, polypropylene dextran microspheres, and polylactic acid microspheres. Solid and hollow microspheres have large variations in their densities and hence different applications (Scarfato et al. 2008). The polymers utilized in synthesis of microspheres can be either biodegradable or non-biodegradable. Different polymers are being used in fabrication of these carriers, including gelatin, albumin, polypropylene, modified starch, polylactic acid, dextran, polylactide, etc. (Burgess and Hickey 2009). The drug release in this type of delivery system is regulated by the dissolution and degradation of encapsulating matrix.

Numerous methods, including evaporation and ionic cross-linking techniques, are being used (Das and Senapati 2008) in preparation of mucoadhesive, buoyant microspheres. These microparticle systems have advantages as compared to other delivery systems as they can be ingested or injected and help in sustained release action and site-specific delivery. Gastroretentive floating microspheres of silymarin have been studied for sustained drug delivery (Garg and Gupta 2010). Microspheres containing turmeric oleoresin have been prepared using a spray-drying technique. The stable emulsion safeguarded the resin from degradation by different conditions and chemical agent, and it also enhanced the therapeutic effect of the drug (Kshirsagar et al. 2009). Targeted delivery of rutin (Xiao et al. 2008) from formulated microspheres (rutin alginate-chitosan) was observed through its targeting on cardiovascular and cerebrovascular regions. Oxidized cellulose microspheres incorporated with camptothecin (Chao et al. 2010) were synthesized by a spray-drying process and have been effectively utilized to enhance the solubility and cytotoxicity of camptothecin.

6.5.6 *Ethosome*

Ethosomes are phospholipid-based pliable nanovesicles with higher content of ethanol (20–45%). Ethanol is observed to be an effectual permeation enhancer. Ethosomes are prepared as unique lipid carriers comprising of phospholipids, ethanol, and water. Ethosomes enable the herbal drugs to reach deeper skin layers and/or systemic circulation. Owing to higher amounts of ethanol, the lipid membrane is packed less compactly as compared to conventional vesicles, with comparable stability (Touitou and Godin 2000). Transdermal drug delivery systems assist the movement of therapeutic drugs through the skin and into the general circulation for systemic effects (Shaik et al. 2011). It is observed that herbal drugs can be employed with superior efficacy by incorporating them into transdermal drug patches alongside penetration enhancers such as terpenes (Rathva et al. 2012). The commonly available transdermal drug delivery patches are the nicotine patches to help people quit smoking (Jatav et al. 2011).

6.5.7 *Solid Lipid Nanoparticles*

Solid lipid nanoparticles were first developed in the 1990s. These are specifically used in the delivery of lipophilic herbal drugs. Solid lipid nanoparticles can be prepared by various methods such as homogenization and warm microemulsion. The typical size of solid lipid nanoparticles lies between 50 and 1000 nm. Solid lipid nanoparticles comprise lipid matrix, which solidifies at room and body temperatures (Pople and Singh 2006). The main advantages of solid lipid nanoparticles (SLNs) as compared to conventional drugs include exceptional physical stability and protection of assimilated drugs from deterioration (Kakkar et al. 2010). Various types of nanodrug delivery systems that have been synthesized for delivering plant-based anticancer drugs are tabulated in the Table 6.1.

Table 6.1 Different types of nanodrug delivery systems synthesized for delivering plant-based anticancer compounds

Drug carrier	Active ingredients	Biological activity	References
Phytosome	Quercetin	Quercetin is a dietary flavonol which shows poor absorption after oral administration. To increase its bioavailabilty, quercetin-loaded phytosome nanoparticles (QP) have been prepared, and its encapsulation efficiency was found to increase. Treatment of ovariectomized model with QP has shown increased level of serum calcium and glutathione as well as improved lipid profile. Therefore, QP emerged as a promising therapy for hormonal replacement	El-Fattah et al. (2017)
	Silybin	In this study, the effect of silybin on the accumulation and metabolism of lipid and oxidative stress was examined on liver steatotic cells by incubating it with phytosome and silybin complex with vitamin E. And, it has been shown that fat accumulation as well as oxidative imbalance was decreased	Vecchione et al. (2016)
	Naringenin	Complex of naringenin-phospholipid has been investigated for its antioxidant activity in carbon tetrachloride-induced rats. After performing the liver function test, naringenin-phospholipid complex was found to have better antioxidant capacity than free form of naringenin; it also enhanced the antioxidant activity of biomolecules for liver protection	Semalty et al. (2010)
Liposome	*Myrtus communis*	In this study, antibacterial activity of *Myrtus communis* has been investigated against several bacterial strains as food preservatives	Bouyahya et al. (2016)
	Diospyrin	Diospyrin is a plant product, i.e., bisnaphthoquinonoid. It shows antitumor activity *in vivo* and cancer cell lines of human *in vitro*. But due to some toxicity, it was encapsulated in lysosome to reduce the toxicity	Hazra et al. (2005)

(continued)

Table 6.1 (continued)

Drug carrier	Active ingredients	Biological activity	References
Nanoparticles	Berberine	Berberine is a plant alkaloid that reduces the *H. pylori* proliferation. This study has developed a unique nanoparticle berberine carrier with a heparin shell for localization of berberine at the infection site of *H. pylori*, and it has been observed that the *in vitro* drug carrier method controlled the berberine release, which interacted to intercellular space at the infection site precisely	Chang et al. (2011)
	Piperine	In this study, the piperine liposome complex formulation has been prepared and investigated by spectroscopic methods, and the formulation was found to be stable for 3 weeks after storing it at 4 °C	Pentak (2016)
	Ginseng	Ginseng, a phytochemical-mediated synthesis of gold nanoparticles have exhibited shieling effect from aggregation alongside leading to the nontoxic nature of ginseng-AuNP conjugates analyzed against normal cervical cell lines	Leonard et al. (2011)
Emulsions	Docetaxel	Microemulsions containing docetaxel as bioactive constituent were prepared, and the increased bioavailability of the drug was observed. Also the increased solubility of the hydrophobic drug was observed	Cui et al. (2009)
Microspheres	Quercetin	The study focused on development of a delivery system which can exhibit controlled release of the drug in treatment of rheumatoid arthritis along with increased biocompatibility of the delivery system	Natrajan et al. (2010)
	Zedoary oil	Zedoary turmeric oil (ZTO) shows broad pharmacological action spectrum including antitumor, antibacterial activity, etc. The ZTO-microsphere formulation increases the bioavailability of drugs in comparison to conventional methods	Ghule et al. (2016)
Ethosomes	Matrine	Ethosome-matrine complex has been found to increase per cutaneous permeation, and it also helps in the improvement of anti-inflammatory effect	Bhokare et al. (2016)

6.6 Advantages of Plant-Based Nanodrug Delivery Systems

Nanosized delivery systems have numerous advantages when compared to conventional drug delivery systems (Kuntal et al. 2005). Some of them are given below:

(1) Capable to deliver high drug concentrations to infected sites due to their distinctive size and high loading capacities.
(2) Deliver the drug in the smaller particle size thus increasing the surface area of the drugs resulting in better dissolution in the blood.
(3) The drug persists at active sites for longer periods, resulting in enhanced permeation and retention effect, which include greater permeation through barriers due to smaller size and better retention owing to poor lymphatic drainage in tumor-affected tissues.
(4) Novel drug delivery systems exhibit passive targeting to the site of action without the need of any particular ligand moiety.
(5) Use of novel drug delivery systems results in lowering of the side effects.
(6) Decrease in the dose of the drug formulation.

6.7 Conclusions and Future Prospects

Herbal medicine is accepted throughout the world as an alternative system of medicine. Yet, the available drug delivery systems for herbal compounds are traditional and outdated. A widespread research is happening in the field of NDDS for herbal medicine. Though, research in the mentioned fields is still at the initial stages. Numerous plant constituents including tannins, flavonoids, terpenoids, etc. exhibited improved therapeutic efficacies after incorporation into NDDS as compared to conventional plant extracts. Henceforth, great potential is available in the development of NDDS. Nanomaterials in drug delivery systems have been incorporated into formulations having specific characteristics both *in vivo* and *in vitro*. Small number of clinical studies have exhibited greater efficacy in animal models, but translation of these results into clinical success is very limited. Successful conversion needs to reconsider the role of nanomaterials in drug delivery, their limitations, and facing inconvenient facts. Nano-based drug delivery systems for herbal drugs exhibit potential to augment the biological activity. Though, noteworthy challenges persist for execution of clinically sustainable therapies. Trials of novel approaches to regulate nanomaterial- biological system interactions signify recent challenges in rendering these technologies into therapies. The challenges in the development of nano-based drug delivery systems include the feasibility of scale-up and the possibility of attaining multifunctional systems to accomplish numerous biological and therapeutic requirements. For the mentioned approach to be fruitful, fine-tuning of the procedures to increase the usefulness of the nanoparticle in novel drug delivery systems is required. Herbal remedies are thriving resources for valuable compounds including antioxidants and other phytoconstituents. This type of two-way approach among the traditional herbal remedies and novel approaches of contemporary drug delivery system such as nanotechnology has established the therapies to the pharmaceutical sciences in the near future that will augment human health. It is estimated that the effective significance of the natural products and herbal remedies being combined with the nanocarriers will improve the importance of existing drug delivery systems.

References

Adams JM, Cory S (2007) The Bcl-2 apoptotic switch in cancer development and therapy. Oncogene 26:1324–3713

Arumugam G, Swamy MK, Sinniah UR (2016) *Plectranthus amboinicus* (Lour.) Spreng: botanical, phytochemical, pharmacological and nutritional significance. Molecules 21:369. https://doi.org/10.3390/molecules21040369

Atmakuri LR, Dathi S (2010) Current trends in herbal medicines. J Pharm Res 3:109–113

Bhati-Kushwaha H, Malik CP (2017) Upcoming impression of nanotechnology in therapeutic medicine. Int J Life Sci 6:86–106

Bhokare SG, Dongaonkar CC, Lahane SV, Salunke PB, Sawale VS, Thombare MS (2016) Herbal novel drug delivery-a review. World J Pharm Pharm Sci 5:593–611

Bhowmick NA, Neilson EG, Moses HL (2004) Stromal fibroblasts in cancer initiation and progression. Nature 432:332–337

Biju SS, Talegaonkar S, Mishra PR, Khar RK (2006) Vesicular system: an overview. Indian J Pharm Sci 68:141–153

Bouyahya A, Abrini J, Khay EO, Charfi S, Boujida N, Harsal AEL, Talbaoui A, Et-Touys A, Bakri Y, Dakka N (2016) *In vitro* antibacterial activity of organic extracts from North-West Moroccan medicinal plant *Myrtus communis* (L.) Br Biotech J 16:1–9

Burgess DJ, Hickey AJ (2009) Microsphere technology and applications. In: Swarbrick J (ed) Encyclopedia of pharmaceutical technology. Informa Healthcare, New York, pp 2328–2338

Burkhart DL, Sage J (2008) Cellular mechanisms of tumour suppression by the retinoblastoma gene. Nat Rev Cancer 8:671–682

Chang CH, Huang WY, Lai CH, Hsu YM, Yao YH, Chen TY, Wu JY, Peng SF, Lin YH (2011) Development of novel nanoparticles shelled with heparin for berberine delivery to treat Helicobacter pylori. Acta Biomater 7:593–603

Chao P, Deshmukh M, Kutscher HL, Gao D, Rajan SS, Hu P (2010) Pulmonary targeting microparticulate camptothecin delivery system: anticancer evaluation in a rat orthotopic lung cancer model. Anti-Cancer Drugs 21:65–76

Conde J, Doria G, Baptista P (2012) Noble metal nanoparticles applications in cancer. J Drug Deliv 2012:751075. https://doi.org/10.1155/2012/751075

Cragg GM, Newman DJ (2005) Plants as a source of anticancer agents. J Ethnopharmacol 100:72–79

Cui FD, Yin Y, Choi MK, Chung S (2009) Docetaxelmicroemulsion for enhanced bioavailability: preparation and *in vitro* and *in vivo* evaluation. J Control Release 140:86–94

Dagg RA, Pickett HA, Neumann AA, Napier CE, Henson JD, Teber ET, Arthur JW, Reynolds CP, Murray J, Haber M, Sobinoff AP (2017) Extensive proliferation of human cancer cells with ever-shorter telomeres. Cell Rep 19:2544–2556

Das MK, Senapati PC (2008) Furosemide loaded alginate microspheres prepared by ionic cross linking technique: morphology and release characteristics. Indian J Pharm Sci 70:77–84

De Mejia EG, Prisecaru VI (2005) Lectins as bioactive plant proteins: a potential in cancer treatment. Crit Rev Food Sci Nutr 45:425–445

Dzutsev A, Badger JH, Perez-Chanona E, Roy S, Salcedo R, Smith CK, Trinchieri G (2017) Microbes and cancer. Annu Rev Immunol 35:199–228
El-Fattah AIA, Fathy MM, Ali ZY, El-Garawany AERA, Mohamed EK (2017) Enhanced therapeutic benefit of quercetin-loaded phytosome nanoparticles in ovariectomized rats. Chem Biol Interact 271:30–38
Evan G, Littlewood T (1998) A matter of life and cell death. Science 281:1317–1322
Flavahan WA, Gaskell E, Bernstein BE (2017) Epigenetic plasticity and the hallmarks of cancer. Science 357:2380
Folkman J, Shing Y (1992) Angiogenesis. J Biol Chem 267:10931–10934
Fricker G, Kromp T, Wendel A, Blume A, Zirkel J, Rebmann H, Setzer C, Quinkert RO, Martin F, Müller-Goymann C (2010) Phospholipids and lipid-based formulations in oral drug delivery. Pharm Res 27:1469–1486
Garg T, Goyal AK (2012) Iontophoresis: drug delivery system by applying an electrical potential across the skin. Drug Deliv Lett 2:270–280
Garg R, Gupta GD (2010) Gastroretentive floating microspheres of silymarin: preparation and *in vitro* evaluation. Trop J Pharm Res 9:59–66
Geldenhuys WJ, Khayat MT, Yun J, Nayeem MA (2017) Drug delivery and nanoformulations for the cardiovascular system. Res Rev Drug Deliv 1:32–40
Ghule PK, Sayyad SF, Chaudhari SR (2016) Self microemulsifying drug delivery system: a method for enhancement of bioavailability. World J Pharm Res 5:578–590
GLOBOCAN (2012) Cancer fact sheet. http://globocan.iarc.fr/Pages/fact_sheets_cancer.aspx. Accessed 16 May 2017
Gortzi O, Lalas S, Chinou L (2008) Re-evaluation of bioactivity and antioxidant activity of *Myrtus communis* extract before and after encapsulation in liposome. Eur Food Res Technol 226:583–590
Goyal G, Garg T, Malik B, Chauhan G, Rath G, Goyal AK (2013) Development and characterization of niosomal gel for topical delivery of benzoyl peroxide. Drug Deliv 22:1027–1042
Goyal G, Garg T, Rath G, Goyal AK (2014) Current nanotechnological strategies for an effective delivery of drugs in treatment of periodontal disease. Crit Rev Ther Drug Carrier Syst 31:89–119
Gunasekaran T, Haile T, Nigusse T, Dhanaraju MD (2014) Nanotechnology: an effective tool for enhancing bioavailability and bioactivity of phytomedicine. Asian Pac J Trop Biomed 4:S1–S7
Hanahan D, Weinberg RA (2011) Hallmarks of cancer: the next generation. Cell 144:646–674
Hazra B, Kumar B, Biswas S, Pandey BN, Mishra KP (2005) Enhancement of the tumour inhibitory activity, *in vivo*, of diospyrin, a plant derived quinonoid, through liposomal encapsulation. Toxicol Lett 157:109–117
van der Heiden MG, De Berardinis RJ (2017) Understanding the intersections between metabolism and cancer biology. Cell 168:657–669
Hnawate RM, Deore P (2017) Nanoparticle-novel drug delivery system: a review. Pharma Tutor 5:9–23
Itokaw H, Wang X (2005) Homoharringtonine and related compounds. In: Cragg GM, Kingston DGI, Newman D (eds) Anticancer agents from natural products. Brunner- Routledge Psychology Press, Taylor and Francis Group, Boca Raton, pp 47–70
Iwaki T, Kakihara Y, Toda T, Abdullah M, Okuyama K (2003) Preparation of high coercivity magnetic FePt nanoparticles by liquid process. J Appl Phys 94:6807–6811
Jatav VS, Saggu JS, Jat RK, Sharma AK, Singh RP (2011) Recent advances in development of transdermal patches. Pharmacophore 2:287–297
Ju Qun XI, Guo R (2007) Studies on molecular interection between puerarin and pc liposome. Chin Sci Bull 52:2612–2617
Kakkar V, Singh S, Singla D, Sahwney S, Chauhan AS, Singh G, Kaur IP (2010) Pharmacokinetic applicability of a validated liquid chromatography tandem mass spectroscopy method for orally administered curcumin loaded solid lipid nanoparticles to rats. J Chromatogr B 878:3427–3431

Kanagesan S, Hashim M, Aziz SAB, Ismail I, Tamilselvan S, Alitheen N, Swamy MK, Purna Chandra Rao B (2016a) Evaluation of antioxidant and cytotoxicity activities of copper ferrite ($CuFe_2O_4$) and zinc ferrite ($ZnFe_2O_4$) nanoparticles synthesized by sol-gel self-combustion method. Appl Sci 6:184

Kanagesan S, Hashim M, Aziz SAB, Ismail I, Tamilselvan S, Alitheen NB, Swamy MK, Rao BPC (2016b) Synthesis, characterization and *in vitro* evaluation of manganese ferrite ($MnFe_2O_4$) nanoparticles for their biocompatibility with murine breast cancer cells (4T1). Molecules 21:312. https://doi.org/10.3390/molecules21030312

Kaur P, Garg T, Rath G, Murthy RS, Goyal AK (2016a) Development, optimization and evaluation of surfactant-based pulmonary nanolipid carrier system of paclitaxel for the management of drug resistance lung cancer using Box-Behnken design. Drug Deliv 23:1912–1925

Kaur P, Garg T, Rath G, Murthy RS, Goyal AK (2016b) Surfactant based drug delivery systems for treating drug-resistant lung cancer. Drug Deliv 23:727–738

Kharisov BI, Dias HR, Kharissova OV, Vázquez A, Pena Y, Gomez I (2014) Solubilization, dispersion and stabilization of magnetic nanoparticles in water and non-aqueous solvents: recent trends. RSC Adv 4:45354–45381

Kshirsagar AC, Yenge VB, Sarkar A, Singhal RS (2009) Efficacy of pullulan in emulsification of turmeric oleoresin and its subsequent microencapsulation. Food Chem 113:1139–1145

Kumar AB, Habbu P, Hullatti P, Kumar RS (2017) Phytosomes as novel drug delivery system for herbal medicine-a review. Syst Rev Pharm 8:5–7

Kuntal M, Mukherjee K, Ahamed H (2005) Enhanced therapeutic benefit of Quercitin- phospholipid complex in carbon tetrachloride induced acute liver injury in rats: a comparative study. Iran J Pharmacol Ther 4:84–90

Lambert WJ (2010) Considerations in developing a target product profile for parenteral pharmaceutical products. AAPS Pharm Sci Tech 11:1476–1481

Leonard K, Ahmmad B, Okamura H, Kurawaki J (2011) In situ green synthesis of biocompatible ginseng capped gold nanoparticles with remarkable stability. Colloids Surf B Biointerfaces 82:391–396

Liong M, Lu J, Kovochich M, Xia T, Ruehm SG, Nel AE, Tamanoi F, Zink JI (2008) Multifunctional inorganic nanoparticles for imaging, targeting, and drug delivery. ACS Nano 2:889–896

Ma J, Lee SM, Yi C, Li CW (2017) Controllable synthesis of functional nanoparticles by microfluidic platforms for biomedical applications-a review. Lab Chip 17:209–226

Mohanty SK, Mallappa KS, Godavarthi A, Subbanarasiman B, Maniyam A (2014) Evaluation of antioxidant, *in vitro* cytotoxicity of micropropagated and naturally grown plants of *Leptadenia reticulata* (Retz.) Wight & Arn.-an endangered medicinal plant. Asian Pac J Trop Med 7:S267–S271

Mohanty SK, Swamy MK, Sinniah UR, Anuradha M (2017) *Leptadenia reticulata* (Retz.) Wight & Arn.(Jivanti): botanical, agronomical, phytochemical, pharmacological, and biotechnological aspects. Molecules 22:1019. https://doi.org/10.3390/molecules22061019

Natrajan V, Madhan B, Sehgal P (2010) Formulation and evaluation of qucertin polycaprolactone microsphere for the treatment of rheumatoid arthritis. J Pharm Sophora Alopecuroides Sci 100:195–205

Nyberg P, Xie L, Kalluri R (2005) Endogenous inhibitors of angiogenesis. Cancer Res 65:3967–3979

Ohsumi K, Nakagawa R, Fukuda Y, Hatanaka T, Morinaga Y, Nihei Y, Ohishi K, Suga Y, Akiyama Y, Tsuji T (1998) Novel combretastatin analogues effective against murine solid tumors: design and structure activity relationships. J Med Chem 41:3022–3032

Ost P, Bossi A, Decaestecker K, De Meerleer G, Giannarini G, Karnes RJ, Roach M, Briganti A (2015) Metastasis-directed therapy of regional and distant recurrences after curative treatment of prostate cancer: a systematic review of the literature. Eur Urol 67:852–863

Parveen S, Sahoo SK (2008) Polymeric nanoparticles for cancer therapy. J Drug Target 16:108–123

Paulis LE, Geelen T, Kuhlmann MT, Coolen BF, Schäfers M, Nicolay K, Strijkers GJ (2012) Distribution of lipid based nanoparticles to infarcted myocardium with potential application for MRI-monitored drug delivery. J Control Release 162:276–285

Pentak D (2016) *In vitro* spectroscopic study of piperine encapsulated nanosize liposomes. Eur Biophys J 45:175–186

Pople PV, Singh KK (2006) Development and evaluation of topical formulation containing solid lipid nanoparticles of vitamin A. AAPS Pharm Sci Tech 7:91

Prabhu N, Gowari K, Raj D (2010) Synthesis of silver phyto nanoparticles and their antibacterial activity. Dig J Nano Biostruct 5:185–189

Rao PV, Nallappan D, Madhavi K, Rahman S, Jun Wei L, Gan SH (2016) Phytochemicals and biogenic metallic nanoparticles as anticancer agents. Oxidative Med Cell Longev 2016:3685671. https://doi.org/10.1155/2016/3685671

Rathva SR, Patel NN, Shah V, Upadhyay UM (2012) Herbal transdermal patches: a review. Int J Drug Discov Herbal Res 2:397–402

Rudramurthy GR, Swamy MK, Sinniah UR, Ghasemzadeh A (2016) Nanoparticles: alternatives against drug-resistant pathogenic microbes. Molecules 21:836

Sangtani A, Nag OK, Field LD, Breger JC, Delehanty JB (2017) Multifunctional nanoparticle composites: progress in the use of soft and hard nanoparticles for drug delivery and imaging. WIREs Nanomed Nanobiotechnol 9:e1466. https://doi.org/10.1002/wnan.1466

Saraf S (2010) Applications of novel drug delivery system for herbal formulations. Fitoterapia 81:680–689

Scarfato P, Avallone E, Iannelli P, Aquino RP, Lauro MR, Rossi A, Acierno D (2008) Qucertin microsphere by solvent evaporation: preparation characterization and release behaviour. J Appl Polym Sci 109:2994–3001

Schwarz JA, Contescu CI, Putyera K (2004) Dekkar encyclopedia of nanoscience and nanotechnology, vol 3. CRC press, Boca Raton

Semalty A, Semalty M, Singh D (2010) Supramolecular phospholipid polyphenolics interection: the phytosome strategy to improve the bioavailabilty of phytochemicals. J Incl Phenom Macrocycl Chem 67:253–260

Shabani A (2016) A review of anticancer properties of herbal medicines. J Pharm Care Health Syst 3:160. https://doi.org/10.4172/2376-0419.1000160

Shaik HR, Haribabu R, Mohiddin K, Vineela J, Ravitejab A, Pathuri RK, Gajavalli SR, Naidu LV (2011) Transdermal drug delivery system simplified medication regimen-a review. Res J Pharm, Biol Chem Sci 2:223–238

Sharma M (2014) Applications of nanotechnology based dosage forms for delivery of herbal drugs. Res Rev Int J Pharm Sci Nanotech 2:23–30

Sharon N (2007) Lectins: carbohydrate-specific reagents and biological recognition molecules. J Biol Chem 282:2753–2764

Sherbet GV (2017) Suppression of angiogenesis and tumour progression by combretastatin and derivatives. Cancer Lett 403:289–295

Singh D, Minz AP, Sahoo SK (2017) Nanomedicine-mediated drug targeting of cancer stem cells. Drug Discov Today 22:952–959

Sre PR, Reka M, Poovazhagi R, Kumar MA, Murugesan K (2015) Antibacterial and cytotoxic effect of biologically synthesized silver nanoparticles using aqueous root extract of *Erythrina indica* Lam. Spectrochim Acta A 135:1137–1144

Stathis A, Hess D, von Moos R, Homicsko K, Griguolo G, Joerger M, Mark M, Ackermann CJ, Allegrini S, Catapano CV, Xyrafas A, Enoiu M, Berardi S, Gargiulo P, Sessa C (2017) Phase I trial of the oral smoothened inhibitor sonidegib in combination with paclitaxel in patients with advanced solid tumors. Invest New Drugs (Online). https://doi.org/10.1007/s10637-017-0454-z

Swamy MK, Sinniah UR (2015) A comprehensive review on the phytochemical constituents and pharmacological activities of *Pogostemon cablin* Benth.: an aromatic medicinal plant of industrial importance. Molecules 20:8521–8547

Swamy MK, Sinniah UR (2016) Patchouli (*Pogostemon cablin* Benth.): botany, agrotechnology and biotechnological aspects. Ind Crop Prod 87:161–176

Swamy MK, Sinniah UR, Akhtar MS (2016) Antimicrobial properties of plant essential oils against human pathogens and their mode of action: an updated review. Evid Based Complement Alternat Med 2016:3012462. https://doi.org/10.1155/2016/3012462

Swamy MK, Arumugam G, Kaur R, Ghasemzadeh A, Yusoff MM, Sinniah UR (2017) GC-MS based metabolite profiling, antioxidant and antimicrobial properties of different solvent extracts of Malaysian *Plectranthus amboinicus* leaves. Evid Based Complement Alternat Med 2017:1517683. https://doi.org/10.1155/2017/1517683

Touitou E, Godin B (2000) Ethosome novel vesicular carrier for enhanced delivery: characterization and skin penetration properties. J Control Release 3:403–418

Tsutsui Y, Tomizawa K, Nagita M, Michiue H, Nishiki T, Ohmori I, Seno M, Matsui H (2007) Development of bionanocapsules targeting brain tumors. J Control Release 122:159–164

Uhumwangho MU, Okor RS (2005) Current trends in the production and biomedical applications of liposomes: a review. J Biomed Sci 4:9–21

Vecchione G, Grasselli E, Voci A, Baldini F, Grattagliano I, Wang DQ, Portincasa P, Vergani L (2016) Silybin counteracts lipid excess and oxidative stress in cultured steatotic hepatic cells. World J Gastroenterol 22:6016–6026

Vyas SP, Khar RK (2002) Targeted and controlled drug delivery novel carrier systems. CBS Publishers and Distributors, New Delhi

Wang W, Zhang J, Chen F, He D, Anpo M (2008) Preparation and photocatalytic properties of Fe^{3+} doped Ag@TiO_2 core-shell nanoparticles. J Colloid Interface Sci 323:182–186

WHO (2015) Cancer factsheets. http://www.who.int/mediacentre/factsheets/fs297/en/. Accessed 16 Oct 2017

Xia D, Yang X, Liu W, Shen F, Pan J, Lin Y, Du N, Sun Y, Xi X (2017) Over-expression of CHAF1A in epithelial ovarian cancer can promote cell proliferation and inhibit cell apoptosis. Biochem Biophys Res Commun 486:191–197

Xiao L, Zhang YH, Xu JC, Jin XH (2008) Preparation of floating rutin-alginate-chitosan microcapsule. Chin Tradit Herb Drugs 2:209–212

Xiong XY, Pan X, Tao L, Cheng F, Li ZL, Gong YC, Li YP (2017) Enhanced effect of folated pluronic f87-pla/tpgs mixed micelles on targeted delivery of paclitaxel. Int J Biol Macromol 103:1011–1018

Yadav D, Suri S, Choudhary AA, Sikender M, Hemant BN, Beg NM (2011) Novel approach: herbal remedies and natural products in pharmaceutical science as nano drug delivery systems. Int J Pharm Tech 3:3092–3116

Zhang Z, Feng SS (2006) The drug encapsulation efficiency, *in vitro* drug release cellular uptake and cytotoxicity of paclitaxel-loaded poly(lactide)tocopheryl polyethylene glycol succinate nanoparticles. Biomaterials 27:4025–4033

Chapter 7
Toxicological and Pharmacological Use of Anticancer Compounds

Charles Gnanaraj, Sivapragasam Gothai, Katyakyini Muniandy, Rengarajan Thamaraiselvan, and Palanisamy Arulselvan

Abstract Knowledge of medicinal plants from traditional medicinal drug therapy has brought limitless implications towards the discovery of novel compounds for modern therapeutic applications particularly for anticancer research. Research approaches have been taken for the development of traditional medicinal products into effective, non-toxic and clinically proven pharmaceuticals. Plant-derived anticancer compounds might exert adverse toxic effects in humans, depending on the species of plant, drug administration route, dose regimens and drug receptiveness. Due to this possibility, it is important to assess the toxicity profiles of new potential anticancer compounds through proper toxicological tests. The importance of preclinical toxicology studies is continually drifting from the use of conventional methods to assess the impending adverse effects of new anticancer agents over the past 10 years. Recent practice encompasses designing and carrying out more personalized agent-directed research within the framework of clinical pharmacology guidelines. Distinctive standpoints have been brought into view by clinical application of anticancer drugs that are apparent in their discovery and development. Evolving

C. Gnanaraj
Biotechnology Research Institute, Universiti Malaysia Sabah, Jalan UMS, Kota Kinabalu, Sabah, Malaysia

Scigen Research and Innovation, Periyar Technology Business Incubator, Thanjavur 613403, Tamil Nadu, India

S. Gothai · K. Muniandy
Laboratory of Vaccines and Immunotherapeutics, Institute of Bioscience, Universiti Putra Malaysia, Serdang, Selangor, Malaysia

R. Thamaraiselvan
Scigen Research and Innovation, Periyar Technology Business Incubator, Thanjavur 613403, Tamil Nadu, India

P. Arulselvan (✉)
Laboratory of Vaccines and Immunotherapeutics, Institute of Bioscience, Universiti Putra Malaysia, Serdang, Selangor, Malaysia

Muthayammal Centre for Advanced Research, Muthayammal College of Arts and Science, Rasipuram, Namakkal, Tamilnadu, India

M. S. Akhtar, M. K. Swamy (eds.), *Anticancer Plants: Clinical Trials and Nanotechnology*, https://doi.org/10.1007/978-981-10-8216-0_7

standards in estimating novel drug candidates through clinical and preclinical trials manifest the cytotoxic compound enhancements over time, with substantial aid from circumstances, and the current interest in discovering drug candidates with specifications on reaching targeted cells. This chapter highlights the relevance of plant-derived anticancer compounds through their pharmacological and toxicological uses.

Keywords Anticancer compounds · Clinical pharmacology · Drug discovery · Toxicity · Polyphenols

7.1 Introduction

Knowledge of medicinal plants from traditional medicinal drug therapy has brought limitless implications towards the discovery of novel compounds for modern therapeutic applications particularly for anticancer research. Plant-derived metabolic compounds have been used in modern medicinal practices for cancer prevention and as a possible curative treatment in current clinical trials (Arulselvan et al. 2016a; Mut-Salud et al. 2016; Redondo-Blanco et al. 2017). Almost 35,000 plant species have been reported to possess medicinal properties that have ethnobotanical values for various ailments (Bao and Fenwick 2004). This number may not be accurate, since most of the ethnobotanical identifications were orally passed on through decades and not all of them have been recorded. It was stated that more than 3000 plant species are involved directly in anticancer drug development process. Over the past 70 years, nearly 40% of anticancer drugs were derivatives of plants, and an additional 8% were analogues of natural products, which might be indirectly attributed to the known medicinal properties of the host plant (Van et al. 2004). There is an improvement in the research areas of anticancer therapeutics as a result of increasing demand among the public for natural products with minimal side effects. Plant-derived anticancer compounds might exert adverse toxic effects in humans, depending on the species of plant, drug administration route, dose regimens and drug receptiveness (Ndhlala et al. 2013). Due to this possibility, it is important to assess the toxicity profiles of new potential anticancer compounds through proper toxicological tests. Recently, naturally derived chemotherapeutic agents such as vincristine (vinca alkaloids) and vinblastine, teniposide (podophyllotoxin analogues) and etoposide, camptothecin-derived topotecan and paclitaxel (Taxol) were included in the medical treatment methods (Jantan et al. 2015). Formerly, these renowned drugs were in the form of tea, tinctures, powders, poultices and other crude drugs and herbal formulations. Evolving standards in estimating novel drug candidates through clinical and preclinical trials manifest the cytotoxic compound enhancements over time, with substantial aid from serendipity, and the current interest in discovering drug candidates with specifications on reaching targeted cells (Mishra and Tiwari 2011).

Pathophysiology of cancer is implemented through oxidative stress, which manipulates each stage of cancer mutagenesis beginning from damage incurred to DNA to the final stage of metastasis (Cook et al. 2004). Free radicals are known to target cellular contents such as carbohydrates, lipids, proteins and most importantly DNA (Bishayee et al. 2011). Free radicals or reactive oxygen species (ROS) are capable of damaging the DNA or other constituents of the cell by abruptly producing massive numbers of pro-oxidants. ROS are able to modify the cell signalling pathways through a cascade of events, causing cancer proliferation and promoting tumour formation. Apoptosis- and cell proliferation-inducing transcription factors and pathways such as NF-kB, MAP-kinase and AP-1 are notably influenced by the ROS causing detrimental effect (Rehman et al. 2013; Ponmari et al. 2014). It is suggested that prevention of DNA damage could be the key to chemoprevention of ROS-related genetic disorders. Plant-derived secondary metabolite compounds with antioxidative properties have been proven to possess efficient ability to scavenge free radicals.

The recent broadening of the drug discovery horizons along with the applications of advanced computational methods including instrumentations and bioinformatics tools has further added to new avenues of drug development (Cragg and Newman 2005). Medicinal plants have provided an opening point for new synthetic compounds. Plant-derived compounds with anticancer properties could serve as a drug lead to be optimized into synthetic drug or as a potential drug candidate by itself. *Podophyllum peltatum* (Berberidaceae)-derived compound, podophyllotoxin, was terminated from clinical trials as a result of inefficiency and intolerable toxicity in mice (Gordaliza et al. 2004). However, the pioneer clinically approved drug etoposide was made out of this compound as a chemical derivative. At times, there might be certain obstacles in finding for novel anticancer compounds from plants, but current research trend allows vital drug leads to be derived from commonly known compounds with prominent biological activity. Camptothecin derived from *Camptotheca acuminata* Decne. (Nyssaceae) tree initially exhibited intolerable myelosuppression. But later, this compound was found to be interesting and revitalized as it was able to cause a selective inhibition of topoisomerase I, which is involved in the cleavage and reorganization of DNA (Cragg and Newman 2004). Thereafter in the year 2002, camptothecin was reckoned for one-third of the worldwide anticancer market valued at over 2.75 billion dollars.

Some plants have medicinal properties and can be used as natural therapeutics; besides this, they could also cause lethal toxicity to humans because of their self-protection mechanism (Wink and van Wyk 2008). Few plant-derived compounds are known to cause mortality to humans even at a small dose. Compounds with known toxic effects were later isolated and utilized for beneficial purposes by humans (Ndhlala et al. 2013). Certain toxic compounds have been reported to have medicinal values as well. For instance, cardiac glycosides which are normally used to poison arrows for hunting by indigenous people have also shown antimicrobial activity (Botha and Penrith 2008). Similarly, plants containing lectins are safely being used in traditional medicine as treatment for bacterial and fungal infections, but lectins are proven to cause cellular necrosis to the gastrointestinal tract cell linings (Gonzalo 2011).

In the past few decades, compounds with cytotoxic properties were identified and developed for anticancer therapeutics. Thousands of cytotoxic compounds are being identified, but only a few of them are used clinically (Narang and Desai 2009) because the discovery of anticancer drug process is unique and has to pass through several challenges at every stage of drug development. Severe toxic effects might be caused by anticancer agents; hence, it should be carefully estimated and examined. Proliferation of normal cells such as hair follicles and cells in the digestive tract as well as blood cell formation in the bone marrow is affected due to therapeutic treatments targeting cell cycle, resulting in gastrointestinal toxicity, neutropenia and hair loss (Shoemaker 2006). These obvious side effects are generally reversible and can be recovered in the posttreatment period depending on the type of therapy. Mild or visible side effects will deteriorate with time; however, certain anticancer agents may cause a permanent toxicity or major dysfunction to vital organs such as the liver, kidney and heart and the nervous system (Narang and Desai 2009). In general, the consequences of therapeutic treatment by anticancer agents will develop immediately or shortly upon treatment, but sometimes they become apparent only after a longer period of time past treatment. If this period of delay is long enough, it may affect the benefit-risk equilibrium, and this interval might correspond to the time of progression-free survival.

Cancer is generally known as the disease of overwhelming cell division; hence all initiatives to overcome this problem are focused towards identifying novel compounds with anti-proliferative properties. Hence, efficiency in reducing the tumour size is focused as the main aim of preclinical and clinical studies. As part of drug discovery programmes, screening of novel compounds is done by testing them on rapidly growing tumour models. Although this screening model is able to identify several clinically important anticancer compounds, the success has been prominent in rapidly proliferating malignant tumours such as germ cell tumours, lymphoma and childhood leukaemia. Adversely, these compounds are not successful for common slowly proliferating epithelial tumours found in adults such as colorectal, breast and lung cancers (Samoylenko et al. 2013). Due to this drawback, researchers are encouraged to upgrade the prescreening and screening methods by including diverse tumour types and cell lines (Sznarkowska et al. 2017).

The global success ratio of assessments and anticancer drug development demonstrate difficulties and challenges involved in the process. Previous studies and analysis show that the percentage of failure for anticancer compounds in clinical trials was at its worst at the time period of 18–27 years ago (Mathijssen et al. 2014). It was recorded that only 1 out of 20 anticancer compounds would proceed from clinical trials to the market upon regulatory approval, and the rest failed. Candidate compounds in other therapeutic areas demonstrated double the percentage of success in reaching the market as compared to anticancer drugs. An analysis carried out in the USA showed similar approval success ratio for anticancer drugs in the past decade. Similarly, studies on the economic advancements of novel anticancer drugs and the priority review ratings given by the Food and Drug Administration have shown that anticancer drugs gain maximum profit (Workman and Collins 2014). It was also found in the study that clinical endorsement scales were comparable for anticancer drugs and non-cancer therapeutic drugs but huge volume of anticancer drug candidates were neglected in advanced stages of clinical tests where failures would be costly (Zhang et al. 2012). Reports from the past decade show that the time frame for anticancer drug development was usually lengthier as compared to non-cancer therapeutics, estimated to be around 7 years. except for central nervous system drugs which averagely take longer time, due to the need for multiple assessments over a wide range of indications. The report also says that the number of endorsed novel anticancer compounds improved significantly and gradually from the past three decades (Alqahtani et al. 2013).

On the whole, it is our vision to watch the success of anticancer drugs to increase in the market and a rapid drug development process to be imposed with a minimal failure. It is important to understand the deteriorating factors of anticancer compounds in the clinical use as it is the only way to focus and restructure problematic areas. Earlier statistics exhibited that in the early 1990s, bioavailability and pharmacokinetic limitations of anticancer compounds are the major known problems (Van et al. 2004). Thus, analytical assays for the absorption, distribution, metabolism and excretion properties were developed to overcome the problem of bioavailability and pharmacokinetics of anticancer compounds. But as time changes, new problems have aroused. The success ratio of clinical evaluation of anticancer drugs can be enhanced through testing them in different animal models shaving a closer physiology with humans. For instance, molecularly labelled human tumour xenografts in immunosuppressed mice are persistently beneficial to discover novel drugs, with elevated prominence on early channel patient-derived xenograft and orthotopic implantation (Guo et al. 2006).

It is clearly seen that by focusing on developments in medicinal chemistry, the eminence of the chemical leads being recognized for the optimization against novel targets has enhanced drastically with the identification of lead-like and drug-like chemical properties. The practice of high-performance, multi-parameter outlining currently permits us to anticipate impending problems such as toxicity and metabolism, which could be highlighted during the early stages of the optimization process (Workman and Collins 2014). Remarkably, natural products still remain to be of great importance to be used as therapeutic agents and possess several biological functions characterized with various inhibitory mechanisms and novel targets regardless of their possible cytotoxic adverse effects. Furthermore, the diverse assortments of natural compound structures relentlessly impress and instigate the medicinal chemistry research. In several circumstances, synthetic alterations of natural products are essential to improve the understanding of structure-activity relationships as well as to progress on physicochemical structures and biological characteristics. This chapter aims to highlight the pharmacological and toxicological uses of plant-derived anticancer compounds and the importance of natural products as potential therapeutic agents against cancer.

7.2 Discovery of Plant-Derived Anticancer Compounds

Plants are rich in antioxidant compounds which are well-known for their free radical scavenging effects and oxidative stress prevention, thus providing chemopreventive measures against chronic ailments (Arzamastseva et al. 2007). Antioxidants are able to restrict free radicals and restore cellular and organ functions by healing the injury done by free radicals. Polyphenols are a major group of antioxidants which are most abundantly found in plants and possess a unique chemical structure having the potential of scavenging free radicals. High reactivity and the potential to donate electron or hydrogen to stabilize the unpaired electron of free radicals attribute to the antioxidative property of polyphenols (Chanda and Dave 2009). Antioxidant compounds are well-known to possess anticancer properties. Natural antioxidant deficiency in humans is the major cause for many clinical conditions. The onset progression of pathological conditions due to oxidative stress could be delayed, reduced or reversed significantly through additional distribution of antioxidant compounds (Gnanaraj et al. 2016). Therefore, scientific measures are currently being carried out to explore and identify plant-derived antioxidant compounds with anticancer potentials along with clinical efficiency (Mut-Salud et al. 2016; Redondo-Blanco et al. 2017). Though microbes and marine products are natural sources of antioxidants, plants are always prioritized, because they are easily accessible and are rich in secondary metabolite compound diversity.

Polyphenols have tremendous medicinal properties, but the metabolism of these compounds in the human body has certain limitations. The absorption and distribution of polyphenols in the human body usually occur during the consumption of fruits and vegetables in the diet. But for a specific antioxidant compound to reach the target cell or organ through diet is a challenging process; hence bioavailability of the specific compound is solely accountable. Upon drug administration into the human body, it is recognized as a xenobiotic and subsequently undergoes biotransformation in the liver. During this process, there are high possibilities for the pure compound drug to lose its structure, thus losing its medicinal properties. This is why most of bioactive compounds isolated or derived from plants do not pass the clinical trials. Drug discovery process is thriving each year with hundreds of compounds being reported with therapeutic properties (Alsarhan et al. 2014). Most of the reported compounds exhibit potential as chemotherapeutic agents in preclinical evaluations but eventually fail in the clinical trials (Androutsopoulos et al. 2010). Scientists came with a solution to overcome this problem by developing drug transporters that will be co-administered along with the original drug. These drug transporters are supposed to protect the drug from losing its medicinal properties by maintaining its chemical structure till it reaches the target cell or organ. The primary technique for structural damage prevention of drug is through modifying the liver phase II metabolizing enzymes associated with biotransformation and excretion and xenobiotics from the body. Though it seems to be successful, it raises certain doubts on the long-term effect of liver enzyme modification which might lead to unwanted mutations or adverse effects and loss of activity of certain classes of enzymes. If the condition prolongs, it might cause the organ to be exposed to greater harms as it will become defenceless against the toxic organic compounds. This debate is opened in a constructive way in order to achieve a winning solution for this problem. Androutsopoulos et al. (2010) reported in a review on bioactive compounds as prospective anticancer agents that methoxylated flavonoids have a better bioavailability compared to hydroxylated flavonoids and could serve as therapeutic cancer drug for the chemoprevention of cancer cells. The metabolic pathways of methoxylated flavonoids need to be studied in depth to ascertain its clinical prospects for safety pharmacological insights for human diet consumption.

Compounds evaluated for possible anticancer properties might be of synthetic or natural origin. Natural products often deliver fresh indications in the discovery of unique chemical structure with potential anticancer properties. A recent study found that 80% of the available small-molecule anticancer drugs are natural products, the compound-related analogues or imitations of natural product compounds (Newman and Cragg 2012). Moreover, this percentage has not significantly changed in the past 70 years. It is noted that most of anticancer drugs approved around the world contain naturally inspired agents. Several naturally derived compounds such as etoposide, a derivative of mandrake plant *P. peltatum*; Taxol, a derivative of the Pacific yew, *Taxus brevifolia*; and vincristine, a derivative from *Vinca rosea*, the periwinkle plant, are known for their anticancer properties. There are also several microbe-derived compounds that are used as anticancer agents such as bleomycin and doxorubicin which are *Streptomyces* bacteria fermentation products, rhizoxin which is a derivative of the fungus *Rhizopus chinensis* and L-asparaginase which is a derivative isolated from *Escherichia coli* or *Erwinia carotovora* broths. Anticancer compounds isolated from natural products other than plants and microbes include bryostatin, a derivative of the sea moss *Bugula neritina*, and cytarabine, derived from the marine sponge, *Cryptotethya crypta* (Sultana and Ata 2008; Narang and Desai 2009; Aung et al. 2017).

7.2.1 Notable Types of Antioxidants

Natural compounds with potent anticancer properties are extensively present in various plant groups. Studies have been conducted in depth on the phytochemical constituents of plants for their therapeutic potentials. There are several methods and techniques to extract phytochemicals or pure compounds from plants, predominantly using chromatography such as high-performance liquid chromatography, microemulsion electrokinetic chromatography and micellar electrokinetic chromatography, and their structures will be elucidated based on nuclear magnetic resonance spectroscopy analysis. In the late 1980s, major successful anticancer compounds of plant origin were discovered, including doxorubicin, vinblastine, irinotecan, paclitaxel, etoposide, methotrexate and gemcitabine, but due to lack of popularity in the 1990s, anticancer compounds from natural products were not given priority. Recent researches have gained interest in using plant derivative anticancer compounds as potent curatives of cancer (Arulselvan et al. 2016b). Interestingly, natural product compounds such as flavonoids, alkaloid, lactones, sesquiterpenes, diterpenoids and polyphenols have been widely studied and are found to exhibit a diverse range of chemopreventive mechanisms against numerous cancer forms in both *in vitro* cell culture and *in vivo* animal models (Gothai et al. 2016). Some of these compound's chemical structures are depicted in Fig. 7.1.

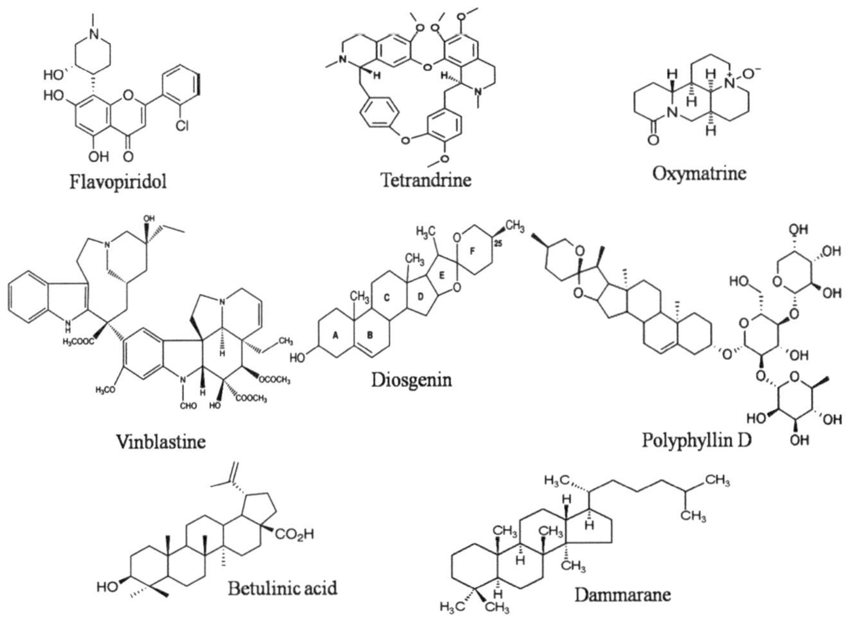

Fig. 7.1 Chemical structures of some of the natural plant-derived anticancer compounds

7.2.1.1 Flavonoids

Flavonoids are the common antioxidants found in plants under six subgroups, namely, anthocyanins, flavones, isoflavonoids, flavonols, flavanones and chalcones. They provide support to plants as secondary metabolites through chemical signalling, pigmentations, microbial and insect repellent, promoting or inhibiting growth and many more. Flavonoids have been reported to have therapeutic potentials; however, at the same time, they could cause detrimental side effects (Kyselova 2011). Previous study has shown that flavonoids impede cell proliferation and angiogenesis, induce cell apoptosis, cause cell cycle arrest and reverse multidrug resistance or a mixture of the previously described mechanisms. Flavonoids with various structures and methods of action are currently under clinical evaluation (Aung et al. 2017). Genistein has been examined on many patients, for preventive as well as curative treatment for prostate, breast and bladder cancers. Flavopiridol, a semisynthetic flavonoid, with anticancer properties is credited to downregulation of cyclins D1 and D3, inhibition of cyclin-dependent kinases (CDKs) and inhibiting phosphorylation of CDKs. Flavopiridol induces apoptosis, cell mortality and cell cycle checkpoint arrest, but the specific pathway mechanisms for anticancer properties remain uncertain. Clinical trials have shown that combination effects of few flavonoids are more encouraging as compared to a single compound effect. Therapeutic effects of flavonoid compounds in combination are assumed to have a promising outcome as a potent anticancer drug for cancer treatment and prevention, with concerns to their safety and accessibility.

7.2.1.2 Alkaloids

Naturally present toxic compounds are known as alkaloids and commonly found in both plants and animals. These compounds have been traditionally used for medicinal purposes as well as other activities like hunting. They function as protective and supportive metabolites as well as neurotransmitters in plants. Though alkaloids are poisonous and could cause lethality, they are still being structurally modified to be used as synthetic drugs in modern medication systems. Some notable alkaloids that are currently in use as medicine are quinidine, codeine, colchicine, quinine, vincamine, emetine and reserpine (Qiu et al. 2014). Certain well-synthesized semisynthetic alkaloid derivative anticancer compounds are vinorelbine, vindesine, vincristine and vinblastine. There are other alkaloid compounds with anticancer properties such as oxymatrine, a major quinolizidine alkaloid found in Aiton plant *Sophora flavescens*, and tetrandrine, a natural alkaloid found in the medicinal plant *Stephania tetrandra*. Oxymatrine is reported to possess cytotoxic effect on human colon cancer cells through downregulation of human telomerase reverse transcriptase (hTERT) gene and upregulation of mad1 gene as well as p53 gene in a dose-dependent manner. Oxymatrine also inhibits the growth of gallbladder cancer cells by activating apoptotic signal Bax along with caspase-3 and suppressing Bcl-2 and NF-kB. Oxymatrine is also reported to have anticancer properties for human lung cancer cells and Hela cells by inducing apoptosis. Therefore, this compound could be useful for cancer therapeutic treatments. Tetrandrine also has anticancer properties such as anti-proliferative and apoptosis-inducing effect on human oesophageal cancer cell lines, ECa109 and ECa109-C3, and human monoblastic leukaemia cells U937. It is also reported to be effective in reversing multidrug resistance in human nasopharyngeal cancer cells (KBv200) and Adriamycin-resistant human breast cancer cells (MCF-7/Adr). Tetrandrine compound's molecular mechanisms of action in cancer cells include upregulating cell death promoters such as Bak, Bax, Bad and apaf-1 and downregulating apoptotic regulators such as Bcl-xl and Bcl-2, discharging cytochrome c and activating caspase-3 and caspase-9 in the apoptotic mitochondrial pathway (Aung et al. 2017). Activation of intrinsic apoptosis pathway highlights the effectiveness of tetrandrine as a potent therapeutic anticancer compound. A semisynthetic analogue of alkaloid compound isolated originally from the periwinkle plant *Catharanthus roseus* known as vinblastine is an antimitotic drug. It was reported that this compound disrupts microtubules resulting in disappearance of mitotic spindles, thus inhibiting cell proliferation.

7.2.1.3 Saponins

Saponins are a type of secondary metabolite compounds of plants and marine organisms. It is easy to detect saponins in plants as they have foaming features. These compounds are usually used in the industry as biochemical reagents. The information regarding the functions of saponins in plants is scarce except for the protection against microbes. They have been reported to possess cytotoxicity against human cancer cell lines as well as being used as nutraceuticals and adjuvants in vaccines due to their membrane-penetrating mechanism (Francis et al. 2002; Nassiri and Hosseinzadeh 2008). Chikusetsusaponin IVa butyl ester (CS-IVa-Be) is an apoptotic triterpenoid saponin extracted from the medicinal plant, *Acanthopanax gracilistylus*. The herbal derivative has been found to cause cell cycle arrest at G0/G1 stage in multiple cancer cell lines. This compound is reported to promote apoptosis in MDA-MB-231 cells through inhibition of IL-6-induced STAT3 activity via IL-6/JAK/STAT3 signalling pathway. It also triggers the tumour necrosis factor (TNF)-related apoptosis-inducing ligand (TRAIL), a particular inducer of cancer cell apoptosis, in TRAIL-resistant MDA-MB-231 cells by upregulating death receptor 5 (DR5) (Aung et al. 2017). Polyphyllin D is a steroidal saponin compound with promising anti-proliferative property, extracted from the medicinal plant, *Paris polyphylla*. The anticancer activity of polyphyllin D was detected through the degeneration of mitochondrial membrane and fragmentation of DNA, leading to mitochondrial dysfunction and loss of membrane integrity in MDA-MB-231 and MCF-7 cells. It has been documented that administration of polyphyllin D for 10 consecutive days reduced up to 50% of tumour growth in mice. Another steroidal saponin compound with anticancer properties is diosgenin which has been reported to promote apoptosis by reducing protein expression and mRNA of 3-hydroxy-3-methylglutaryl-CoA reductase in HCT-116 human colon cancer cells. Similarly, macranthoside B, a triterpenoid saponin extracted from *Lonicera macranthoids*, has proven to induce mitochondrial-mediated apoptosis in numerous tumour cells via increased Bax/Bcl-2 ratio (Millimouno et al. 2014; Aung et al. 2017). A wide range of saponins have been evaluated for their cytotoxic properties, and a number of compounds have shown to be a potent inducer of apoptosis. However, the prospective of this type of compound remains to be explored thoroughly as they might be efficient in combination with other classes of anticancer compounds or drugs to synergistically boost their therapeutic action in cancer.

7.2.1.4 Terpenoids

Terpenoids are large multicyclic lipid compounds which are almost the same as terpenes and are found mostly in all living organisms. Terpenoid derivatives of plants have a strong aromatic quality and have been documented in traditional medicinal practices for their pharmacological benefits (Ayoola 2008). Derivatives of triterpene molecules, isopentenyl pyrophosphate oligomers, are known as triterpenoids (Bishayee et al. 2011). These compounds have been reported to possess much therapeutic potential and, most importantly, exhibit cytotoxic effects in cancer cells without causing toxicity to healthy cells (Rabi and Bishayee 2009). Preclinical evaluation of triterpenoids has proven antitumour effect in animal models (Petronelli et al. 2009). Till date, many triterpenoids have been derived and modified from natural compounds to optimize their anticancer properties and anti-inflammatory effects for clinical trials (Mullauer et al. 2010). Triterpenes such as betulinic acid, dammarane, squalene, lupine, ursane, ursolic acid and oleanolic acid have been reported to possess cytotoxic and antineoplastic activities against cancer cell lines (Chudzik et al. 2015). A recent report by Borkova et al. (2017) analysed the cytotoxic activity of triterpenoid thiazoles against eight cancer cell lines and two non-cancer fibroblasts and found that they possessed effective anti-proliferative properties.

7.2.2 Advances in Plant-Derived Anticancer Compounds

Plant-derived anticancer compounds often target tumour cells through the regulation of cell death pathways such as intrinsic and extrinsic apoptosis and autophagic pathways. Furthermore, it has been reported that cell signalling pathways activated by natural anticancer compounds are diverse and distinct for different targets. It was found that different signalling pathways could be activated by the same compound depending on the target cell types. Many studies have been conducted to proof the pathways, and one of them is the *in vitro* and *in vivo* test for prostate cancer treatment using isoflavones and phytoestrogens from soy which exhibited deactivation of NF-kB signalling, inhibition of angiogenesis and induced apoptosis. NF-kB pathway is known to initiate cancer cell proliferation and also causes inflammation (Nair et al. 2006). Induction of NF-kB activation has been reported to cause resistance to chemotherapy. Constitutively activated NF-kB stimulates the proliferation of cancer cells thus saving them from apoptosis or autophagy. Fundamentally activated NF-kB can often be seen in various cell lines, cancer cells, xenograft animal models or clinical sites (Hassanzadeh 2011). In recent studies, NF-kB signalling is being deeply analysed as a favourite target of cancer cell proliferation. It seems to be an encouraging target and is worth for further research to find out the possible therapeutic mechanisms for cancer prevention (Arulselvan et al. 2016b). Though there is a clinical use of NF-kB inhibitors, they seem to be distant from model drugs due to their detrimental side effects. Therefore, it is recommended to perform extensive research on natural product-derived anticancer compounds for their optimal therapeutic mechanism on NF-kB signalling pathway and preventive measures of cancer cell proliferation.

Analogues of anticancer compounds derived from natural products have often been synthesized as a step to increase anticancer efficiency or reduce toxicity (Aung et al. 2017). For instance, daunomycin was found to be causing cardiotoxicity; therefore, an analogue doxorubicin was created. Similarly, an analogue of cisplatin was synthetized, carboplatin was able to reduce renal toxicity, and an analogue of camptothecin, topotecan, also has reduced toxic effect. Analogues of prevailing anticancer agents have also been developed to increase the pharmacokinetic availability and improve tumour targeting accuracy of compounds. Examples of such anticancer agents are 9-alkylmorpholinyl, an analogue of doxorubicin, synthesized to reduce drug affinity to P-glycoprotein which is the cellular efflux protein, and another agent is tauromustine, a nitrosourea anticancer compound joined to the brain targeting peptide taurine to target central nervous system tumours. Anticancer compounds of natural origin sometimes could not compensate the excessive demand for drug supply (Sultana and Ata 2008). To overcome this problem, the original compound-related analogues have to be developed to continuously supply therapeutic drugs. An example of such drug is taxotere, an analogue of Taxol which is a plant-derived anticancer compound with extremely low yields. Synthetic compounds have been developed in recent years to be used as analogues of natural anticancer compounds with novel structures (Redondo-Blanco et al. 2017). Selection process for a suitable synthetic drug candidate is developed through combinational chemistry and computer modelling of drug-receptor interactions. Laboratory synthesis-based development of potential anticancer compounds has advanced from analogues, and improvisations have led to coherent design established for drug-enzyme or drug-receptor relations. Prominent models of synthetic anticancer drugs developed from analogues of natural products are 5-fluorouracil, fluorinated pyrimidine base and methotrexate, an analogue of folic acid. Other examples of anticancer drugs discovered via coherent designing method by exploring drug-target interactions and molecular mechanisms include mitomycin C-related indolequinone EO9 which is actively used against hypoxic tumours and edelfosine, an ether lipid which targets cellular membranes (Mathijssen et al. 2014; Aung et al. 2017). Developments in anticancer drug discovery process have found numerous compounds subsequently requiring a thorough screening process to shortlist potential compounds to be subjected to preclinical tests. The selected compounds will undergo tests such as *in vitro* cell line and *in vivo* animal assays for toxicological classifications, efficiency quantification and bioavailability through pharmacokinetics and pharmacological potentials prior to clinical trials.

7.3 Toxicological Measures of Anticancer Compounds

In the early years of the 1950s, almost all cancer drugs were known to be cytotoxic agents. They commonly react by damaging and inhibiting DNA synthesis pathways or by disrupting the mechanism of cell proliferation, for instance, via binding to microtubules or blocking topoisomerases. Most anticancer compounds were found via screening for possible compounds with the ability to destroy cancer cells, such as paclitaxel, a plant-derived compound with microtubule inhibition property. Based on nitrogen and sulphur mustards, DNA-alkylating agents were structurally reformed to control their chemical reactivity rates thus prompting the formation of drugs like ifosfamide and cyclophosphamide. It was later found that these drugs could cause adverse effects to other vital organs and some effects were irreversible and could cause mortality (Workman and Collins 2014). Therefore, toxicity assessment for anticancer compounds is an important process. An approximation of the clinical schedule of anticancer drugs should be assessed by toxicology tests as the toxicity of the drug could be highly influenced by its administration schedule (Philomena 2011). General toxicities encountered during cancer treatments are skin and hair follicle toxicity; gastrointestinal, haematological and urinary tract toxicity; cardiac toxicity; pulmonary toxicity; metabolic abnormalities; nervous system toxicity; gonadal toxicity; and hepatic toxicity.

Estimation of the possibility to recover from the toxicity caused by therapeutic agents must be given in order to determine the seriousness of contrary effects and to understand whether it is reversible or not. Whenever there is severe toxicity occurrence during a clinical exposure of cancer treatment and the period of recovery could not be determined by scientific measures, a terminal non-dosing period study will be imposed. The degree and severity of pathologic injury and the recovering capability of the target organ system could be included in this study to show the effect of the drug (Remesh 2012). When a study on recovery from toxicity has been initiated, it should be presentable to assist the clinical development of therapeutic drugs. However, it is not necessary to exhibit the complete recovery process as time is crucial for cancer treatments.

7.3.1 Approaches in Toxicological Assessments

Toxicity and therapeutic effects caused by cytotoxic compounds are the results from similar molecular mechanisms, and they are often correlated directly to the dose regimens. In clinical studies, patients are usually dosed to a maximum concentration of drug compounds to achieve maximal medical benefits, provided the dosage does not cause serious side effects (Mathijssen et al. 2014). The serious ill situation of the patients, disparity in individual drug reaction and harmfulness along with the limitations of the therapeutic anticancer compounds constitute the major challenge for their clinical practice and development. Considering these factors, novel approaches have been made for the development of instruments for specifying a drug therapy for individual patients and to monitor the drug response and toxicity by means of alternate indicators. As an example, tumour treatment has been specialized for individual patients depending on the nature of tumour, state of disease and histology of tumour cells.

Toxicological studies of anticancer compounds involving animal models are targeted to assess the toxicities of the drug, possibilities for reversibility and severity of the compound toxicities prior to calculating a safe entry dose and dosage schedule for clinical trials on humans. Toxicity is estimated at a regular and therapeutically active dose for most of the cytotoxic anticancer compounds (Shoemaker 2006). This is due to the same mechanism attributing to toxicity and therapeutic effect. The common harmfulness of cytotoxic compounds includes the suppression of the bone marrow which can be clearly witnessed in rodent species, such as mice and rat, and therefore toxicological evaluation is mostly conducted on smaller animals under this principle. Considering the cost and ethical reasons, higher animals such as apes have been avoided in the routine toxicological tests. The severity and occurrence of acute toxicities are determined in the targeted organs through biochemical, histopathological and haematological evaluations soon after the dosage administration, while the long-term or chronic toxicities are determined by means of sacrificing and investigating the animals few weeks upon the dosage administration.

7.3.2 *Effective Dosage Estimation of Anticancer Compounds*

The drug to dosage correlation varies among species and is normally determined based on the body surface area, though other factors such as body weight and age have also been considered. Dosage levels that cause mortality to 10%, 50% and 90% of the animals (LD_{10}, LD_{50} and LD_{90}; LD, lethal dose) will be subjected through the same route of administration to determine the phase I starting dosage of a cytotoxic compound. These dosage parameters could also be referred to the terms of dosages that cause severe lethal toxicity or severe toxic dose (STD) instead of counting mortality as an end point (Narang and Desai 2009). The proposed phase I starting dosage is normally 1/10th of the LD_{10} or STD_{10}.The proposed phase I starting dosage is usually examined in other animal models, usually dogs or rats, to determine the presence of substantial toxicity as well as to reduce the risk related with human trial of a new anticancer compound (Samoylenko et al. 2013). Therefore, the preclinical animal toxicology practice normally comprises of single-dose and multiple-dose fatality or lethal toxicity tests in mice and subsequently single-dose and multiple-dose toxicity verification tests in rats or dogs. In case of severity, permanent toxicities are demonstrated in the animal model at the proposed phase I entry dosage, then the human entry dosage will be decreased to 1/6th of the highest dosage given to the animal model, provided the dosage did not cause lethal or permanent injury. The steps involved in the toxicity assessment for potential anticancer compounds are represented in Fig. 7.2.

Assessment of anticancer compound toxicities in animals has certain restrictions, since anticancer compounds are fundamentally harmful with normally a dose-dependent expression of symptoms. The assessment of dosage regimens related to the clinical doses for human using animal model toxicology is difficult to attain within a therapeutic range, causing the risk of lower or higher approximation of the compound's toxicological outline. This situation could be attributed to the genomic variations in the pharmacodynamic and pharmacokinetic reactions of the compounds which might be due to dissimilarities in protein binding, the metabolic and excretion pathways and the sensitivity of target cells. Moreover, uncommon toxicities such as those of neuromuscular or cardiovascular basis are challenging to be detected in animal models. Previous studies have witnessed about 0.5% of mortality due to toxicity in the phase I clinical tests of anticancer drugs among approximately

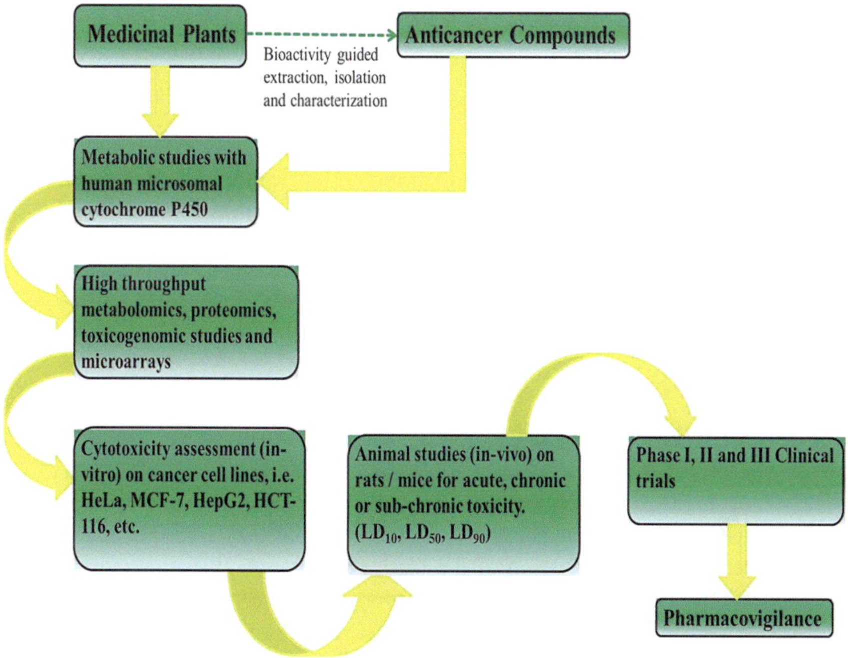

Fig. 7.2 Schematic diagram of toxicity assessment for potential anticancer compounds

6000 volunteers (Grunwald 2007).Underestimation of starting dosage could increase the time and procedures in the dosage acceleration experiments, hence raising the costs and time for development and causing unethical administration of unproductive dosages to a big number of extremely sick patients. On the other hand, higher estimation of starting dosage is very much unwanted due to safety precautions. Both under- and higher estimations could cause a novel compound to be rejected from further development due to these reasons. Higher estimation example was observed in the anticancer drug fludarabine where it caused substantial bone marrow suppression at the 1/10th mice LD_{10} dosage at the phase I studies but did not cause any substantial bone marrow suppression in dogs at dosage levels up to 20 times higher than the LD_{10} dosage (Narang and Desai 2009). This discrepancy has been related to the differences in species towards the drug pharmacodynamics, where the efficacy of the drug phosphorylation is higher in human bone marrow cells as compared to dog bone marrow cells. Underprediction of dosage is demonstrated by brequinar sodium through increased clinical trial period, which required 19 dose acceleration steps over a time of approximately 3 years or more to attain the maximum tolerated dose (MTD), since the MTD was 40-folds increased than that of the mice LD_{10} dosage (Eisenhauer et al. 2000). Another notable example is the flavone acetic acid, a synthetic flavonoid, for which the single dosage LD_{10} in mice was showing similar results to that in rats, but testing on dogs and humans showed tolerance level up to fourfold increased dosages. The reason for this is rapid drug excretion in the higher species, hence causing underestimation of the clinical starting dose. These cases prove the importance of a better estimation of cytotoxic compound toxicities in humans to avoid prolonged phase I trials other than lethal drug toxicities. Currently, pharmacokinetic experiments are often being added in the toxicology assessment methods. Microdosing of drug has been recommended for human trials to comprehend the drug pharmacokinetic properties before the proposed dosages could be administered (Narang and Desai 2009). Microdosing is a new method to learn the behaviour of drugs in humans through administration of very low doses but high enough to study cellular response without producing whole body effects.

7.4 The Evolving Role of Natural Products as a Renewable Source of New Drug Leads

A very common method to identify bioactive compounds in plants is through pharmacological testing of crude extracts from the potential medicinal plants followed by isolation and characterization of the bioactive compounds attributable for the therapeutic effect of the extract. The extraction method of plant extracts is rather easy and low in time investment and cost. However, to ensure reproducible data, it is critical to cautiously record the processing of the plant material beginning from the collection, identification and extraction, since the technique of extraction can highly impact on the chemical composition and consequently influence the bioactivity of the plant extract (Atanas et al. 2015). Therefore, selecting an appropriate extraction method involving different solvents is very crucial and should be carefully well-thought-out. To maximize the yield, the extraction and fractionation are needed to be optimized for medicinal plants with identified target compound classes and chemical structures. In cases with unknown bioactive compounds in the investigated plant, extracting diverse classes of compounds is valuable. Commonly used solvents in extraction method are 70% ethanol or methanol, as these solvents have been found to extract a maximum yield of bioactive molecules.

There are two principal approaches in the isolation of new lead compounds which are bioactivity-guided isolation and metabolic profiling (Halket et al. 2005). Plant crude extracts demonstrating bioactivity are usually subjected to the identification of pure bioactive compounds. To define the chemical traits of plant-derived compounds, isolation of the pure compounds is mandatory using stepwise separation methods and spectral analysis to inaugurate the exact chemical structure of the compound. Novelty of structure and new action methods are the normal characteristics of plant drugs. Bioassay-guided isolation involves the assessment of crude extracts biological activity, followed by fractionation and biological testing until pure active compounds are obtained. Technologies used in the fractionation process are liquid-liquid extraction, using organic solvents of opposing polarity such as ethyl acetate, n-butanol and chloroform. Following each fractionation step, bioactive fractions are guided to deeper separation, generally by column chromatography, flash chromatography, vacuum liquid chromatography and thin-layer chromatography (Luc and Arnold 2005). However, the traditional bioactivity-guided isolation is time-consuming, costly and exhausting. Furthermore, plant crude extracts represent complex mixtures interactions of active and inactive compounds which may result in strong bioactivity in a crude extract from additively or synergistically weak active compound action. Thus, fractionations may break this collective bioactivity, resulting in a trustable bioactivity (Park et al. 2007).

Metabolic profiling is an evolving substitute to bioactivity-guided fractionation. As for metabolic profiling, the metabolites present in the natural sources provide information on the chemical composition, thus allowing for the detection and isolation. This type of profiling can be obtained by using liquid chromatography-mass spectrometry (LC-MS) and gas chromatography-mass spectrometry (GC-MS) (Berkov et al. 2008). Pattern of the compound bioactivity and chemical profile of plant extracts correlate in guiding the isolation and identification of compounds at an earlier stage. Quantification and characterization of plant-derived secondary metabolites require extremely sensitive and reproducible analytical techniques such as the nuclear magnetic resonance (NMR) spectroscopy or high-performance liquid chromatography (HPLC). This method permits pointed isolation of specific bioactive compounds and dereplication of already known compounds. It is recommended to use bioactive fractions with fewer compounds for screening and quantification purposes (Kawaguchi et al. 2010).

7.4.1 Primary Screening of Anticancer Drug Discovery

The assortment of plants based on molecule entities serves as the initial biological screening in a drug discovery programme. The discovery of drug from plant necessitates a multidisciplinary field in which the accomplishment is mostly related to a relevant set of *in vitro* and *in vivo* assays. However, the choice of the bioassay largely depends on the objectives of a study (Lee et al. 2012). The examination of plant derivative compounds starts with the screening of numerous bioactive compounds using *in vivo* organ, tissue or animal models followed by *in vitro* investigation of mechanistic underpinnings (forward pharmacology approach). Although screening of random compounds in animal models is still a useful method, the disadvantages are apparent. Further, it is laborious and necessitates the involvement of large amount of compounds with low sensitivity. In most situations, it is impractical to supply adequate amounts of pure compound isolates for animal trialling, since the amount of bioactive compounds present in plants is relatively low (Ming-Wei et al. 2007).

Remarkable improvements in the pharmacological research have led to modern approaches involving various cellular and molecular bioassays known as the reverse pharmacology approaches (Alqahtani et al. 2013). It begins with screening of numerous compounds isolated from plants against predesignated ailment-relevant targeted protein, with the objective to recognize promising hit bioactive constituents with the preferred activity that will be studied in detail using relevant *in vivo* models with the purpose to validate them. Animal testing could be reduced through this method, but it must be understood that promising hits might accidentally get rejected upon exerting toxic effects on cell-based screening, which in fact might be successful in animal models due to biotransformation in the liver. It generally needs great time and effort during the primary phases with complex and huge library of unique bioactive chemical constituents. In this sense, optimization of the lead is needed in order to guarantee *in vivo* efficacy. Screening of thousands of compounds within a short time period was made possible through the development of combinatorial chemistry (high-throughput technologies), thus accelerating the pace of isolating target molecules which could be developed further to potential drug candidates with preferred therapeutic activities.

Molecular target assays depend on the evaluation of the physical interaction of the investigated target protein with the desired compound. The biggest group of target proteins approved by the FDA is receptors (e.g. G-protein-coupled receptors (GPCRs), followed by enzymes (e.g. hydrolases) and transporter proteins (e.g. voltage-gated ion channels) (Alkhalfioui et al. 2009). These assays reflect the protein-protein interaction essential for the practical activity or protein initiation upon binding with the test compound and provide important preliminary data about the mechanism of action binding test compound to the active centre of the target protein early in the discovery process. This preliminary information on how a compound may affect the progress of a disease process can be valuable asset in determining the status of the compound for further development. Thus, rapid biochemical screening test of large libraries of pure compounds is being employed as targets for plant-derived lead discovery (Klekota et al. 2006; Swinney and Anthony 2011).

Assays with biological systems such as mammalian cell cultures (cell-based assay) are an alternate to *in vivo* assays for the primary evaluation of pharmacological activity due to the ease of the assay design. These assays deliver cellular level information on bioactivity compounds depending on the objective of the research and the expected outcome (Cragg and Newman 2004). For example, studies of angiogenesis most appropriately require endothelial cells and primary mammalian cells, or immortalized cell lines can be used for this purpose. However, immortalized cell line results are often less relevant because prolonged *in vitro* maintenance permits build-up of phenotypic changes and mutations. Cell-based assays can be phenotypic or phenotypic oriented. Phenotypic models provide results on an altered cellular phenotype with entangled regulation, for instance, cell migration, while target-oriented assays aim on the inter-reaction of test compounds with the role of a particular protein or pathway such as caspase activity (Atanasov et al. 2013). The relevance of *in vitro* assays is however restricted as compared to *in vivo*, where they are unable to deliver knowledge on the regulating aspects of the compound pharmacokinetic activities (absorption, distribution, metabolism and excretion). Yet, cell-based assay signifies imperative tools to isolate and characterize active constituents.

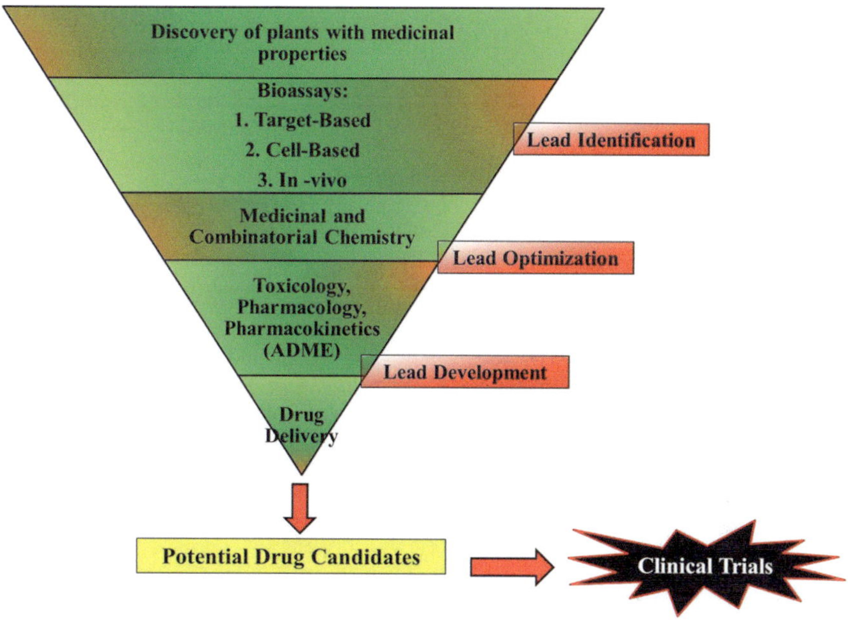

Fig. 7.3 Schematic diagram of plant-derived drug lead development

According to the reverse pharmacology method, isolated bioactive compounds from traditional cell-based assays or molecular target are further evaluated for their anticancer potential using *in vivo* assays using likely animal models to provide toxicological and pharmacological results prior to clinical trials on humans. Animal models such as mice and rat are crucial for drug validation and evaluation due to the sensibly high similarity and homology among mammalian physiology and genomes. Moreover, information on the bioavailability, efficacy, toxicity and side effects of the drug could be gathered through animal models that will be basically used for safety and pharmacokinetic studies that are imperative for clinical trials (Zhang et al. 2012; Alqahtani et al. 2013). It is a common pharmacological practice to assess candidate drug compounds on larger animals such as dogs and swine. Due to high cost and ethical issues, drug testing on larger animals is only carried out at the final stage prior to human tests. Indeed, such regulations require the pharmaceutical industry to employ both rodent and non-rodent models. And so, preclinical animal models will continue to be essential for assessing efficacy and side effects, and their predictive values will be revealed when more compounds enter human clinical trials (Leonard and Randall 2005; Parasuram 2011). Figure 7.3 represents the scheme of developing plant-derived drug lead molecules.

7.4.2 The Notion of Synergy and Mechanisms of Actions

At present, 62% of the newly developed anticancer agents have been from natural products or based on natural product models. All of these natural products brought substantial biological findings associated with their distinctive mechanisms of action with most cases evidenced in synergistic role (Marcy and Douglas 2005). Plants develop a great complexity of phytochemicals resulting from plant's defences to produce chemical defence to predatory micro- and macro-organisms. These defence chemicals exist as small molecules that apt into predator's enzymes or receptor, thus interfering with its immune system. However, the predators over time habitually acquire resistance to plant's chemical defence, and ergo, plants co-evolve to produce resistance inhibitors, and these plant defences serve as a great source of therapeutic molecules. In contrary to pharmaceutical models which act on the basis of single compound, often culminating in resistance and failure in drug response synergistic effect of multiple bioactive compounds is considered to have better therapeutic potentials (Rasoanaivo et al. 2011). However, scientific efforts to understand the complexity of interactions among the constituents in medicine plants with clinical benefit can be problematic because of the lack of sufficient information, documentation and standardization. Herbalists justifiably point out most medicinal plants are categorized as an idiosyncratic herb which has been alleged to specific kinds of toxicity. The best known are dryopteris, viscum, and corynanthe with potential hepatotoxicity (Hamid and Hedayatollah 2013).

Though validation of traditional remedies can be challenging, a great prospective exists to investigate the possible therapeutic values of medicinal plants. For reaping such benefit, the medical and pharmaceutical communities have adopted the concept and philosophy of the "silver bullet" approach therapy, where pharmaceutical industries have been very much geared to the "single-ingredient" and "one-drug-one-target" paradigm (Chun et al. 2011). Therefore, new approaches like metabolomics aid to formulate novel strategies for recognizing valid targets and discovering leads from medicinal plants against them. Thousands of novel molecular targets have been identified as implications of various ailments, since human genome was sequenced. Plant-derived compounds with known bioactivity might have targeted and promising effect in the modern screening assay directed towards the disease-causing molecular targets. Several reports have been made on the reactivity of compounds isolated from traditional medicinal plants on validated molecular targets, hence resurrecting the interest in regularly isolated compound classes in plants (Jeffrey and Michal 2009; Philomena 2011).

Yet a relatively new-fangled field, medicinal plant-based drug discovery, has played a crucial part in most new clinical applications for the past 500 years towards battling cancer. Cancer chemoprevention is a strategy to control cancer by administration of synthetic or natural compounds to inverse or suppress the process of carcinogenesis (Butler 2004). Since cancer therapy is intended to befall preceding the onset of cancer diagnosis, toxicity is mostly unexpected. In the case of prophylactic treatment and inhibiting cancer reappearance, tolerable toxicity levels are acceptable. As such, edible plants, dietary supplements, botanicals and all medicinal constituents were recommended as potential chemopreventive agents for cancer. This is attributable to well-documented ethnopharmacological information about possible drug candidate compounds which were used traditionally with therapeutic potentials for humans (Park and Pezzuto 2002). Currently, plant-derived anticancer agents in clinical use can be grouped into four main classes: vinca (or *Catharanthus*) alkaloids, epipodophyllotoxins, taxanes and camptothecins. These compounds have been found to act on two important biochemical targets, tubulin and topoisomerase (Marcy and Douglas 2005).

Vinca alkaloids such as vinblastine and vincristine were isolated from *Catharanthus roseus* (Madagascar periwinkle plant). Medicinal applications of this plant lead to the discovery of major vinca alkaloids in clinical use: vinblastine (VBL), vinorelbine (VRL), vincristine (VCL) and vindesine (VDS). Only VBL and VCL are approved for clinical use. The main mechanisms of vinca alkaloid cytotoxicity are their interactions with tubulin and disruption of microtubule "vinca domain" function, disrupting the assembly of microtubules (essential to cell division) in mitosis and leading to cell apoptosis (Moudi et al. 2013). Epipodophyllotoxins are substances naturally occurring in the root of American Mayapple plant (*P. peltatum*). Etoposide and teniposide are semisynthetic derivatives of epipodophyllotoxins. These are topoisomerase II (enzyme essential for cell growth by regulating levels of DNA supercoiling) inhibitors which bind tubulin, causing DNA strand breaks during the G2 phase of the cell cycle by irreversibly inhibiting DNA topoisomerase II (Demain and Vaishnav 2011).

Taxanes are a class of diterpenes. They were originally identified from *T. Brevifolia* Nutt. The taxanes, including paclitaxel derivatives, act as a microtubule stabilizer, thereby blocking cell cycle progression as depolymerization is prevented or the tubulin assembly is interfered (Jordan and Wilson 2004). Camptothecin is isolated from *C. Acuminate* Decne. Camptothecin and its clinically used analogues, irinotecan and topotecan, arrest the cell cycle at the S-phase by affecting the activity of topoisomerase I, leading to the inhibition of DNA replication and transcription (Kingston 2009).

7.4.3 Bioavailability of Plant-Derived Anticancer Agent in Drug Discovery and Development: The Barrier of Oral Administration

Chemo drugs are mainly designed in a way it can disrupt the cell cycle of the cells, so it may not be able to distinguish between the malignant and the healthy cells in the body, i.e. interfere with both of them. However, there is a gap in between them to achieve anticancer effects or life-threatening cytotoxicity, known as the therapeutic index. Unfortunately, most of the anticancer drugs are listed under the narrow therapeutic index (NTI) drugs. Therefore, they are either prescribed closely or even at the maximum tolerated dose (Mathijssen et al. 2014). Despite optimizing the dosage according to the NTIs, another major problem faced by the drug development company is to overcome the low bioavailability (Godugu et al. 2014). Bioavailability is the relative absorption of drug to the targeted site of action from the total amount administrated dosage of drug into the system. This is important to ensure an adequate amount of drug reaches the action site, essential for an efficient pharmacodynamics to establish an effective therapeutic achievement.

In cancer chemotherapies especially, a preparation of a wide range of dosage to cater for a better bioavailability is almost impossible as it may cause adverse drug reactions resulting in more DNA damage and more cell killing including the normal fast-growing healthy cells. Blood cells, digestive and reproductive system-associated cells and hair follicles are likely to be affected by chemotherapy causing side effects such as fatigue, nausea, indigestion, diarrhoea, hair loss and anaemia (Nonnekens and Hoeijmakers 2017). Therefore, to reduce the likelihood and the severity of the side effects caused, it is important to study the whole mechanism of the drug as well as its therapeutic actions in order to optimize the bioavailability of the drug.

Orally administered drugs remain as one of the patients' preferences for chemo as they are cheaper and involve low technical maintenance compared to the injection-related therapy (Arber et al. 2017). Anticancer drugs via oral administration shows three-quarters of all the cancer therapies available and being developed (Colomer et al. 2010). However, limited bioavailability from this route of administration is the chief challenge faced, and they are mainly due to the activity of the cytochrome P450 (CYP) enzyme such as CYP 34 and the drug transporters, such as P-glycoprotein (P-gp), deposited in the intestinal wall and the liver (Leeuwen et al. 2013). Upon oral delivery, the drug undergo first-pass metabolism in the liver by the enzyme CYP 34 before being released to the systemic circulation. Moreover, the efflux pump P-gp that is also present in the gut lining plays its role in the drug expulsion back to the intestine rather than releasing to the systemic blood circulation. These factors affect the bioavailability of the drug negatively by influencing the achievement of desired plasma drug concentration for an efficient therapeutic effect. Due to the high susceptibility of metabolism in the gastrointestinal tract, the dosage needed to cause a response will be significantly higher than the intravenous dose, but this may cause untoward effects. However, the functionality of the enzyme as well as the transporters varies among patients which further contribute to the limitation for oral drugs. On the other hand, intravenous route assured 100% bioavailability compared to other routes of administration. Yet, this venipuncture method is less preferred compared to oral route due to its inconveniences as it is exposed to many risks including the development of cannula-related infections and thrombophlebitis and a frequent visit to the clinic (Cyriac and James 2014).

As each and every patient is made up of uniquely different set of a genome, the susceptibility of therapy will be slightly different from one individual to another. Besides genomic variance, other inter- and/or intra-individual varying factors such as the age, gender, body mass and health conditions influence the drug bioavailability level (Almeida et al. 2009). Drug prescriptions should be incorporated with the mentioned factors in order to enhance the patients' compliance with the medication. An individual's body composition, blood flow and plasma protein concentration also affect drug distribution. As ageing proceeds, human physiological functions undergo changes which usually reduces the absorption rate of the drug thereby affecting the drug distribution process (He et al. 2011). Furthermore, older patients tend to have comorbid medical conditions such as diabetes, hypertension, arthritis and heart complication that affect the overall rate of drug delivery mechanism. Prior knowledge about a patient's medical background is vital, because certain drug, for example, fluoropyrimidines, increases the risk of occurrence of cardiotoxicity in coronary artery disease patients compared to patients without the disease (Albini et al. 2010). Dose reduction or intensification is usually recommended to cater for different conditions of the patients, but it may reduce the therapeutic effect or cause lethal cytotoxicity, respectively. Therefore, to tailor for inter- and/or intra-individual variances, individualized chemotherapy involving the study of the individual's cellular, molecular or genetic contents in selecting an optimized chemotherapy provides a better chemotherapy outcome and survival.

Another approach to increase the drug bioavailability is the modification in the galenic formulation of the drug. Adapting prodrugs into drug design is one of the versatile approaches to increase its bioavailability. Upon administration, the prodrugs undergo some biological or chemical modification by the CYP450 enzyme, before being pharmacologically active (Tsume et al. 2014). Therefore, the pharmacodynamic structures of the parent drug are preserved from being altered by the enzymes. Prodrug-derived ideas have been amended into the development of anticancer drugs, such as cyclophosphamide, dacarbazine, tegafur, flutamide and duocarmycin (Montellano 2013).

7.5 Obstacles in Medicinal Plant-Based Drug Discovery: Renewal of the Interest in Drug Discovery

Compounds derived from medicinal plants have unceasingly fascinated researchers globally for being biocompatible and are regarded as potentially safe and effective. Although medicinal plant-based drug discovery has been proven successful, future undertakings certainly have constant challenges as the drug discovery from plants is a lengthy and complicated process. Natural product chemists need to prove their quality and results consistently in drug discovery and development in order to keep them in level or ahead of other drug discovery efforts (Marcy and Douglas 2005). Lead identification, optimization and development and clinical trials take significantly a long time period of an average 10 years upwards and require a lot of money (>800 million US$) to be spent in obtaining numerous leads that are cast-off during the drug discovery process. Indeed, a ratio of only 1 in 5000 lead drug candidates will effectively progress through clinical trials and be allowed pharmaceutically. Although the trend towards a loss of the development of new bioactive agents from the medicinal plants is evident, several recent positive strategies can be adopted towards reversing this trend (Veeresham 2012).

It is apparent that at present, drug-based therapeutic strategies will predominate in the twenty-first century. Acceptance of the synthetic products increased due to its production cost, time effectiveness, easy quality control, stringent regulation and quick effects. But, their safety and efficacy always remain questionable. The growing capability of cancer cells to resist the chemotherapeutic drug is one of the main drawbacks of conventional chemotherapy. Besides, the effort to maximize the drug efficacy and minimize cytotoxic effects has been a constant struggle in chemotherapy as overdosage can produce many harmful effects in the patients (Aung et al. 2017).The side effects arise typically when the chemotherapy damages the healthy cells that maintain the body's function and appearance depending on the nature of the drug. Cardiotoxicity, hepatotoxicity, nephrotoxicity and ototoxicity are some of the leading problems generating from the non-specific cytotoxic nature of these chemotherapeutic drugs. Although chemotherapy is the key way to control cancer, the dreadful effects of conventional chemotherapeutics result in the dependence on alternative perspective that sets the stage for a future platform of natural products (Mondal et al. 2014).

The advent of conventional cancer chemotherapy limitations has led to a booming of searching for anticancer drugs from natural products, where people began to pay consideration to the herbal ingredients. Furthermore, high selectivity, strong activity, low side effects and a wide range of antitumour drugs from natural resources have led people to focus on cancer prevention research (Butler 2004).Compounds extracted from natural resources can enhance immune cells or body immunity and antitumour effects and inhibit the proliferation of cancer cells. Not just biologically toxic compounds have attracted people's attention but also more non-toxic ingredients that have healthcare effects. Disease prevention is more important than treatment, so is cancer. With the progress of science and technology, people now have deep understanding of tumour, making the prevention of cancer possible. The development of cancer prevention and healthcare products from natural products has broader prospects and greater economic and social benefits when compared with chemically derived anticancer drugs (Veeresham 2012).

7.6 Conclusions and Future Prospects

The importance of natural products for therapeutic purposes has gained public attention in recent years. Plant-derived antioxidant compounds with known anticancer properties are potential candidates for clinical trials and to be developed as possible therapeutic drugs. This is due to minimal side effect of plant products as compared to synthetic drugs. Although plant-derived anticancer compounds might exert toxicity to humans, they are mostly less harmful than other drugs. The toxicological properties of plant-derived compounds are screened with current methods prior to clinical trials. This procedure ensures the safety of the anticancer compounds for human consumption. Ethnobotanical claims and antioxidant activities of several medicinal plants can be taken into consideration for anticancer drug development purposes, implementing reactive oxygen species as the etiological factor for cancer. Preventive measures are always better than curative treatments; hence, it is emphasized to increase the intake of exogenous natural antioxidants in our diet. Additional intake of natural antioxidants will not cause any detrimental effect to the human body system as the antioxidants will be biotransformed and balanced by the liver. Plant-derived anticancer compounds might exert toxic effects if administered at extreme doses; therefore, it is important to standardize the effective dose of each compound as well as the formulated drug. There is a rise in problems related to drug resistance in cancer patients who are undergoing chemotherapy, due to the unsuccessful drug formulations in chemically synthesized anticancer drugs. We believe that plant-derived anticancer compounds will be able to overcome the problem of drug resistance among cancer patients. Drug delivery methods should be enhanced in a way to use natural products alone as drug vessels instead of using synthetic vessels. Pharmacological evidences of anticancer compounds have risen in a steady pace, but yet to attain success. This is due to the limitations of plant-derived pure compounds described earlier in this chapter. We strongly support the use of multiple compounds in chemoprevention for better pharmacological results. Drug discovery and formulation process should include multiple antioxidant compounds from the same plant to provide a synergistic effect towards treating cancer. Through this procedure, the bioavailability and other setbacks of anticancer compounds could be breached, and at the same time, pharmacological values of plant-derived compounds and pharmaceutical needs for novel anticancer drugs could be compensated.

References

Albini A, Pennesi G, Donatelli F, Cammarota R, De Flora S, Noonan DM (2010) Cardiotoxicity of anticancer drugs: the need for cardio-oncology and cardio-oncological prevention. JNCI: Journal of the National Cancer Institute 102(1):14–25

Alkhalfioui F, Magnin T, Wagner R (2009) From purified GPCRs to drug discovery: the promise of protein-based methodologies. Curr Opin Pharmacol 9:629–635

Almeida DA, Piazza JR, Stawski RS (2009) Inter-individual differences and intra-individual variability in the cortisol awakening response: an examination of age and gender. Psychol Aging 24:819–827

Alqahtani S, Mohamed LA, Kaddoumi A (2013) Experimental models for predicting drug absorption and metabolism. Expert Opin Drug Metab Toxicol 9:1241–1254

Alsarhan A, Sultana M, Khatib AA, Kadir MRA (2014) Review on some Malaysian traditional medicinal plants with therapeutic properties. J Basic Appl Sci 10:149–159

Androutsopoulos VP, Papakyriakou A, Vourloumis D, Tsatsakis AM, Spandidos DA (2010) Dietary flavonoids in cancer therapy and prevention: substrates and inhibitors of cytochrome P450 CYP1 enzymes. Pharmacol Ther 126:9–20

Arber A, Odelius A, Williams P, Lemanska A, Faithfull S (2017) Do patients on oral chemotherapy have sufficient knowledge for optimal adherence. A mixed methods study. Eur J Can Care 26:e12413

Arulselvan P, Fard MT, Tan WS, Gothai S, Fakurazi S, Norhaizan ME, Kumar SS (2016a) Role of antioxidants and natural products in inflammation. Oxidative Med Cell Longev 2016:5276130

Arulselvan P, Tan WS, Gothai S, Muniandy K, Fakurazi S, Esa NM, Alarfaj AA, Kumar SS (2016b) Anti-inflammatory potential of ethyl acetate fraction of *Moringa oleifera* in downregulating the NF-kB signaling pathway in lipopolysaccharide-stimulated macrophages. Molecules 21:E1452

Arzamastseva N, Lapina Y, Ageev F, Lankin V, Mareev V, Belenkov Y (2007) Influence of metformin on the parameters of oxidative stress in patients with combined heart failure and diabetes mellitus. Eur J Heart Fail Suppl 6:123–134

Atanas GA, Birgit W, Eva-Maria PW, Thomas L, Christoph W, Pavel U, Eronika T, Limei W, Stefan S, Ele H (2015) Discovery and resupply of pharmacologically active plant-derived natural products: a review. Biotechnol Adv 33:1582–1614

Atanasov AG, Wang JN, Gu SP, Bu J, Kramer MP (2013) Honokiol: a non-adipogenic PPAR gamma agonist from nature. Biochim Biophys Acta 1830:4813–4819

Aung TN, Qu Z, Kortschak RD, Adelson DL (2017) Understanding the effectiveness of natural compound mixtures in cancer through their molecular mode of action. Int J Mol Sci 18:1–20

Ayoola GA (2008) Phytochemical screening and antioxidant activities of some selected medicinal plants used for malaria therapy in southwestern Nigeria. Trop J Pharm Res 7:1019–1024

Bao Y, Fenwick R (2004) Phytochemicals in health and disease. CRC Press, New York

Berkov S, Bastida NM, Viladomat CC (2008) Rapid TLC/GC-MS identification of acetylcholinesterase inhibitors in alkaloid extracts. Phytochem Anal 19:411–419

Bishayee A, Ahmed S, Brankov N, Perloff M (2011) Triterpenoids as potential agents for the chemoprevention and therapy of breast cancer. Front Biosci 16:980–996

Borkova L, Adamek R, Kalina P, Drasar P, Dzubak P, Gurska S, Rehulka J, Hajduch M, Urban M, Sarek J (2017) Synthesis and cytotoxic activity of triterpenoid thiazoles derived from allobetulin, methyl betulonate, methyl oleanonate, and oleanonic acid. Chem Med Chem 12:390–398

Botha CJ, Penrith ML (2008) Poisonous plants of veterinary and human importance in southern Africa. J Ethnopharmacol 119:549–558

Butler MS (2004) The role of natural product chemistry in drug discovery. J Nat Prod 67:2141–2153

Chanda S, Dave R (2009) In-vitro models for antioxidant activity evaluation and some medicinal plants possessing antioxidant properties: an overview. Afr J Microbiol Res 3:981–996

Chudzik M, Szlacheta IK, Krol W (2015) Triterpenes as potentially cytotoxic compounds. Molecules 20:1610–1625

Chun TC, Zhi JW, Moses SC, Christopher WL (2011) Herb-herb combination for therapeutic enhancement and advancement: theory, practice and future perspectives. Molecules 18:5125–5141

Colomer R, Alba E, González-Martin A, Paz-Ares L, Martin M, Llombart A, Lescure AR, Salvador J, Albanell J, Isla D, Lomas M, Rodríguez CA, Trigo JM, Germa JR, Bellmunt J, Tabernero J, Rosell R, Aranda E, Cubedo R, Baselga J (2010) Treatment of cancer with oral drugs: a position statement by the Spanish Society of Medical Oncology (SEOM). Ann Oncol 21:195–198

Cook JA, Gius D, Wink DA, Krishna MC, Russo A, Mitchell JB (2004) Oxidative stress, redox and tumor micro environment. Semin Radiat Oncol 14:256–266

Cragg GM, Newman DJ (2004) A tale of two tumor targets: topoisomerase I and tubulin. The Wall and Wani contribution to cancer chemotherapy. J Nat Prod 67:232–244

Cragg GM, Newman DJ (2005) Plants as sources of anticancer agents. J Ethnopharmacol 100:72–79

Cyriac JM, James E (2014) Switch over from intravenous to oral therapy: a concise overview. J Pharmacol Pharmacother 5:83–87

Demain AL, Vaishnav P (2011) Natural products for cancer chemotherapy. Microb Biotechnol 4:687–699

Eisenhauer EA, O'Dwyer PJ, Christian M, Humphrey JS (2000) Phase I clinical trial design in cancer drug development. J Clin Oncol 18:684–692

Francis G, Kerem Z, Makkar HPS, Becker K (2002) The biological action of saponins in animal systems: a review. Brit J Nut 88:587–605

Gnanaraj C, Shah MD, Makki JS, Iqbal M (2016) Hepatoprotective effects of *Flagellaria indica* L. are mediated through the suppression of pro-inflammatory cytokines and oxidative stress markers in rats. Pharm Biol 54:1420–1433

Godugu C, Patel AR, Doddapaneni R, Somagoni J, Singh M (2014) Approaches to improve the oral bioavailability and effects of novel anticancer drugs berberine and betulinic acid. PLoS One 9:e89919

Gonzalo JD (2011) Toxic plants of veterinary and agricultural interest in Colombia. Int J Poison Plant Res 1:1–15

Gordaliza M, Garcia PA, Del Corral JM, Castro MA, Gomez-Zurita MA (2004) Podophyllotoxin: distribution, sources, applications and new cytotoxic derivatives. Toxicon 44:441–459
Gothai S, Ganesan P, Park SY, Fakurazi S, Choi DK, Arulselvan P (2016) Natural phyto-bioactive compounds for the treatment of type 2 diabetes: inflammation as a target. Nutrition 8:E461
Grunwald HW (2007) Ethical and design issues of phase I clinical trials in cancer patients. Cancer Investig 25:124–126
Guo J, Zhou AW, Fu YC, Verma UN, Tripathy D, Pfrenkel U, Becerra CR (2006) Efficacy of sequential treatment of HCT116 colon cancer monolayers and xenografts with docetaxel, flavopiridol, and 5-fluorouracil. Acta Pharmacol Sin 27:1375–1381
Halket JM, Daniel W, Anna MP, Raj KP, Paul DF, Peter MB (2005) Chemical derivatization and mass spectral libraries in metabolic profiling by GC/MS and LC/MS/MS. J Exp Bot 56:219–243
Hamid N, Hedayatollah S (2013) Toxicity and safety of medicinal plants. J Herb Med Pharmacol 2:21–22
Hassanzadeh P (2011) Colorectal cancer and NF-kB signalling pathway. Gasteroenterol Hepatol 4:127–132
He X, Clarke SJ, McLachlan AJ (2011) Clinical pharmacology of chemotherapy agents in older people with cancer. Curr Gerontol Geriat Res 2011:628670
Jantan I, Ahmad W, Bukhari SNA (2015) Plant-derived immunomodulators: an insight on their preclinical evaluation and clinical trials. Front Plant Sci 6:655
Jeffrey S, Michal B (2009) A combined evolution/structure-based approach to protein function prediction. Brief Bioinform 2:1–14
Jordan MA, Wilson L (2004) Microtubules as a target for anticancer drugs. Nat Rev 4:253–265
Kawaguchi H, Hirakawa K, Miyauchi K, Koike K, Ohno Y, Sakamoto A (2010) Pattern recognition analysis of proton nuclear magnetic resonance spectra of brain tissue extracts from rats anesthetized with propofol or isoflurane. PLoS One 5:e11172
Kingston GI (2009) Tubulin-interactive natural products as anticancer agents. J Nat Prod 72:507–515
Klekota J, Brauner E, Roth FP, Schreiber SL (2006) Using high-throughput screening data to discriminate compounds with single-target effects from those with side effects. J Chem Inf Model 46:1549
Kyselova Z (2011) Toxicological aspects of the use of phenolic compounds in disease prevention. Interdiscip Toxicol 4:173–183
Lee JA, Uhlik MT, Moxham CM, Tomandl D, Sall DJ (2012) Modern phenotypic drug discovery is a viable, neoclassic pharma strategy. J Med Pharm Chem 55:4527–4538
Leeuwen RWF, Brundel DHS, Gelder TV, Mathijssen RHJ, Burger DM, Jansman FGA (2013) Prevalence of potential drug-drug interactions in cancer patients treated with oral anticancer drugs. Brit J Can 108:1071–1078
Leonard IZ, Randall TP (2005) *In vivo* drug discovery in the zebra fish. Nat Rev Drug Discov 4:35–44
Luc P, Arnold JV (2005) Bio-guided isolation of pharmacologically active plant components, still a valuable strategy for the finding of new lead compounds. J Ethnopharmacol 100:57–60
Marcy JB, Douglas KB (2005) Drug discovery from medicinal plants. Life Sci 78:431–441
Mathijssen RHJ, Sparreboom A, Verweij J (2014) Determining the optimal dose in the development of anticancer agents. Nat Rev Clin Oncol 11:272–281
Millimouno FM, Dong J, Yang L, Li J, Li X (2014) Targeting apoptosis pathways in cancer and perspectives with natural compounds from mother nature. Cancer Prev Res 7:1081–1107
Ming-Wei W, Xiaojiang H, Kaixian C (2007) Biological screening of natural products and drug innovation in China. Phil Trans Royal Soc B 362:1093–1105
Mishra BB, Tiwari VK (2011) Natural products: an evolving role in future drug discovery. Eur J Med Chem 46:4769–4807
Mondal J, Panigrahi AK, Khuda-Bukhsh AR (2014) Conventional chemotherapy: problems and scope for combined therapies with certain herbal products and dietary supplements. Aus J Mol Cell Biol 1:10

Montellano PROD (2013) Cytochrome P450-activated prodrugs. Future Med Chem 5:213–228
Moudi MGR, Yien CYS, Nazre M (2013) Vinca alkaloids. Int J Prev Med 4:1231–1235
Mullauer FB, Kessler JH, Madema JP (2010) Betulinic acid, a natural compound with potent anti-cancer effects. Anti-Can Drugs 21:215–227
Mut-Salud N, Alvarez PJ, Garrido JM, Carrasco E, Aranega A, Rodriguez-Serano F (2016) Antioxidant intake and antitumor therapy: toward nutritional recommendations for optimal results. Oxidative Med Cell Longev 2016:6719534
Nair MP, Mahajan S, Reynolds JL, Aalinkeel R, Nair H, Schwartz SA, Kandaswami C (2006) The flavonoid quercetin inhibits proinflammatory cytokine (tumor necrosis factor-alpha) gene expression in normal peripheral blood mononuclear cells via modulation of the NF-kβ system. Clin Vaccine Immunol 13:319–328
Narang AS, Desai DS (2009) Anticancer drug development: unique aspects of pharmaceutical development. In: Lu Y, Mahato RI (eds) Pharmaceutical perspectives of cancer therapeutics. Bristol-Myers Squibb, New Brunswick, pp 49–92
Nassiri AM, Hosseinzadeh H (2008) Review of pharmacological effects of *Glycyrrhiza sp.* and its bioactive compounds. Phytother Res 22:709–724
Ndhlala AR, Ncube B, Okem A, Mulaudzi RB, Staden JV (2013) Toxicology of some important medicinal plants in southern Africa. Food Chem Toxicol 62:609–621
Newman DJ, Cragg GM (2012) Natural products as sources of new drugs over the 30 years from 1981 to 2010. J Nat Prod 75:311–335
Nonnekens J, Hoeijmakers JHJ (2017) After surviving cancer, what about late life effects of the cure. EMBO Mol Med 9:4–6
Parasuram S (2011) Toxicological screening. J Pharmacol Pharmacother 2:74–79
Park EJ, Pezzuto JM (2002) Botanicals in cancer chemoprevention. Can Met Rev 21:231–255
Park JY, Park CM, Kim JJ, Noh KH, Cho CW, Song YS (2007) The protective effect of chlorophyll against oxidative stress and inflammatory processes in LPS-stimulated macrophages. Food Sci Biotechnol 16:205–211
Petronelli A, Pannitteri G, Testa U (2009) Triterpenoids as new promising anticancer drugs. Anti-Can Drugs 20:880–892
Philomena G (2011) Concerns regarding the safety and toxicity of medicinal plants-an overview. J App Pharm Sci 1:40–44
Ponmari G, Annamalai A, Gopalakrishnan VK, Lakshmi PTV, Guruvayoorappan C (2014) NF-kB activation and proinflammatory cytokines mediated protective effect of *Indigofera caerulea* Roxb. on CCl_4 induced liver damage in rats. Int Immunopharmacol 23:672–680
Qiu S, Sun H, Zhang AH, Xu HY, Yan GL, Han Y et al (2014) Natural alkaloids: basic aspects, biological roles, and future perspectives. Chin J Nat Med 12:401–406
Rabi T, Bishayee A (2009) Terpenoids and breast cancer prevention. Breast Cancer Res Treat 115:223–239
Rasoanaivo P, Wright CW, Willcox ML, Gilbert B (2011) Whole plant extracts versus single compounds for the treatment of malaria: synergy and positive interactions. Malar J 10:S4
Redondo-Blanco S, Fernandez J, Gutierrez-del-Rio I, Villar CJ, Lombo F (2017) New insights toward colorectal cancer chemotherapy using natural bioactive compounds. Front Pharmacol 8:109
Rehman MU, Tahir M, Khan AQ, Khan R, Lateef A, Hamiza OO, Qamar W, Ali F, Sultana S (2013) Chrysin suppresses renal carcinogenesis via amelioration of hyperproliferation, oxidative stress and inflammation: plausible role of NF-kβ. Toxicol Lett 216:146–158
Remesh A (2012) Toxicities of anticancer drugs and its management. Int J Basic Clin Pharmacol 1:2–12
Samoylenko A, Hossain JA, Mennerich D, Kellokumpu S, Hiltunen JK, Kietzmann T (2013) Nutritional countermeasures targeting reactive oxygen species in cancer: from mechanisms to biomarkers and clinical evidence. Antioxid Redox Signal 19:2157–2196
Shoemaker RH (2006) The NCI60 human tumour cell line anticancer drug screen. Nat Rev Cancer 6:813–823

Sultana N, Ata A (2008) Oleanolic acid and related derivatives as medicinally important compounds. J Enz Inhibit Med Chem 23:739–756

Swinney DC, Anthony J (2011) How were new medicines discovered. Nat Rev Drug Discov 10:507–519

Sznarkowska A, Kostecka A, Meller K, Bielawski KP (2017) Inhibition of cancer antioxidant defense by natural compounds. Oncotarget 8:15996–16016

Tsume Y, Incecayir T, Song X, Hilfinger JM, Amidon GL (2014) The development of orally administrable gemcitabine prodrugs with D-enantiomer amino acids: enhanced membrane permeability and enzymatic stability. Eur J Pharm Biopharm 86:514–523

Van DH, Jacobs DI, Snoeijer W, Hallard D, Verpoorte R (2004) The Catharanthus alkaloids: pharmacognosy and biotechnology. Curr Med Chem 11:607–628

Veeresham C (2012) Natural products derived from plants as a source of drugs. J Adv Pharm Tech Res 3:200–201

Wink M, Van Wyk BE (2008) Mind-altering and poisonous plants of the world. Timber Press, Portland

Workman P, Collins I (2014) Modern cancer drug discovery: integrating targets, technologies, and treatments for personalized medicine. In: Neidle S (ed) Cancer drug design and discovery, 2nd edn. Academic, San Diego, pp 3–53

Zhang D, Luo G, Ding X, Lu C (2012) Preclinical experimental models of drug metabolism and disposition in drug discovery and development. Acta Pharm Sin B 2:549–561

Chapter 8
Recent Advancements in the Clinical Evaluation of Plant-Derived Anticancer Compounds

Jayabalan Shilpha, Lakkakula Satish, and Manikandan Ramesh

Abstract Cancer is a chief global health burden and the leading cause of human death worldwide next to heart diseases. The alarming rise in the mortality rate owing to cancer has driven the chase for anticancer agents to effectively combat this disease. Searching for novel and efficient compounds of natural origin has been a major aspect of concerns because they exhibit less toxic side effects. Numerous secondary metabolites from plants and their semisynthetic analogs have been identified as an excellent, novel lead structures in developing promising anticancer agents. In the current scenario, several successful anticancer drugs and their derivatives have been obtained from plant sources, and many of them are in clinical trials. Phytocompounds such as vinca alkaloids, taxanes, podophyllotoxin, camptothecin, homoharringtonine, and their derivatives have appreciably influenced cancer research on many facets. Likewise, some of the other plant-derived anticancer agents including omacetaxine mepesuccinate, ingenol mebutate, β-lapachone, flavopiridol, curcumin, etc. are currently being under phase I and II clinical trials, either individually or in concert with other anticancer agents for the treatment of a broad range of tumors like lymphomas, leukemias, and solid tumors. Customary anticancer drug discovery has targeted mainly on the cytotoxic agents that hamper metabolic pathways critical to cell division. However, during recent years, several molecular target-based compounds have been emerged concentrating on other cellular process of cancer cells such as apoptosis, metastasis, angiogenesis, etc. Hence,

J. Shilpha · M. Ramesh (✉)
Department of Biotechnology, Science Campus, Alagappa University,
Karaikudi 630 004, Tamil Nadu, India

L. Satish
Department of Biotechnology, Science Campus, Alagappa University,
Karaikudi 630 004, Tamil Nadu, India

Department of Biotechnology Engineering, Ben-Gurion University of Negev,
Beer Sheva 84105, Israel

The Jacob Blaustein Institutes for Desert Research, Ben-Gurion University of the Negev,
Bergman Campus, Beer Sheva 84105, Israel
e-mail: lsatish@post.bgu.ac.il

M. S. Akhtar, M. K. Swamy (eds.), *Anticancer Plants: Clinical Trials and Nanotechnology*, https://doi.org/10.1007/978-981-10-8216-0_8

the present anticancer drug discovery involves high-throughput screening of phytocompounds against a series of such molecular targets. The present chapter discusses the clinical evidences of some important phytocompounds of anticancer plants, overview of their current clinical status, and recent advances in their molecular mechanism of action.

Keywords Anticancer agents · Cytotoxic agents · Clinical trials · Natural compounds · Secondary metabolites

8.1 Introduction

Cancer is the foremost public health problem that represents the world's primary causes of death with an episode rate of 2.6 million and more cases per annum (May 2014; Siegel et al. 2015). Cancer, a set of devastating diseases and major cause of mortality in the United States of America, next only to heart diseases, is responsible for nearly one of every four deaths. The World Health Organization (WHO) has speculated that without immediate measures, the worldwide deaths due to cancer will elevate to nearly 80% by 2030, with more frequent occurrence in the middle- and low-income countries. Given the morbidity and mortality rate linked to the disease, along with the considerable economic burden, there persists to be a vital necessitate for more effective approaches. Despite the remarkable efforts over the past few decades to develop the existing therapeutic choices, and a vast number of efficient chemotherapeutic anticancer agents have been successfully identified and exploited in clinical practice, cancer still appears to be the major cause of fatality in most countries. The raise in the frequency of cancer, as well as the detrimental side effects of cancer chemotherapy, demands the discovery of novel therapeutic agents from natural sources.

Throughout the history, nature has provided an infinite resource of medicinal plants that have occupied a dominant role in treating human ailments and are used as their crude forms including infusions, syrups, and ointments (Ansari and Inamdar 2010). Plant-derived secondary metabolites significantly represent a boundless source of chemicals for new drug discovery. The compounds obtained from plants as such may not serve as the drug but may form the basis for screening, selecting, and developing many potential novel therapeutic agents (Dias et al. 2012). In reality, the idea of chemoprevention involving chemical drugs is commonly employed for treating cancers, however, with a limited therapeutic success due to toxicity issues (Pan et al. 2013). The chemoprevention of cancer using phytochemicals has revealed effective outcomes against a range of malignancies (Shukla et al. 2014). Thus, natural products as cancer therapeutic agents are of a particular interest among therapist due to their low toxicity profiles and potential effectiveness. Nearly 60% of the existing anticancer drugs are mainly the plant-derived pure compounds or their chemical analogs (Newman and Cragg 2012). Nature exists as an abundant source of biologically diverse and active chemotypes, though relatively a few of the originally isolated natural compounds are clinically tested as effective drugs. Researchers and clinicians are now focusing on search of natural products with multi-targeted potentials for effective chemoprevention (Millimouno et al. 2014; Cragg and Pezzuto 2016). Moreover, it offers a platform for the development of novel and safe drugs by means of appropriate understanding of the intricate synergistic interaction between varieties of constituents of anticancer herbs (Khazir et al. 2014).

Today's cancer therapy involves majorly of natural products as anticancer agents from various sources such as plants, animals, and microorganisms. With the advent of new technologies like high-throughput screening and combinatorial synthesis, plant-based anticancer drug discovery has been now expanding as evidenced by substantial increase of natural anticancer agents under clinical trials. A detailed collection of natural anticancer compounds from plants can be observed at Naturally Occurring Plant-Based Anticancer Compound-Activity-Target database (http://crdd.osdd.net/raghava/npact/) which comprises of about 1980 experimentally demonstrated compound-target interactions. Some of the major classes of plant-derived anticancer compounds that have drawn attention in current clinical trials include vinca alkaloids, taxanes, podophyllotoxin, camptothecin, homoharringtonine, and their derivatives. The present chapter discusses the clinical evidences of some important phytocompounds of anticancer plants, overview of their current clinical status, and recent advances in their molecular mechanism of action.

8.2 Clinical Evidences of Anticancer Potential from Traditional Medicinal Plants

For millennia, plant products have served as a thriving source of therapeutically valued compounds and presently gained significant applications in the fields of pharmacy, medicine, and biology (Swamy and Sinniah 2016; Arumugam et al. 2016). Nature has been proven to be the eternal reservoir of bioactive molecules for drug discovery, offering structural diversity and a broad-spectrum novel and exciting medical activities (Mohanty et al. 2017; Swamy et al. 2017). Besides phytocompounds have a long history of treating several human illnesses and occupy a major fraction of the current pharmacopeia. According to the World Health Organization (WHO), plant-based medicine is still the mainstay of about ~80% of populations in many developing countries (Swamy et al. 2011, 2016; Akhtar et al. 2014; Gezahegn et al. 2015). For the past two decades, almost 50% of new drugs introduced in the market had a natural product origin, and this rate is even high for anticancer drugs, wherein almost two thirds of such drugs were of natural origin (Newman and Cragg 2007). Nearly more than 3000 plant species have been reported to possess anticancer activity and utilized in cancer therapeutics. In addition to the presence of unique pharmacophores and a maximum degree of stereochemistry, phytocompounds are better than synthetic compounds due to their efficient delivery to the intracellular site of action as they commonly belong to the category of active biological metabolites and the real substrates for membrane transport mechanisms (Hert et al. 2009). Furthermore, when compared to synthetic ones, natural products have superior bioavailability which gives more advantage in the early stage of drug discovery by preventing false positives. The intricacy in the development of anticancer drug is very well understood from the fact that merely 5% of the candidates reaching clinical trials get the approval for marketing (Choi et al. 2017). The extent of clinical potency and toxicity of several anticancer agents obtained from traditional medicinal plants is indefinite and elusive. For instance, research on greater part of traditional anticancer plants is in the preclinical stage or is not being actively practiced. Many of the ayurvedic anticancer plants such as *Allium sativum*, *Aloe vera*, *Andrographis paniculata*, *Annona atemoya/muricata*, *Asparagus racemosus*, *Berberis aristata*, *Curcuma longa*, *Calotropis gigantea*, *Datura metel*, *Euphorbia hirta*, *Moringa oleifera*, *Nigella sativa*, *Phyllanthus niruri/amarus*, *Picrorhiza kurroa*, *Piper longum*, *Plumbago zeylanica*, *Podophyllum hexandrum*, *Rubia cordifolia*, *Tinospora cordifolia*, *Semecarpus anacardium*, *Taxus baccata*, *Vinca rosea*, and *Withania somnifera* have scientifically demonstrated anticancer properties and are now actively exploited for the treatment of different cancer types (Balachandran and Govindarajan 2005). In saffron (stigmata of *Crocus sativus*) selective and assuring anticancer properties have been observed at both *in vitro* and *in vivo* conditions but not hitherto in clinical trials (Schmidt et al. 2007). The various forms of pomegranate such as juice peel and oil were found to exert potent anticancer activity, by causing hindrance to cell cycle, tumor cell proliferation, invasion, and angiogenesis (Lansky and Newman 2007). Thus, future research on these anticancer plants would aid in identifying safe and competent anticancer drugs and in furtherance the exploration of their mode of action.

8.3 Clinical Uses of Major Plant-Derived Anticancer Compounds

There are four main structural classifications of anticancer compounds derived from plants, namely, alkaloids, terpenoids, polyphenols, and lignans. These groups of compounds include some of the most active novel chemotherapeutic agents for current use in a clinical setting. One of the major setbacks associated with the usage of plant-based compounds is poor bioavailability or meager solubility that poses complications in manufacturing of bulk amounts required to serve as medicines (Fridlender et al. 2015). The major strategy adopted for diverse plant-derived substances is the utilization of semisynthetic or synthetic analogs. Some of the major plant-derived anticancer compounds in clinical use and under different phases of clinical trials are discussed below.

8.3.1 Alkaloids

8.3.1.1 Vinca alkaloids

The epoch of using plant source as anticancer agents began with the identification of two important alkaloids from the Madagascar periwinkle, *Catharanthus roseus* G. Don (Apocynaceae), such as vinblastine and vincristine (Verma and Singh 2010). For almost a half century, these two compounds have been employed in clinical oncology, which act by inhibiting the polymerization of tubulin molecules and the formation of the mitotic spindle resulting in metaphase arrest and apoptosis (Khazir et al. 2014). Later, a number of semisynthetic analogs of these two important drug molecules have been developed. The first semisynthetic vinca alkaloid to enter into human clinical trials was vindesine (Jordan and Wilson 2004), which is primarily used in the treatment of acute lymphocytic leukemia (ALL). Occasionally, it is used in the treatment of breast cancer, colorectal cancer, renal cell cancer, blast crisis of chronic myelocytic leukemia (CML), and non-small-cell lung cancer (NSCLC). Vinorelbine is another semisynthetic derivative of vinblastine which gained approval to treat metastatic breast cancer (MBC) in 1991. Vinflunine, a dihydro-fluoro derivative of vinorelbine was approved as second-line chemotherapy in metastatic urothelial cancer (Bachner and De Santis 2008; Mamtani and Vaughn 2011). Vinflunine is actively used in phase II/III clinical trials to treat advanced stage of cancers like MBC and NSCLC. In addition, many other structural analogs of vindesine are undergoing clinical trials including anhydrovinblastine (Hydravin), vinglycinate, and vintripol.

8.3.1.2 Homoharringtonine

Cephalotaxus alkaloids including harringtonine, isoharringtonine, and homoharringtonine (HHT) were acquired from *Cephalotaxus harringtonia* var. *drupacea* (Cephalotaxaceae). Cephalotaxus alkaloids are inhibitors of protein synthesis. Homoharringtonine in its purified form has been found effective against a range of leukemias, especially those which are resistant to standard treatment, and has been observed with complete hematologic remission capacity in late chronic-phase chronic myelogenous leukemia (CML) patients. Homoharringtonine (Omacetaxine mepesuccinate) has been clinically tested with CML in myeloid blast crisis (CML-MBC) patients (Cragg and Pezzuto 2016).

8.3.1.3 Colchicine

Colchicine, a nitrogenous alkaloid isolated from *Colchicum autumnale* is one of the oldest known antimitotic drugs, available till now due to its potent antitumor properties. It attaches irreversibly to tubulin and thereby blocks microtubule formation, inhibits cell cycle progression, and subsequently induces apoptosis. The major setback of colchicine in clinical usage is its nonspecific action on both actively dividing tumor cells and normal cells. However, derivatives of colchicine, namely, colchicinamide and deacetylcolchicine were made with minimum toxicity and greater specificity to tumor cells combined with effective drug delivery approaches (Sivakumar 2013). Colchicinamide has been targeted toward the treatment of breast cancer and various other solid tumors. Deacetylcolchicine has been now subjected to phase II clinical trials for its effectiveness against melanoma, Hodgkin's lymphoma, and chronic granulocytic leukemia. Recently, a series of novel deacetamidothiocolchicine-7-ols and their ester exhibited potent inhibitory activity against tubulin polymerization and selective cytotoxicities against colon and CNS cancers as well as melanoma *in vitro* (Muhtasib et al. 2015).

8.3.1.4 Ellipticine

Ellipticine, an indole alkaloid isolated from *Ochrosia elliptica* is a topoisomerase II inhibitor and showed efficient anticancer effect in many humans and animal tumor systems (Marie et al. 2012). The analogs of ellipticine such as DHE (N-2-(diethylaminoethyl)-9-hydroxy-ellipticinium chloride, NMHE (2-N-methyl 9-hydroxy-ellipticine), and retelliptine dihydrochloride (SR 95325 B) were at phase II and phase I clinical trials on metastatic breast cancer and murine solid tumor models, respectively (Iveta et al. 2013).

8.3.1.5 Camptothecin Alkaloids

Camptothecin (CPT) was isolated from *Camptotheca acuminata* which is a quinoline alkaloid, with topoisomerase-I inhibitor activity and stimulates cell death by DNA damage (Khazir et al. 2014). Nevertheless, it was withdrawn from clinical trials owing to its low aqueous solubility and severe toxicity. To overcome these problems, a sequence of novel CPT analogs were developed and approved for use in clinical setting such as topotecan, irinotecan, and belatecan which are effective inhibitors of DNA topoisomerase-I, a crucial component in DNA replication and transcription (Oberlies and Kroll 2004). Irinotecan is used in the treatment of colorectal cancers, while topotecan is actively used for treating ovarian and small-cell lung cancers. In addition, there are other camptothecin derivatives which are in preclinical development stage (Rahier et al. 2005; Khazir et al. 2014) (Table 8.1).

Table 8.1 Semisynthetic analogs and prodrugs of camptothecin under clinical trials

CPT semisynthetic analogs	Clinical status	Target
9-Aminocamptothecin (9-AC)	Phase II	Ovarian and malignant lymphoma
Karenitecin (BNP-1350)	Phase II	Malignant melanoma and brain tumors
Diflomotecan (BN-80915)	Phase I	Solid tumors
Gimatecan (ST-1481)	Phase I and phase II	Advanced solid tumors, epithelial ovarian, and fallopian tube cancers
Elomotecan (BN-80927)	Phase I	Advanced solid tumors
Exatecan mesylate	Phase II	Gastric cancers
Rubitecan	Phase III	Pancreatic cancer
CZ-48	Phase I	Solid tumors
XMT-1001	Phase I b	Lung cancers
EZN-2208	Phase I	Advanced malignancies

8.3.2 *Terpenoids*

8.3.2.1 Taxol and Its Semisynthetic Analogs

Taxol, (paclitaxel) a highly effective chemotherapeutic diterpenoid, isolated from the bark of *Taxus brevifolia* (Pacific yew tree) is active against ovarian and breast cancers (Newman et al. 2003). Paclitaxel was the first compound discovered to support microtubule formation and has been administered in the treatment of different types of cancers such as ovarian, breast, NSCLC, and Kaposi sarcoma (Newman and Cragg 2012). A number of its semisynthetic derivatives such as docetaxel, cabazitaxel, and abraxane (ABI-007) have been developed and proficiently utilized in the treatment of broad range of tumors, prostate cancer, and breast cancer, respectively. There are several other novel semisynthetic analogs in development which are being designed to improve the pharmacology and therapeutic index (Saklani and Kutty 2008; Khazir et al. 2014) (Table 8.2).

8.3.2.2 Betulinic Acid

Betulinic acid, a pentacyclic triterpenoid and potent inhibitor of topoisomerase, is commonly occurring in Betula species (Betulaceae) (Cichewicz and Kouzi 2004; Fulda 2009; Mullauer et al. 2010). Betulinic acid has been reported to induce apoptosis in human melanoma cell lines by selectively arresting their growth and to exhibit *in vivo* activity in athymic mice possessing human melanoma cells (Pan et al. 2010). Betulinic acid has also shown significant cytotoxic effect against neuroectodermal and malignant brain tumor cell lines (Cichewicz and Kouzi 2004). So far, a greater number of betulinic acid derivatives have been synthesized to enhance its therapeutic activity. NVX-207 is one of the novels and well-tolerated Beta derivatives that has shown prominent antitumor activity in treatment-resistant canine

Table 8.2 Semisynthetic analogs of paclitaxel in clinical trials

Paclitaxel semisynthetic analogs	Clinical status	Target
Paclitaxel poliglumex (CT-2103, Xyotav)	Phase III	Non-small-cell lung carcinoma (NSCLC)
7-DHA-Taxol (Taxoprexin 13)	Phase II	Advanced NSCLC, eye and skin melanoma
	Phase III	Advanced lung cancers
Larotaxel (XPR9881)	Phase II	Breast cancer
	Phase III	Metastatic urothelial tract or bladder cancer
Ortataxel (BAY-59-862)	Phase II	Breast cancer
TPI-287	Phase I/II	Melanoma, neuroblastoma, and medulloblastoma
Milataxel (MAC-321)	Phase II	Colorectal cancer
Tesetaxel (DJ-927)	Phase I/II	Advanced solid tumors
BMS-184476	Phase I	Human tumor xenografts and advanced solid tumors
RPR 109881A	Phase III	Breast cancer

cancer patients. NVX-207 has also shown excellent responses in phase I/II study of dogs afflicted with naturally occurring cancer undergoing local treatment of NVX-207 (Willmann et al. 2009). The topical application of 20% of betulinic acid ointment has been evaluated under phase I/II clinical trials for the treatment of dysplastic melanocytic nevi (NIH 2013). Recently, Kutkowska et al. (2017) have showed the synergistic capacity of betulinic acid with sorafenib on induction of apoptosis and inhibition of clonogenic activity in various non-small-cell lung cancer cell lines.

8.3.3 Polyphenols

8.3.3.1 Resveratrol

Resveratrol (trans-3,5,4′-trihydroxystilbene) is a phytoalexin that occurs in various plant species such as mulberries, peanuts, and grapes. Resveratrol has been recognized as a useful candidate for cancer chemoprevention owing to its remarkable inhibitory effects on sequential cellular actions such as tumor initiation, promotion, and progression (Zhu 2011). In all three stages of carcinogenesis (initiation, promotion, and progression), resveratrol has exhibited chemotherapeutic and chemopreventive activities, as demonstrated in both UVB and chemically induced skin carcinogenesis in mice and various other murine models of human cancers (Nobili et al. 2009). Several *in vivo* and *in vitro* studies have confirmed its capacity to modulate and control diverse signaling pathways and various targets (Athar et al. 2007). Being a chemopreventive agent, resveratrol is now undergoing phase I studies in patients with colorectal cancer as well as in healthy subjects who are highly susceptible to the development of melanoma (Nobili et al. 2009). Recently, Singh et al.

(2016) have summarized the various ongoing clinical trials of resveratrol against human cancer. Resveratrol has been evaluated for its role in treating colorectal cancer, gastrointestinal tumors, and in follicular lymphoma.

8.3.3.2 Combretastatin and Its Analogs

Combretastatin, a stilbenoid separated from the bark of *Combretum caffrum* Kuntze (Combretaceae) is a vascular disrupting agent. It restrains the growth of blood vessel in tumors, resulting in necrosis and cell death of tumors. The combretastatin A series are well known for their tubulin assemblage inhibitory and cytotoxicity activities against *in vitro* human cancer cell lines. Combretastatin A4 is active against colon and lung cancers and leukemia which are now at phase I/II clinical trials (Banerjee et al. 2008). Combretastatin A4 phosphate (CA4P) is the phosphate-bearing prodrug of combretastatin A4, which is capable of inducing morphological alterations within endothelial cells and cause selective and prompt vascular dysfunction of tumors (Delmonte and Sessa 2009; Pan et al. 2010). CA4P is currently examined in phase II clinical studies for treating anaplastic thyroid carcinoma. The other phosphate prodrug of combretastatin A-1, OXi4503 has been found outstanding in preclinical studies and has been investigated in phase IIb clinical trial to study its safety and efficacy against solid tumor in the liver (Patterson et al. 2012). Moreover, combretastatin CA4P has been actively investigated for several types of thyroid cancer including anaplastic, medullary, and stage four follicular thyroid cancer (Jaroch et al. 2016; Xie and Zhou 2017).

8.3.4 Lignans

8.3.4.1 Podophyllotoxins

Podophyllotoxin is an important anticancer lignan obtained from *Podophyllum emodi* or *P. peltatum* (Berberidaceae). The two most important and successful analogs of podophyllotoxin are etoposide (VP16 or VP16-213) and teniposide (VM26). Etoposide is one of the most dynamic anticancer agents administered in the treatment of Hodgkin's and non-Hodgkin's lymphomas, testicular teratoma, small-cell lung cancer, and a variety of other malignancies. Due to limitations, such as reduced water solubility, the development of resistance, and toxicity by tumor cells, investigators are exploring new synthetic analogs for new lead compounds with superior therapeutic index and extensive therapeutic scope. As a result of these several semi-synthetic analogs such as tafluposide, etoposide phosphate, NKH-611, GL331, azatoxin, and Top-53 have gone through clinical trials for treating a variety of cancers (Liu et al. 2002; Sargent et al. 2003; Lee and Xiao 2012). Very recently, deoxypodophyllotoxin (DPT) has been reported to have strong antitumor effect on lung cancer, breast cancer, and gastric cancer (Khaled et al. 2017).

8.4 Other Anticancer Compounds in Clinical Trials

Curcumin, a yellow polyphenol obtained from *Curcuma longa* L. (Zingiberaceae), is the most important dietary supplement and chemopreventive agent (Singh 2007). Curcumin arrests multiplication of a broad range of malignant cells via modulating several intracellular signaling pathways, by exerting its action on various targets, such as growth factor receptors, cell surface adhesion molecules, transcription factors, and protein kinases (Aggarwal et al. 2007). Curcumin is now undergoing phase II clinical trials to treat pancreatic cancer at advanced stage as well as colorectal cancer (Johnson and Mukhtar 2007; Dhillon et al. 2008). Recently, Panda et al. (2017) have extensively reviewed the promising role of curcumin as an adjuvant in combinatorial therapy for treating several cancer types. Currently, many clinical trials involving curcumin as an anticancer agent have been ongoing, and few have been completed including breast cancer, lung cancer, colon cancer, oral cancer, prostate cancer, and cervical and gastric cancers.

Indirubin, an isomer of indigo is an alkaloid present in "Qingdai," a Chinese traditional medicine that has been reported to possess broad-spectrum anticancer activity (Eisenbrand et al. 2004). A derivative of indirubin, namely, meisoindigo, has significant ability to hinder cyclin-dependent kinases, arrest cell cycle at G0/G1 phase, promote cell differentiation and apoptosis and has also been found to inhibit growth of tumor in xenografts of HT-29 colon cancer (Zuo et al. 2008). Meisoindigo is currently under phase III stage of clinical trial as a chemotherapeutic agent against chronic myelogenous leukemia (CML) (Pan et al. 2010). Perillyl alcohol, a limonene type of monoterpenoid, occurring in essential oils of aromatic plants like lavender (*Lavandula intermedia*), peppermint (*Mentha piperita*), etc., has been reported to stimulate apoptosis, cellular differentiation, and inhibition of cell cycle at G1 phase (Yeruva et al. 2007). Recently, perillyl alcohol has been involved in phase II clinical trial for treating metastatic breast cancer patients (Bailey et al. 2008).

Protopanaxadiol (20S-protopanaxadiol) is a dammarane group of triterpene aglycone obtained by hydrolysis of ginseng (*Panax ginseng* C.A. Mey.; Araliaceae) saponins. The mode of action of protopanaxadiol was found to be inducing apoptosis in cancer cells and arresting cell cycle by means of regulating different signaling pathways involving caspases. It has also been observed to possess cytotoxic activity in multidrug-resistant tumors by exhibiting P-glycoprotein blocking activity (Liu et al. 2007). Protopanaxadiol has now been taken to phase I clinical trial for the chemotherapy of lung cancer and other solid tumors (Saklani and Kutty 2008). Flavopiridol is a synthetic compound whose chemical structure is similar to rohitukine a compound isolated from *Dysoxylum binectariferum*, an indigenous plant of India (Khazir et al. 2014). Flavopiridol being a potent inhibitor of cyclin-dependent kinases (cdks) was the first compound to reach clinical trial (Shapiro 2004). It is currently undergoing a number of phase I and phase II clinical trials for leukemias, lymphomas, and solid tumors.

Ingenol mebutate (ingenol-3-angelate 44) is a substance identified from the sap of *Euphorbia peplus* (Fallen and Gooderham 2012). It plays a prominent role in skin

Table 8.3 Other anticancer agents in clinical trials

Name	Source	Functions	Clinical status	References
Bruceantin	*Brucea antidysenterica*	Inhibitor of peptidyl transferase in elongation reaction	Phase II	Da Rocha et al. (2001)
4-Ipomeanol	*Ipomoea batatas*	DNA binder	Phase II	Lakhanpal et al. (2001)
Pervilleine A	*Erythroxylum pervillei*	P-glycoprotein inhibitor	Lead molecule	Mi et al. (2003)
Dimethyl xanthene-9-one-4-acetic acid	Flavone-8-acetic acid analog	TNF-α inducer	Phase II	Westwell (2003)
NM-3	Isocoumarin derivative	Inhibitor of angiogenesis by VEGF expression inhibition	Phase I complete	Quesada et al. (2006)
Silvestrol	*Aglaia foveolata*	Mitochondrial apoptotic pathway promoter	Lead molecule	Kim et al. (2007)
Salvicine	*Salvia prionitis*	Non-intercalative topoisomerase II inhibitor	Phase II	Cai et al. (2008)
Triptolide	*Tripterygium wilfordii*	Inducer of apoptosis by modulating p53	Phase I	Saklani and Kutty (2008)
Phenoxodiol	Genistein derivative	Topoisomerase II inhibitor and apoptosis promoter	Lead molecule	Silasi et al. (2009)
Quercetin	Onions, apples, red wine, and tea	Inducer of apoptosis, androgen receptor inhibitor, and angiogenesis inhibitor	Phase I	
Berbamine	*Berberis amurensis*	Caspase-3-dependent apoptosis	Lead molecule	Kapoor (2012)
Roscovitine (CYC 202)	(*Raphanus sativus*)	CDK inhibitor	Phase II	Slovackova et al. (2012)
Kanglaite	*Coix lacryma-jobi*	Inhibitor of mitosis of tumor cells during G2/M phase	Lead molecule	Zhan et al. (2012)
Santonin	*Artemisia maritima*	NF-Kb inhibitor	Lead molecule	Khazir et al. (2013)

chemotherapy; particularly, a gel form of the compound has been used for the treatment of acid keratosis. Beta-lapachone a quinone obtained from lapacho tree (*Tabebuia avellanedae*) bark is a powerful DNA topoisomerase I inhibitor that acts by causing delay in cell cycle at G1 or S phase prior to inducing either apoptosis or necrotic cell death of human carcinoma cells, including colon, lung, ovary, breast, and prostate (Nobili et al. 2009). At present, it is investigated in a phase I and II study (Anonymous 2014). Moreover, a large number of other anticancer agents in clinical trials are listed in Table 8.3.

8.5 Recent Advancements in Mechanism of Action of Natural Anticancer Drugs

The conventional screening of plant-based anticancer agents was principally based on examining the cytotoxic activity either *in vitro* using cancer cell lines or *in vivo* using animal models. Several natural anticancer agents discovered actually through such assays have been found to exhibit their cytotoxic activity by interacting with tubulin. But the major limitations of cytotoxic agents are attributed to their toxicity and tumor regression effects. As recent researches have revealed a number of molecular targets associated with specific cancers, anticancer drug discovery is currently focused on high-throughput screening of phytocompounds against a series of such targets. The major advantage of using molecularly targeted agents is that their mechanism of action may not exhibit significant or direct toxicity. Their actions are mainly targeted toward transmembrane, nonnuclear intracellular or extracellular processes and are illustrated by matrix metalloproteinase (MMP) inhibitors, farnesyl transferase inhibitors, receptor tyrosine kinase inhibitors, and angiogenesis inhibitors. There are various modes of action by which phyto-anticancer compounds exert their action on tumor cells such as facilitation of oxidative stress, inflammation, apoptosis, autophagy, and inhibition of tumor cell survival, cell division and differentiation, metastasis, and angiogenesis (Fig. 8.1). So far, the major molecular mechanisms such as facilitation of apoptosis, inhibition of metastasis, inhibition of angiogenesis, and using antibodies against tumor-specific antigens have been investigated for the development of molecular target-based anticancer agents (Narang and Desai 2009; Singh et al. 2016).

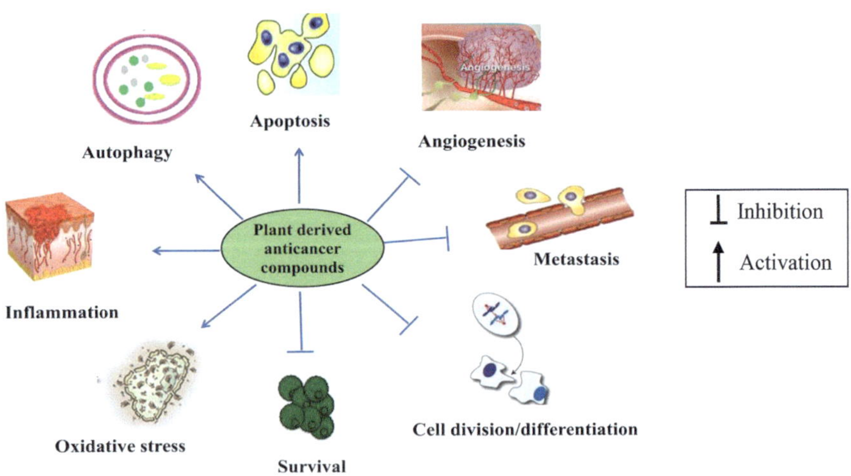

Fig. 8.1 Mechanism of action of plant-derived anticancer compounds on tumor cells. ⊥ = Inhibition; ↑ = Activation

8.5.1 Facilitation of Apoptosis

Apoptosis is a physiological intracellular process mediated by a network of organized signaling pathways ultimately leading to cell death and removal of dead cells by means of phagocytosis. DNA damage caused by cytotoxic drugs usually triggers apoptosis in the cell via two signaling mechanisms – Bcl-2 family proteins mediated activation and release of mitochondrial pro-apoptotic proteins known as caspases or upregulation of pro-apoptotic receptors on cancer cells, which further activates apoptotic signaling pathway by consequent interaction with their ligands. These pro-apoptotic receptors comprise the tumor necrosis factor-related apoptosis-inducing ligand (TRAIL) receptors Fas (CD95 or APO-1). Moreover, anticancer drugs may also activate lipid-based signaling pathways resulting in reduced apoptosis threshold or alter other cytoprotective pathways involving heat shock proteins, nuclear factor-kB, and cell cycle regulatory pathways. Muhtasib et al. (2015) have briefly reviewed the apoptosis-mediated cell death mechanisms of some important plant-derived anticancer drugs such as resveratrol, genistein, berberine, triptolide, and betulinic acid. Some brassinosteroids such as homocatasterone and epibrassinolide were also found to exert anticancer effects through apoptosis. Also, apigenin was observed to induce apoptosis in lung adenocarcinoma cells by targeting leptin receptor pathway (Singh et al. 2016). Harmine, a -carboline alkaloid isolated from the seeds of *Peganum harmala* was found to exhibit *in vitro* and *in vivo* anticancer activity against thyroid cancer by means of inducing apoptosis of TPC-1 (thyroid cancer cell line) cells via regulating the ratio of Bcl-2/Bax in a dose-dependent manner (Ruan et al. 2017). Similarly, curcumin was found to induce an endoplasmic reticulum stress-mediated apoptosis in prostate cancer (Rivera et al. 2017).

8.5.2 Inhibition of Metastasis

Metastasis is the process where the tumor cells spread from region of the body to another which is facilitated by cellular movement and remodeling of the extracellular matrix (ECM), which gives shape to the tissue. Several extracellular enzymes such as plasminogen activators, matrix metalloproteinases, and proteases have been involved in the remodeling of extracellular matrix resulting in metastasis of cancer. Thus, compounds targeting these proteases and matrix metalloproteinases have been now developed as potential anticancer agents. Nimbolide from *Azadirachta indica* functions as a potent anticancer agent by inhibiting proliferation and metastasis of tumor cells in pancreatic cancer (Subramani et al. 2016). Chanvorachote et al. (2016) have reviewed a number of potential anti-metastatic plant compounds against lung cancer including imperatorin, artonin E, oroxylin A, gigantol, moscatilin, baicalein, etc.

8.5.3 Inhibition of Angiogenesis

Recent strategies for cancer therapy, other than just tumor cytotoxicity, have been under intensive development, such as anti-angiogenic therapeutic approach. The process of formation of new blood vessels called angiogenesis is an indispensable prerequisite for the growth and metastasis of tumor cells. Thus, inhibition of angiogenesis is considered as the ideal way to block oxygen and nutrient supply to the tumors, causing tumor deterioration. Tumor angiogenesis involves three major events such as switch, proliferation, followed by endothelial cell migration to initiate the formation of new blood vessels as well as remodeling of extracellular matrix. There are a number of signal transducers that take part in angiogenesis including the vascular endothelial growth factor (VEGF), the integrins, plasminogen activation system, and the matrix metalloproteinases. Hence, nowadays, drug targets have been developed to prohibit one or many aspects of these pathways, e.g., VEGF antibodies and VEGF receptor antagonists. Zhang et al. (2014) have elaborated the various anticarcinogenic actions of ellagic acid including angiogenesis via VEGFR-2 signaling pathway in breast cancer and Akt, Shh, and Notch pathways in pancreatic cancer. Epigallocatechin-3-gallate (EGCG) from green tea was found to exhibit anti-angiogenesis effect by downregulating the expression of VEGF in various cancer types (Min and Kwon 2014). On contrary, ferulic acid has been reported to inhibit fibroblast growth factor receptor 1(FGFR1) in a non-VEGF angiogenic pathway (Yang et al. 2015). Platycodin D from *Platycodon grandiflorum*, a Chinese medicinal plant, has been shown to exhibit angio-suppressive activity by regulating VEGF expression (Khan et al. 2016).

8.5.4 Targeting of Tumor-Specific Antigens by Antibodies

Induction of immune responses by employing antigens that are specific to each type of tumor is a much appreciated objective in cancer therapy, due to its exception from dose-limiting toxicity. On the other hand, usage of antibodies against antigens of specific tumor can also be done to target tumors by involving radio isotopes, prodrug-converting enzymes, and toxins. Tumor regression through antibodies involves complement fixation or antibody-dependent cellular toxicity (ADCC) mediated via monocytes, natural killer cells, and granulocytes. In addition, some novel strategies including the expression of target antigens on the antigen-presenting cells (APCs) or dendritic cells have also been exploited to activate T-cell immune response. In recent times, anticancer immunotherapy involving monoclonal antibodies targeting a number of antigens such as PD-1, VEGF, EGFR, PAP, etc. has gained much clinical significance. Currently, the monoclonal antibodies like ipilimumab (metastatic melanoma), nivolumab (NSCLC), pembrolizumab (metastatic breast cancer), trastuzumab (breast cancer), etc. are in phase III clinical trials (Wurz et al. 2016).

8.6 Conclusions and Future Prospects

Natural anticancer drug discovery is a dominant strategy to identify not only new classes of anticancer agents but also their novel mechanisms of action. Having known the bottomless diversity of nature, it is worth mentioning that chemical lead molecules can be created which are capable of interacting with the majority or nearly all therapeutic targets. Moreover, perhaps only less than 20% of the plants identified on earth have been taken up for investigating their therapeutic potential, thus indicating that relatively a smaller percentage is still available for screening anticancer activity. Recently, with the arrival of high-throughput screening, a huge number of effective experimental agents can be readily screened and examined, and choice can be made for discovering prototype ligands valuable of future development as therapeutic substances. While drug-associated toxicity continues to be a considerable obstacle for presently available chemopreventive and chemotherapeutic drugs, the usage of natural compounds in combination with current chemotherapeutic agents as adjuvant therapy may aid to alleviate drug-associated toxicities. Hence, it is apparent that efficient drug discovery and development will entail multidisciplinary collaborative approaches including the discovery of natural product lead, integrated with combinatorial and medicinal chemistry.

However, one of the major hurdles faced in the isolation of anticancer compounds from plants is the limited accessibility of the starting material. Even though many plant-derived compounds have previously been isolated and characterized, often the available compound quantities are inadequate for evaluating a wide range of anticancer activities. Hence, supply constraints remain a major hindrance for the successful clinical research and commercialization of plant anticancer compounds. Also, correct identification of anticancer plants is yet another challenge in anticancer drug discovery. Hence, the usage of combination of methods like genetic and biochemical analysis along with morphological and anatomical characterization is essential for an unambiguous identification of anticancer plants. *In vitro* production of potent anticancer compounds through various biotechnological approaches such as *in vitro* propagation, plant cell cultures, hairy root cultures, elicitation, metabolic engineering, and heterologous expression would be a promising strategy to meet the pharmaceutical demand of anticancer plants. The conduct of meticulous clinical trials desirable for approval of anticancer drugs signifies another major intricacy. High-throughput screening of phytochemicals for identifying anticancer activity in turn requires high robustness, reliability, reproducibility, and accuracy, as upholding the stability of plant extracts might be difficult, owing to their complexity. Therefore, a highly sophisticated experimental design is indispensable for the anticancer drug discovery. Moreover, computational approaches are powerful knowledge-based method that acts as an efficient filter tool in the search of anticancer activities from plants. Hence, the above-described challenges have reduced interest in plant-based anticancer drug discovery nowadays, which can be rejuvenated through the use of synergistic approaches including in silico methods, combinatorial synthesis, ethnopharmacological knowledge, plant biotechnology, and a wide range of cell-based, *in vitro* and *in vivo* models for testing bioactivity.

References

Aggarwal BB, Sundaram C, Malani N, Ichikawa H (2007) Curcumin: the Indian solid gold. Adv Exp Med Biol 595:1–75

Akhtar MS, Birhanu G, Demisse S (2014) Antimicrobial activity of *Piper nigrum* L. and *Cassia didymobotyra* L. leaf extract on selected food borne pathogens. Asian Pac J Trop Dis 4:S911–S919

Anonymous (2014) A study to assess the effectiveness of the combination of carboplatin, paclitaxel, bevacizumab and combretastatin (CA4P) in patients with chemotherapy naive lung cancer. BioPortfolio 3:31–32

Ansari J, Inamdar N (2010) The promise of traditional medicines. Int J Pharmacol 6:808–812

Arumugam G, Swamy MK, Sinniah UR (2016) *Plectranthus amboinicus* (Lour.) Spreng: botanical, phytochemical, pharmacological and nutritional significance. Molecules 21:369. https://doi.org/10.3390/molecules21040369

Athar M, Back JH, Tang X, Kim KH, Kopelovich L, Bickers DR, Kim AL (2007) Resveratrol: a review of preclinical studies for human cancer prevention. Toxicol Appl Pharmacol 224:274–283

Bachner M, De Santis M (2008) Vinflunine in the treatment of bladder cancer. Ther Clin Risk Manag 4:1243–2125

Bailey HH, Attia S, Love RR, Fass T, Chappell R, Tutsch K, Harris L, Jumonville A, Hansen R, Shapiro GR, Stewart JA (2008) Phase II trial of daily oral perillyl alcohol (NSC 641066) in treatment-refractory metastatic breast cancer. Cancer Chemother Pharmacol 62:149–157

Balachandran P, Govindarajan R (2005) Cancer-An ayurvedic perspective. Pharmacol Res 51:19–30

Banerjee S, Wang Z, Mohammad M, Sarkar FH, Mohammad RM (2008) Efficacy of selected natural products as therapeutic agents against cancer. J Nat Prod 71:492–496

Cai YJ, Lu JJ, Zhu H, Xie H, Huang M, Lin LP, Zhang XW, Ding J (2008) Salvicine triggers DNA double-strand breaks and apoptosis by GSH-depletion driven H_2O_2 generation and topoisomerase II inhibition. Free Radic Biol Med 45:627–635

Chanvorachote P, Chamni S, Ninsontia C, Phiboonchaiyanan PP (2016) Potential anti-metastasis natural compounds for lung cancer. Anticancer Res 36:5707–5718

Choi H, Cho SY, Pak HJ, Kim Y, Choi J, Lee YJ, Gong BH, Kang YS, Han T, Choi G, Cho Y, Lee S, Ryoo D, Park H (2017) NPCARE: database of natural products and fractional extracts for cancer regulation. J Cheminform 9:2. https://doi.org/10.1186/s13321-016-0188-5

Cichewicz RH, Kouzi SA (2004) Chemistry, biological activity, and chemotherapeutic potential of betulinic acid for the prevention and treatment of cancer and HIV infection. Med Res Rev 24:90–114

Cragg GM, Pezzuto JM (2016) Natural products as a vital source for the discovery of cancer chemotherapeutic and chemopreventive agents. Med Princ Pract 25:41–59

Da Rocha AB, Lopes RM, Schwartsmann G (2001) Natural products in anticancer therapy. Curr Opin Pharmacol 1:364–369

Delmonte A, Sessa C (2009) AVE8062: a new combretastatin derivative vascular disrupting agent. Expert Opin Investig Drugs 18:1541–1548

Dhillon N, Aggarwal BB, Newman RA, Wolff RA, Kunnumakkara AB, Abbruzzese JL, Ng CS, Badmaev V, Kurzrock R (2008) Phase II trial of curcumin in patients with advanced pancreatic cancer. Clin Cancer Res 14:4491–4499

Dias DA, Urban S, Roessner U (2012) A historical overview of natural products in drug discovery. Metabolites 2:303–336

Eisenbrand G, Hippe F, Jakobs S, Muehlbeyer S (2004) Molecular mechanisms of indirubin and its derivatives: novel anticancer molecules with their origin in traditional Chinese phytomedicine. J Cancer Res Clin Oncol 130:627–635

Fallen RS, Gooderham M (2012) Ingenol mebutate: an introduction. Skin Ther Lett 17:1–3

Fridlender M, Kapulnik Y, Koltai H (2015) Plant derived substances with anticancer activity: from folklore to practice. Front Plant Sci 6:799. https://doi.org/10.3389/fpls.2015.00799

Fulda S (2009) Betulinic acid: a natural product with anticancer activity. Mol Nutr Food Res 53:140–146

Gezahegn Z, Akhtar MS, Woyessa D, Tariku Y (2015) Antibacterial potential of *Thevetia peruviana* leaf extracts against food associated pathogens. J Coast Life Med 3:150–157

Hert J, Irwin JJ, Laggner C, Keiser MJ, Shoichet BK (2009) Quantifying biogenic bias in screening libraries. Nat Chem Biol 5:479–483

Iveta V, Michaela M, Petr H, Rene K, Eva F, Marie S (2013) The anticancer drug ellipticine induces cytochromes P450 1A1, 1A2 and 3A, cytochrome b5 and NADPH: cytochrome P450 oxidoreductase in rat liver, kidney and lung. Int J Electrochem Sci 8:1586–1597

Jaroch K, Karolak M, Górski P, Jaroch A, Krajewski A, Ilnicka A, Sloderbach A, Stefanski T, Sobiak S (2016) Combretastatins: *in vitro* structure-activity relationship, mode of action and current clinical status. Pharmacol Rep 68:1266–1275

Johnson JJ, Mukhtar H (2007) Curcumin for chemoprevention of colon cancer. Cancer Lett 255:170–181

Jordan MA, Wilson L (2004) Microtubules as a target for anticancer drugs. Nat Rev Cancer 4:253–265

Kapoor S (2012) Emerging role of berbamine as an anti-cancer agent in systemic malignancies besides chronic myeloid leukemia. J Zhejiang Univ Sci B 13:761–762

Khaled M, Belaaloui G, Jiang ZZ, Zhu X, Zhang LY (2017) Deoxypodophyllotoxin, a semi-synthetic compound from *Dysosma versipellis*, induces selective cell death in human breast cancer cell lines. Med Chem Res 26:1241–1258

Khan M, Maryam A, Zhang H, Mehmood T, Ma T (2016) Killing cancer with platycodin D through multiple mechanisms. J Cell Mol Med 20:389–402

Khazir J, Mir BA, Pilcher L, Riley DL (2014) Role of plants in anticancer drug discovery. Phytochem Lett 7:173–181

Khazir J, Singh PP, Doma MR, Syed S, Hyder I, Gousia C, Ajay M, Alam MS, Saxena AK, Arvinda S, Gupta BD, Kumar HMS (2013) Synthesis and anticancer activity of novel spiro-isoxazoline and isoxazolidine derivatives of α-santonin. Eur J Med Chem 63:279–289

Kim S, Hwang BY, Su BN, Chai H, Mi Q, Kinghorn AD, Wild R, Swanson SM (2007) Silvestrol, a potential anticancer rocaglate derivative from *Aglaia foveolata*, induces apoptosis in LNCaP cells through the mitochondrial/apoptosome pathway without activation of executioner caspase-3 or −7. Anticancer Res 27:2175–2183

Kutkowska J, Strzadala L, Rapak A (2017) Synergistic activity of sorafenib and betulinic acid against clonogenic activity of non-small cell lung cancer cells. Cancer Sci 108:2265–2272

Lakhanpal S, Donehower RC, Rowinsky EK (2001) Phase II study of 4-ipomeanol, a naturally occurring alkylating furan, in patients with advanced hepatocellular carcinoma. Investig New Drugs 19:69–76

Lansky EP, Newman RA (2007) *Punica granatum* (pomegranate) and its potential for prevention and treatment of inflammation and cancer. J Ethnopharmacol 19:177–206

Lee KH, Xiao Z (2012) Podophyllotoxins and analogs. In: Cragg GM, Kingston DGI, Newman DJ (eds) Anticancer agents from natural products, 2nd edn. CRC Taylor and Francis, Boca Raton, pp 95–122

Liu GY, Bu X, Yan H, Jia WWG (2007) 20S-Protopanaxadiol-induced programmed cell death in glioma cells through caspase-dependent and independent pathways. J Nat Prod 70:259–265

Liu JM, Chen LT, Chao Y, Li AF, Wu CW, Liu TS, Shiah HS, Chang JY, Chen JD, Wu HW, Lin WC, Lan C, Whang-Peng J (2002) Phase II and pharmacokinetic study of GL331 in previously treated Chinese gastric cancer patients. Cancer Chemother Pharmacol 49:425–428

Mamtani R, Vaughn DJ (2011) Vinflunine in the treatment of advanced bladder cancer. Exp Rev Anticancer Ther 11:13–20

Marie S, Jitka P, Eva M, Jitka U, Vilim S, Zdenek D, Eva F (2012) Ellipticine oxidation and DNA adduct formation in human hepatocytes is catalyzed by human cytochromes P450 and enhanced by cytochrome b5. Toxicology 302:233–241

May M (2014) Statistics: attacking an epidemic. Nature 509:S50–S51

Mi Q, Lantvit D, Reyes-Lim E, Chai H, Pezzuto JM, Kinghorn AD, Swanson SM (2003) Pervilleine F, a new tropane alkaloid aromatic ester that reverses multidrug resistance. Anticancer Res 23:3607–3615

Millimouno FM, Dong J, Yang L, Li J, Li X (2014) Targeting apoptosis pathways in cancer and perspectives with natural compounds from mother nature. Cancer Prev Res 7:1081–1107

Min KJ, Kwon TK (2014) Anticancer effects and molecular mechanisms of epigallocatechin-3-gallate. Integr Med Res 3:16–24

Mohanty SK, Swamy MK, Sinniah UR, Anuradha M (2017) *Leptadenia reticulata* (Retz.) Wight & Arn. (Jivanti): botanical, agronomical, phytochemical, pharmacological, and biotechnological aspects. Molecules 22:1019. https://doi.org/10.3390/molecules22061019

Muhtasib GH, Hmadi R, Kareh M, Tohme R, Darwiche N (2015) Cell death mechanisms of plant-derived anticancer drugs: beyond apoptosis. Apoptosis 20:1531–1562

Mullauer FB, Kessler JH, Medema JP (2010) Betulinic acid, a natural compound with potent anticancer effects. Anti-Cancer Drugs 21:215–227

Narang AS, Desai DS (2009) Anticancer drug development. In: Lu Y, Mahato RI (eds) Pharmaceutical perspectives of cancer therapeutics. Springer, New York, pp 65–66

Newman DJ, Cragg GM (2012) Natural products as sources of new drugs over the 30 years from 1981 to 2010. J Nat Prod 75:311–335

Newman DJ, Cragg GM, Snader KM (2003) Natural products as sources of new drugs over the period 1981–2002. J Nat Prod 66:1022–1037

Newman DJ, Cragg GM (2007) Natural products as sources of new drugs over the last 25 years. J Nat Prod 70:461–477

NIH (2013) Evaluation of 20% betulinic acid ointment for treatment of *Dysplastic nevi* (moderate to severe dysplasia). https://clinicaltrials.gov/ct2/show/NCT00346502. Assessed 5 Oct 2017

Nobili S, Lippi D, Witortd E, Donninic M, Bausi L, Minia E (2009) Sergio capacciolic natural compounds for cancer treatment and prevention. Pharmacol Res 59:365–378

Oberlies NH, Kroll DJ (2004) Camptothecin and taxol: historic achievements in natural products research. J Nat Prod 67:129–135

Pan L, Chai H, Kinghorn AD (2010) The continuing search for antitumor agents from higher plants. Phytochem Lett 3:1–8

Pan S, Zhou S, Gao S, Yu Z, Zhang S, Tang M, Sun J, Ma D, Han Y, Fong W, Ko K (2013) New perspectives on how to discover drugs from herbal medicines: CAM's outstanding contribution to modern therapeutics. Evidence-Based Compl Altern Med 2013:627375. https://doi.org/10.1155/2013/627375

Panda AK, Chakraborty D, Sarkar I, Khan T, Sa G (2017) New insights into therapeutic activity and anticancer properties of curcumin. J Exp Pharmacol 9:31–45

Patterson DM, Zweifel M, Middleton MR, Price PM, Folkes LK, Stratford MR, Ross P, Halford S, Peters J, Balkissoon J, Chaplin DJ, Padhani AR, Rustin GJ (2012) Phase I clinical and pharmacokinetic evaluation of the vascular disrupting agent OXi4503 in patients with advanced solid. Clin Cancer Res 18:415–1425

Quesada AR, Munoz-Chapuli R, Medina MA (2006) Anti-angiogenic drugs: from bench to clinical trials. Med Res Rev 26:483–530

Rahier NJ, Thomas CJ, Hecht SM (2005) Camptothecin and its analogs. In: Cragg GM, DGI K, Newman DJ (eds) Anticancer agents from natural products. Brunner-Routledge Psychology Press, Taylor and Francis Group, Boca Raton, pp 5–21

Rivera M, Ramos Y, Rodroaguez-Valentõan M, Loapez-Acevedo S, Cubano LA, Zou J, Zhang Q, Wang G, Boukli NM (2017) Targeting multiple pro-apoptotic signaling pathways with curcumin in prostate cancer cells. PLoS One 12:e0179587

Ruan S, Jia F, Li J (2017) Potential antitumor effect of harmine in the treatment of thyroid cancer. Evidence-Based Compl Altern Med 2017:9402615. https://doi.org/10.1155/2017/9402615

Saklani A, Kutty SK (2008) Plant-derived compounds in clinical trials. Drug Discov Today 13:161–171

Sargent JM, Elgie AW, Williamson CJ, Hill BT (2003) Ex-vivo effects of the dual topoisomerase inhibitor tafluposide (F11782) on cells isolated from fresh tumor samples taken from patients with cancer. Anti-Cancer Drugs 14:467–473

Schmidt M, Betti G, Hensel A (2007) Saffron in phytotherapy: pharmacology and clinical use. Wien Med Wochenschr 157:315–409

Shapiro GI (2004) Preclinical and clinical development of the cyclin-dependent kinase inhibitor flavopiridol. Clin Cancer Res 10:4270–4275

Shukla S, Meeran SM, Katiyar SK (2014) Epigenetic regulation by selected dietary phytochemicals in cancer chemoprevention. Cancer Lett 355:9–17

Siegel RL, Miller KD, Jemal A (2015) Cancer statistics, 2015. CA Cancer J Clin 65:5–29

Silasi DA, Alvero AB, Rutherford TJ, Brown D, Mor G (2009) Phenoxodiol: pharmacology and clinical experience in cancer monotherapy and in combination with chemotherapeutic drugs. Expert Opin Pharmacother 10:1059–1067

Singh S (2007) From exotic spice to modern drug. Cell 130:765–776

Singh S, Sharma B, Kanwar SS, Kumar A (2016) Lead phytochemicals for anticancer drug development. Front Plant Sci 7:1667. https://doi.org/10.3389/fpls.2016.01667

Sivakumar G (2013) Colchicine semisynthetics: chemotherapeutics for cancer. Curr Med Chem 20:892–898

Slovackova J, Smarda J, Smardova J (2012) Roscovitine-induced apoptosis of H1299 cells depends on functional status of p53. Neoplasma 59:606–612

Subramani R, Gonzalez E, Arumugam A, Nandy S, Gonzalez V, Medel J, Camacho F, Ortega A, Bonkoungou S, Narayan M, Ak D, Lakshmanaswamy R (2016) Nimbolide inhibits pancreatic cancer growth and metastasis through ROS-mediated apoptosis and inhibition of epithelial-to-mesenchymal transition. Sci Rep 6:19819

Swamy MK, Akhtar MS, Sinniah UR (2016) Antimicrobial properties of plant essential oils against human pathogens and their mode of action: an updated review. Evidence-Based Compl Altern Med 2016:21. https://doi.org/10.1155/2016/3012462

Swamy MK, Arumugam G, Kaur R, Ghasemzadeh A, Yusoff MM, Sinniah UR (2017) GC-MS based metabolite profiling, antioxidant and antimicrobial properties of different solvent extracts of Malaysian *Plectranthus amboinicus* leaves. Evidence-Based Compl Altern Med 2017:1517683. https://doi.org/10.1155/2017/1517683

Swamy MK, Pokharen N, Dahal S, Anuradha M (2011) Phytochemical and antimicrobial studies of leaf extract of *Euphorbia neriifolia*. J Med Plant Res 5:5785–5788

Swamy MK, Sinniah UR (2016) Patchouli (*Pogostemon cablin* Benth.): botany, agrotechnology and biotechnological aspects. Ind Crop Prod 87:161–176

Verma AK, Singh RR (2010) Induced dwarf mutant in *Catharanthus roseus* with enhanced antibacterial activity. Indian J Pharm Sci 72:655–657

Westwell AD (2003) Novel antitumour molecules. Drug Discov Today 8:47–50

Willmann M, Wacheck V, Buckley J, Nagy K, Thalhammer J, Paschke R, Triche T, Jansen B, Selzer E (2009) Characterization of NVX-207, a novel betulinic acid-derived anticancer compound. Eur J Clin Investig 39:384–394

Wurz GT, Kao CJ, DeGregorio MW (2016) Novel cancer antigens for personalized immunotherapies: latest evidence and clinical potential. Ther Adv Med Oncol 8:4–31

Xie S, Zhou J (2017) Harnessing plant biodiversity for the discovery of novel anticancer drugs targeting microtubules. Front Plant Sci 8:720. https://doi.org/10.3389/fpls.2017.00720

Yang GW, Jiang JS, Lu WQ (2015) Ferulic acid exerts anti-angiogenic and anti-tumor activity by targeting fibroblast growth factor receptor 1-mediated angiogenesis. Int J Mol Sci 16:24011–24031

Yeruva L, Pierre KJ, Elegbede A, Wang RC, Carper SW (2007) Perillyl alcohol and perillic acid induced cell cycle arrest and apoptosis in non-small cell lung cancer cells. Cancer Lett 257:216–226

Zhan YP, Hyang XE, Cao J, Lu YY, Wu XY, Liu J, Xu X, Xiang J, Ye LH (2012) Clinical safety and efficacy of kanglaite (coix seed oil) injection combined with chemotherapy in treating patients with gastric cancer. Asian Pac J Cancer Prev 13:5319–5321
Zhang HM, Zhao L, Li H, Xu H, Chen WW, Tao L (2014) Research progress on the anticarcinogenic actions and mechanisms of ellagic acid. Cancer Biol Med 11:92–100
Zhu HL (2011) Resveratrol and its analogues: promising antitumor agents. Anti Cancer Agents Med Chem 11:479–490
Zuo M, Li Y, Wang H, Zhou J, Li H, Liu H, Liu H, Xin H, Zhang S, Chen X (2008) The antitumor activity of meisoindigo against human colorectal cancer HT-29 cells *in vitro* and *in vivo*. J Chemother 20:728–733

Chapter 9
Organosulfur Compounds of Garlic as Potent Chemotherapeutic Agents Against Cancer

Irfan Ahmad Ansari, Imran Khan, Mohd Salman Khan, and Mohd Sayeed Akhtar

Abstract Experimental and epidemiological studies over the past few decades have provided ample evidences with regard to the association amid plant food consumption and decreased cancer risk. Many phytochemicals have proved their anticancer potential to be used as therapeutics against cancer. Among them garlic (*Allium sativum*) has been of much interest, mostly due to the epidemiological reports, which linked the increased garlic consumption with reduced prevalence of many human diseases. Garlic and their constituent organosulfur compounds (OSCs) have been attributed to several medicinal properties like hypocholesterolemic, fibrinolytic, immunostimulatory, antimicrobial, antiviral, antifungal, and anticancer effects. The OSCs have been shown to have strong anticarcinogenic property against a variety of chemical carcinogens in different preclinical animal model studies. Extensive researches are still being carried out to elucidate the molecular mechanism of action of OSCs. The main focus of the present chapter is to give an overview of the past and present studies undergoing on organosulfur compounds for exploring their potential as an adjunct in cancer chemotherapeutics and research.

Keywords Allicin · Carcinogenesis · Chemoprevention · Organosulfur compounds · Phytochemicals

I. A. Ansari (✉) · I. Khan · M. S. Khan (✉)
Department of Biosciences, Integral University, Lucknow, Uttar Pradesh, India
e-mail: iaansari@iul.ac.in; skhan@iul.ac.in

M. S. Akhtar (✉)
Department of Botany, Gandhi Faiz-e-Aam College, Shahjahanpur, Uttar Pradesh, India

M. S. Akhtar, M. K. Swamy (eds.), *Anticancer Plants: Clinical Trials and Nanotechnology*, https://doi.org/10.1007/978-981-10-8216-0_9

9.1 Introduction

The experimental and epidemiological studies over the past few decades have provided ample evidences for the association amid plant food consumptions and decreased cancer risk. Many phytochemicals have proven their anticancer activities and now been used as an effective therapeutics against the cancer. Among them garlic (*Allium sativum* L.) has received much attention because of the epidemiological reports, which may linked the increased garlic consumption with reduced prevalence of several human diseases. The protective and beneficial role of garlic is known from ancient times since 1400 BC (Rivlin 2001). Garlic and their organosulfur compounds (OSCs) have been attributed to have several medicinally important properties like hypocholesterolemic (Yeh and Liu 2001; Mikaili et al. 2013), hypoglycemic (Sheela et al. 1995; Augusti and Sheela 1996), immunostimulatory (Lau et al. 1991; Lamm and Riggs 2000), antimicrobial (Lu et al. 2011), antiviral (Shojai et al. 2016), antifungal (Cellini et al. 1996; Avato et al. 2000; Goncagul and Ayaz 2010), and anticancer effects (Thomson and Ali 2003). Early population-based studies have provided strong proof in support of the anticancer property of *Allium* vegetables (You et al. 1989; Steinmetz et al. 1994; Hsing et al. 2002). The OSCs have been shown to have strong anticarcinogenic property against a variety of chemical carcinogens in different preclinical animal model studies (Reddy et al. 1993; Suzui et al. 1997).

The epidemiological investigations from past researchers have shown an inverse correlation between garlic intake and cancer incidences (Kim and Kwon 2009; Salem et al. 2011). For example, a lower incidence of esophageal and stomach cancers was found to be inversely correlated with garlic consumption (Munoz et al. 2001; Pelucchi et al. 2009). Similarly, the reduced risk of colorectal and prostate cancer has been well documented and linked with increased garlic intake (Hsing et al. 2002; Tanaka et al. 2004, 2006). Similarly, Satia et al. (2009) have also reported an inverse correlation between regular garlic intake and incidence of breast, endometrium, and lung cancer. Despite of extensive researches in the past two decades, the scientists are still working to elucidate the molecular mechanism of anticancer property of OSCs. Thus, the main focus of the present chapter is to give an overview of the past and present studies undergoing on organosulfur compounds for exploring their potential chemotherapeutic agents against cancer.

9.2 Chemical Constituents of Garlic

Chemical analyses have indicated that garlic bulbs, leaf, shoot, flower, and roots are the source of many compounds having medicinal properties with varied pharmacological functions (Ross 1999). Fresh garlic has been shown to have 65% water, 28% carbohydrates, 2% proteins (mainly alliinase), 1.5% fiber, 1.2% essential amino acids, vitamins (β-carotene, ascorbic acid, biotin, thiamin, riboflavin, niacin, nicotinic acid, etc.), minerals (Al, Fe, Mg, Co, Cu, Mn, Ni, P, K, Na, Sn, Zn, Se, Cr. etc.), organosulfur compounds (alliin, allicin, Z-ajoene, diallylsulfide (DAS),

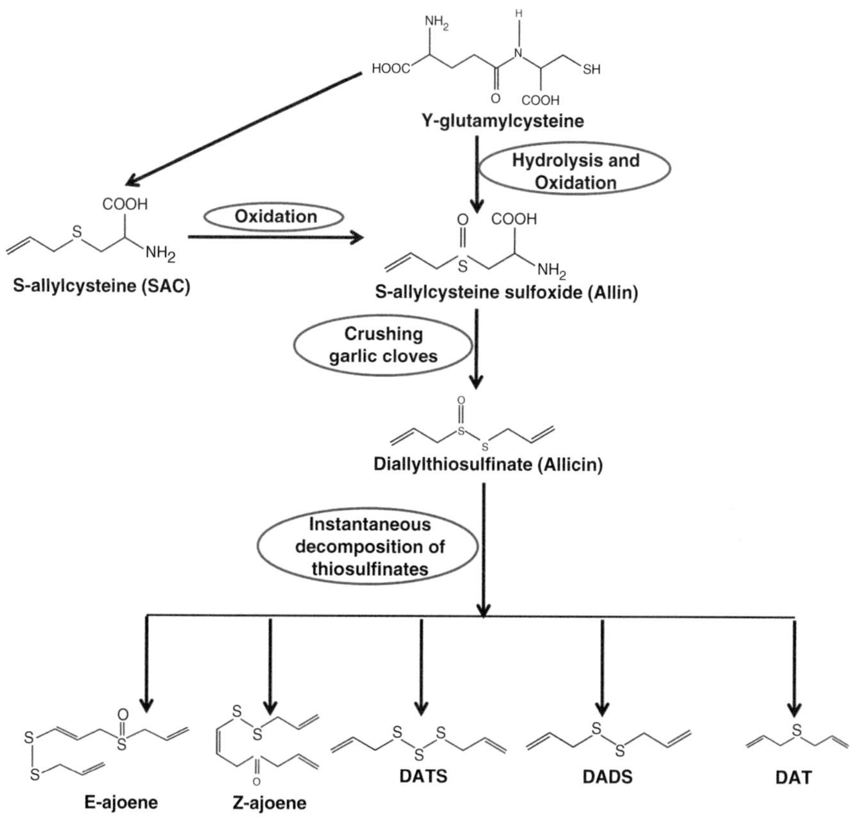

Fig. 9.1 Chemical reactions in processed garlic and generation of some important organosulfur compounds

diallyldisulfide (DADS), diallyltrisulfide (DATS), methyl-allyldisulfide (MADS), methylallylsulfide (MAS), methylallyltrisulfide (MATS), methyl-propyldisulfide (MPDS), dimethyltrisulfide (DMTS), dimethyldisulfide (DMDS), allyl-propyldisulfide (APDS), S-(2-carboxy-propyl)-glutathione, S-allylmercaptocysteine (SAMC), S-allylcysteine (SAC), and many other chemical compounds like p-coumaric acid, caffeic acid, ferulic acid, quercetin, sativosides, scordines, scordinines, degalactotigonin, gitonin, dithiins, dithiane thione, thiazole, etc. (Abdullah et al. 1988; Li 2000; Ejaz et al. 2003).

Block (1985) has nicely explored the chemistry of organosulfur compounds of garlic. γ-glutamyl-S-alk(en)yl-L-cysteine is the main sulfur compound of garlic, and it is hydrolyzed and oxidized into alliin (Fig. 9.1). Alliin is an important precursor of the wide array of organosulfur compounds of garlic, which have now been attributed for their anticancer effect (Reddy et al. 1993; Suzui et al. 1997). An enzyme alliinase is released when the garlic cloves are processed which acts on alliin to give rise to very unstable compounds called thiosulfinates. Allicin is one of the major thiosulfinates generated during the processing of garlic. These thiosulfinates, including allicin, which may be degraded or oxidized to oil-soluble OSCs like DAS, DADS, DATS, dithiins, and ajoene (Munchberg et al. 2007) (Fig. 9.1). These OSCs can undergo further chemical transformations after reacting with free sulfhydryl groups (-SH) of cysteine, glutathione, or proteins producing S-allylcysteine, S-allylmercapto-cysteine, S-(2-carboxy-propyl)-glutathione, etc. (Lanzotti 2006; Corzo-Martinez and Villamiel 2007; Verma et al. 2008).

9.3 Mode of Action of OSCs in Cancer Prevention

OSCs of garlic have been shown to exhibit anticancer activity against many cancer cell lines via various mechanisms like inhibition of mutagen/carcinogen activation via modulation of cytochrome P450 (CYP)-dependent enzymes (phase I detoxification enzyme) and phase II detoxification enzymes, neutralization of free radicals, inhibition of the formation of DNA adducts and effects on cancer cell proliferation and tumor growth like disruption of the cell cycle progression, induction of apoptosis, modification of histone proteins, inhibition of metastasis and angiogenesis, and reversal of drug resistance (Fig. 9.2). The mechanism of action of OSCs has been established by extensive research on several types of cancer cell lines and various experimental animal models (Thomson and Ali 2003; Sengupta et al. 2004a, b; Velmurugan et al. 2005; Kalra et al. 2006; Prasad et al. 2008; Howard et al. 2008).

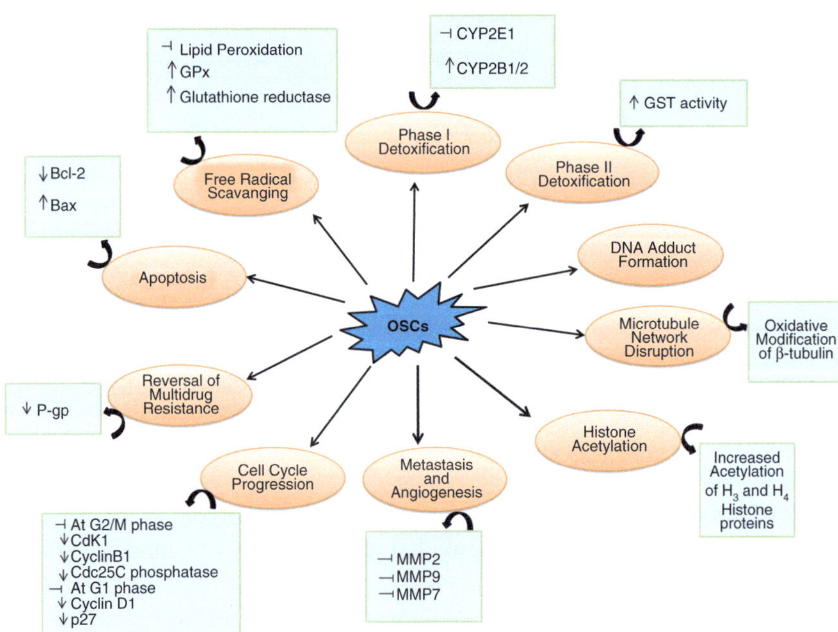

Fig. 9.2 Diagrammatic representation of mechanism of action of organosulfur compounds (OSCs) of garlic. ⊥ indicates inhibition at that step; ↑ indicates upregulation; ↓ indicates downregulation

9.3.1 Alterations in the Gene Expression of Phase I Detoxification Enzymes

The cytochrome P450-dependent monooxygenase has been considered to be responsible for the carcinogenic activation of many environmental pollutants. Garlic-derived OSCs have been extensively studied for their potential to reduce tumor growth in many animal models, primarily through alteration of expression of cytochrome P450 (CYP) enzymes (Ho et al. 2010; Melino et al. 2011; Wanwimolruk and Prachayasittikul 2014). For example, DAS-induced suppression of mutagenesis, caused by vinyl carbamate and N-nitrosodimethyl amine in S *almonella typhimurium* TA100, was found to be correlated with the inhibition of CYP2E1 enzyme-mediated phenol hydroxylation and N-nitrosodimethylamine demethylation (Surh et al. 1995). N-dimethylnitrosamine demethylase activity was also competitively inhibited by DAS (Brady et al. 1988). In another study, intake of garlic powder or oil has been reported to inhibit the hepatic microsomal CYP2E1 activity in mouse and rat (Park et al. 2002a; Zeng et al. 2009). A time- and dose-dependent decrease in the activity of CYP2E1 and increase in CYP2B1 by DAS were observed in hepatic microsomes (Brady et al. 1991; Jin and Baillie 1997). A significant decrease in the CYP2E1 activity by DAS, DADS, and AMS in rat liver was also observed (Reicks and Crankshaw 1996; Davenport and Wargovich 2005; Wargovich 2006). Expression and activities of several monooxygenases and transferases including epoxide hydrolase and CYP2B1/2 in rat liver and intestine were found to be increased by DADS (Haber et al. 1995; Wu et al. 2002; Davenport and Wargovich 2005). An increase in the CYP2B1 expression and decrease in the CYP2E1 expression were also observed by using garlic oil (Sheen et al. 1999). An increase in the activities of ethoxyresorufin O-deethylase, methoxyresorufin O-demethylase, and pentoxyresorufin O-depentylase and decrease in the activities of nitrosodimethylamine N-demethylase and erythromycin N-demethylase were observed in rats treated with dimethyl sulfide (DMS), methylpropyl disulfide (MPDS), dipropyl sulfide (DPS), dipropyl disulfide (DPDS), and DADS. The above observation was accompanied by an increase in the expression of CYP2B1/2 and decrease in the expression of CYP2E1 (Siess et al. 1997). The above observation clearly indicated that inhibition of the carcinogen activation might be one of the mechanisms by which OSCs provided protection against chemically induced cancers.

9.3.2 Induction of Phase II Detoxification Enzymes

The OSCs of garlic have been shown to behave as a double-edged sword in the chemoprevention of chemically induced cancers by suppressing carcinogen activation and augmenting degradation of activated carcinogenic species through the activation of phase II detoxification enzymes like glutathione-S-transferase and quinone reductase, etc. (Sparnins et al. 1988; Hu and Singh 1997; Hu et al. 1997; Ho et al.

2010; Wanwimolruk and Prachayasittikul 2014). The OSCs of garlic have been shown to prevent B(a)P-induced forestomach and lung cancer in mice, and this was found to be correlated with an increase in the hepatic GST activity (Sparnins et al. 1988). In another study feeding of garlic powder increased the mammary and liver GST in rats (Liu et al. 1992). Expression of different classes of GST like alpha (mGSTA3-3, mGSTA1-2, mGSTA4-4), mu (mGSTM1-1), and pi class GST (mGSTP1-1) in the liver, lung, and forestomach was found to be induced by DAS, DADS, and DATS treatment to A/J mice (Hu et al. 1997; Bose et al. 2002; Andorfer et al. 2004). However, chemoprevention of B(a)P-induced forestomach cancer in A/J mice by OSCs was found to be correlated with the induction of mGSTP1-1 (Hu and Singh 1997; Andorfer et al. 2004). In other interesting studies, OSCs of garlic like DAS, DADS, DPS, and DPDS have been shown to induce the expression of NAD(P)H: quinone oxidoreductase enzyme, implicated in the degradation of activated quinone metabolite of B(a)P, in the lung and forestomach of A/J mice (Singh et al. 1998; Fukao et al. 2004). DADS has also been shown to augment the expression of quinone reductase, GST, and uridine diphosphate (UDP)-glucuronosyl transferase in different organs of rats, and induction of these enzymes was attributed to the protective activity of OSCs against gastrointestinal tract (GIT) cancers (Munday and Munday 1999). In general, all the above studies suggested that the activation of phase II detoxification enzymes represents another mechanism to explain OSC-induced chemoprevention of experimental cancers.

9.3.3 Inhibition of the Formation of DNA Adducts

Formation of DNA adducts is supposed to be a preliminary step in carcinogenesis. The anticarcinogenic action of garlic can be attributed to its role in preventing DNA-carcinogen adduct formation and activation of carcinogen (Liu et al. 1992; Lin et al. 2002). Formation of dimethylbenz[a]anthracene (DMBA)-DNA adduct has been found to be decreased by different garlic preparations in rat mammary gland, and these adducts were positively correlated with mammary tumor incidence (Amagase and Milner 1993). N-acetyl-2-aminofluorene (2-AF)-induced DNA adduct formation was found to be inhibited by DAS and DADS in human bladder tumor cells (Chung et al. 2004). Further, benzo(a)pyrene[B(a)P]-DNA adduct formation in human peripheral blood lymphocytes was significantly reduced by a water extract of raw garlic and SAC (Hageman et al. 1997). The *in vitro* nitrosation reactions were inhibited by onion and garlic juices in a dose-dependent manner (Shenoy and Choughuley 1992). The water extract of garlic and deodorized garlic powder have been reported to reduce N-nitrosomorpholine (NMOR) production, a liver carcinogen (Dion et al. 1997). The garlic powder has also been shown to decrease the formation of 7-methyldeoxyguanosine (7-MedG) and O6-methyldeoxyguanosine (O6-MedG) in rat liver (Lin et al. 1994).

9.3.4 *Scavenging of Free Radicals*

Free radicals have been implicated in several age-related diseases, including cancer (Borek 2001; Chung 2006; Iciek et al. 2009). Garlic possesses strong antioxidant activity due to its organosulfur compounds (Shobana and Naidu 2000). The free radical scavenging property of garlic has been attributed to allicin, a major thiosulfinate of crushed garlic (Siegers et al. 1999). Allicin has also been shown to prevent lipid peroxidation in liver homogenates (Prasad et al. 1995). N-methyl-N′-nitro-N-nitrosoguanidine (MNNG)-induced genotoxicity and oxidative stress were also reduced by garlic preparations in Swiss mice (Kumaraguruparan et al. 2005). In an interesting study, glutathione peroxidase (GPx) activity was increased by garlic and onion oils, and ratio of reduced/oxidized GSH was inhibited (Perchellet et al. 1986). DAS treatment also leads to increased GPx activity in mouse forestomach tumor induced by B(a)P. Glutathione reductase activity was also increased by DAS and DADS administration, while garlic oil augmented SOD activity (Gudi and Singh 1991). Similarly, catalase activity was decreased by DAS and crushed garlic preparations in the liver of rat and mice (Sheen et al. 1996). Aged garlic extract (AGE), SAC, and SAMC have been shown to have free radical scavenging property against t-butyl hydroperoxide-induced ROS in microsomal fraction of rat (Imai et al. 1994). DAS, DADS, and AMS have also been shown to neutralize free radicals generated by CCl_4 (Fanelli et al. 1998). AGE, containing both water- and lipid-soluble organosulfur components including S-allylcysteine (SAC) and S-allylmercaptocysteine (SAMC), has been reported to have antioxidant property by augmenting cellular antioxidant enzyme like superoxide dismutase (SOD), catalase (CAT), and glutathione peroxidase (GPx) (Borek 2001). SAC and SAMC have been attributed for the increased biosynthesis of glutathione in human prostate cancer cells (Pinto et al. 1997).

9.3.5 *Disruption of Cell Cycle Progression*

Cell cycle is a complex control system which ensures precise sequence of events like growth, replication, and division of a cell (Molinari 2000; Murray 2004). Cell cycle arrest is induced due to cellular stresses like DNA damage and microtubule network disruption via activation of the cell cycle checkpoints (Molinari 2000; Murray 2004). The G_1/M checkpoint prevents replication of damaged DNA, and G_2/M checkpoint blocks mitosis until the replication is accomplished correctly. Several studies have shown that disruption of cell cycle progression is the primary mechanism of action of many anticancer agents. OSCs of garlic have also been shown to inhibit the growth of many cancer cell lines *in vitro* through inducing cell cycle arrest, mainly in G_2/M phase of the cell cycle (Arunkumar et al. 2006; Omar and Al-Wabel 2010; Wang et al. 2012; Scherer et al. 2009). For the first time, DADS was shown to cause dose- and time-dependent cell cycle arrest in G_2/M phase of human colon cancer cells (Knowles and Milner 1998, 2000). The DADS-induced G_2/M arrest in colon cancer cells was accompanied with decreased kinase activity of Cdk1/cyclinB1 complex, reduced complex formation between Cdk1 and cyclinB1, and decreased Cdc25C phosphatase level (Knowles and Milner 2000). Later on, other OSCs of garlic were also shown to exert similar effects on different *in vitro* and *in vivo* cancer models (Tan et al. 2004; Wu et al. 2004; Xiao et al. 2005; Herman-Antosiewicz and Singh 2005; Arunkumar et al. 2006; Herman-Antosiewicz et al. 2007). For example, DADS was shown to inhibit Cdk1 activity by inducing its phosphorylation in HL-60 cells and also reduced Cdk1 expression in human prostate cancer PC-3 cells in a dose-dependent manner (Tan et al. 2004; Arunkumar et al. 2006). DATS was found to be more potent inducer of cell cycle arrest at G_2/M phase in comparison to DAS or DADS (Xiao et al. 2005). This finding supports the fact that a change in the oligosulfide chain length of OSCs could alter its biological property. Induction of G_2/M phase arrest in PC-3 cells by DATS was shown to be due to increased inhibitory phosphorylation of Cdk1, reduction in Cdk1/cyclinB1 activity, augmented inhibitory phosphorylation of Cdc25C, and reduced expression of Cdc25C (Xiao et al. 2005). Garlic OSCs have also been known to cause cell cycle arrest in phases other than G_2/M. For example, DADS was shown to induce S-phase arrest in human nasopharyngeal carcinoma cells (Zhang et al. 2006). Similarly, allitridi was shown to cause G_1 phase arrest in human gastric cancer cells BGC823 via downregulation of cyclin D1 and upregulation of p27 protein (Lan and Lu 2003). Therefore, the above evidences indicate that disruption of cell cycle progression is the major and common mechanism of action of many structurally different OSCs of garlic.

9.3.6 Disruption of Microtubule Network

OSCs of garlic have been shown to affect the microtubule network in different cancer cells that could lead to mitotic arrest or apoptosis (Xiao et al. 2005; Hosono et al. 2008). For instance, SAMC was shown to cause microtubule depolymerization and cytoskeleton disruption in human colon cancer SW480 cells and NIH3T3 mouse fibroblasts (Xiao et al. 2003). DATS was also shown to disrupt microtubule network and inhibit spindle formation in human colon cancer DLD-1 and HCT-15 cells causing mitotic arrest (Hosono et al. 2005). Interestingly, DATS caused oxidative modification of tubulin-β at Cys12 and Cys354 residues (Hosono et al. 2005). Similarly, Z-ajoene was shown to cause inhibition of tubulin polymerization *in vitro* and disruption of microtubule network leading to G_2/M phase arrest in normal marsupial kidney cells (Li et al. 2002).

9.3.7 Induction of Apoptosis

Apoptosis is an evolutionary conserved process of cell death vital for normal embryonic development and maintenance of tissue homeostasis. A number of pathological conditions including cancer have been attributed to dysregulated programmed cell death, and therefore, apoptosis is considered as a potential target in cancer prevention and therapy (Kaufmann and Gores 2000; Ghobrial et al. 2005). Numerous studies have shown that OSCs of garlic could affect various cellular signaling pathways linked to apoptosis. Majority of these OSCs induce mitochondrial pathway of apoptosis which involves disruption of mitochondrial membrane potential and release of apoptotic factors from the mitochondria to the cytosol (Thornberry and Lazebnick 1998; Hengartner 2000; Xiao et al. 2005; Caro et al. 2012; Wang et al. 2012). Mitochondrial apoptotic pathway is tightly regulated by the Bcl-2 family of anti-apoptotic proteins (Bcl-2, Bcl-xL) and pro-apoptotic proteins (Bax and Bak) (Chao and Korsmeyer 1998). OSCs of garlic have been shown to induce apoptosis by altering the cellular levels of the Bcl-2 family proteins. For example, the ratio of Bax/Bcl-2 was found to be increased in neuroblastoma cells SH-SY5Y and lung cancer cells H460 and H1299 by DAS and DADS treatment (Hong et al. 2000; Karmakar et al. 2007). Likewise, DADS was also shown to induce a time-dependent increase in Bax level and simultaneous reduction in Bcl-xL level in breast cancer cell line MDA-MB-231 (Nakagawa et al. 2001). Z-ajoene was also shown to induce programmed cell death in HL-60 cells in association with caspase-mediated degradation of Bcl-2 protein (Li et al. 2002). DATS was also shown to cause a decrease in Bcl-2 expression with a simultaneous hyperphosphorylation of this protein resulting in an increased Bax/Bcl-2 ratio ultimately leading to induction of mitochondrial pathway of apoptosis (Xiao et al. 2004). The hyperphosphorylation of Bcl-2 in DU145 and PC-3 cells was shown to occur due to activation of JNK and ERK1/2 kinases. Moreover, DATS was shown to be more potent inducer of apoptosis in comparison to DAS and DADS in human prostate cancer PC-3 and DU145 cells (Xiao et al. 2004). Similarly, a significant increase in the Bax and Bak along with apoptosis was observed in prostate cancer LNCaP cells treated with DATS (Kim et al. 2007). In an *in vivo* study, DATS was shown to inhibit PC-3 xenograft tumor growth in nude mice which was correlated with increased apoptosis along with the augmentation of Bax and Bak proteins in tumor tissue (Xiao et al. 2006). In an earlier study, Akt-Bad pathway was attributed, up to some extent, to regulate apoptosis induced by DATS treatment in human prostate cancer PC-3 and DU145 cells (Xiao et al. 2006). A marked reduction in Akt activity was observed in DU145 and PC-3 cells due to DATS treatment, and consequently the phosphorylation of Bad at Ser155 and Ser136 was inhibited, which reduced interaction of Bad with 14–3-3β protein (Xiao et al. 2006).

In several previous studies, reactive oxygen species (ROS) have been shown to play critical role in OSC-induced apoptosis in cancer cells. For example, DADS was shown to induce apoptosis in HL-60 cells which was correlated with augmented intracellular ROS production (Kwon et al. 2002). Moreover, DADS was also shown to induce intracellular ROS generation in neuroblastoma cells SH-SY5Y which was apparent as early as 15 min of treatment and leads to oxidation of cellular lipids and proteins (Filomeni et al. 2003). Interestingly, production of ROS in DADS-treated neuroblastoma cells was found to be due to activation of JNK pathway (Filomeni et al. 2003). Additionally, an increase in the free intracellular Ca^{2++} ion has been attributed to the apoptosis-inducing property of OSCs (Sakamoto et al. 1997; Karmakar et al. 2007; Sundaram and Milner 1996a, b; Park et al. 2002a, b). For example, DADS was shown to augment intracellular calcium ion level with concomitant activation of capase-3 (Park et al. 2002b). Likewise, DAS and DADS were also shown to raise intracellular calcium ion level in SH-SY5Y neuroblastoma cells followed by activation of a calpain, a non-caspase cysteine protease, which triggers caspase-independent pathway of apoptosis (Karmakar et al. 2007).

Interestingly, OSCs have been observed to exert differential effect on normal and cancer cells. Surprisingly, malignant cells were found to be more sensitive to apoptosis induced by OSCs in comparison to normal cells. For instance, DAS and DADS, at the dose of 50 or 100 μM, minimally affected the viability of primary neurons but exerted marked reduction in the viability of neuroblastoma cells at the same concentration (Karmakar et al. 2007). Likewise, DATS did not affect the viability of normal prostate epithelial cells PrEC event at a dose that was cytotoxic to prostate cancer cells (Xiao et al. 2005; Kim et al. 2007). Z-ajoene was also shown to exhibit insignificant effect on peripheral mononuclear blood cells from healthy individuals but induced apoptosis in human leukemia cells (Dirsch et al. 1998). The exact mechanism regarding the differential effect of OSCs on cancer cells and normal cells still remains to be elaborated.

9.3.8 Modification of Histone Acetylation

Modification of histone acetylation has been suggested as one of the mechanisms attributed to the anticancer property of garlic OSCs (Druesne-Pecollo et al. 2004a, 2006; Nian et al. 2009). For the first time, DADS and its metabolite allyl mercaptan were shown to increase acetylation of H_3 and H_4 histone proteins in DS19 mouse erythroleukemia and K562 human leukemia cells (Lea et al. 1999). Similar effect of DADS and allyl mercaptan was shown *in vivo* by the same group in rats (Lea and Randolph 2001). Moreover, increased histone acetylation was also correlated with antiproliferative property of allicin, SAMC, and SAC on human colon cancer cell line Caco-2 and human breast cancer cell line T47D (Lea et al. 2002). DADS was also shown to induce hyperacetylation of histone proteins H3 and H4 in human colon cancer cell line Caco-2 and HT-29 via inhibiting histone deacetylase (Druesne-Pecollo et al. 2004a, b). Moreover, histone deacetylase inhibition was also shown to be responsible for the hyperacetylation of the promoter of p21 gene (Nian et al. 2008, 2009).

9.3.9 Inhibition of Metastasis and Angiogenesis

Metastasis, a primary cause of morbidity in patients with cancer, is a multistep process involving the detachment of cancer cells from the primary tumor site, entry into systemic circulation, and invasion at new sites. The whole metastatic process requires multiple molecular events including angiogenesis (Folkman 2003). Several *in vitro* and *in vivo* studies have shown that garlic and its OSCs can affect metastatic and angiogenic processes (Powolny and Singh 2008; Ng et al. 2012). It was reported for the first time that AGE could suppress migration of rat sarcoma cells *in vitro* (Hu et al. 2002). On the basis of this finding, various OSCs of garlic were then studied for their antimetastatic property. In a recent study, DADS was shown to reduce the migration of prostate cancer cell line LNCaP by inhibiting the expression of MMP2 and MMP9 (Shin et al. 2010). SAC was also shown to inhibit the invasion of breast cancer cell line MDA-MB 231 via augmentation of E-cadherin expression and reduction of MMP2 expression. Recently, DAS, DADS, and DATS were also shown to reduce the expression of MMP2, MMP7, and MMP9 in colon cancer colo 2005 cells (Lai et al. 2011). Furthermore, OSCs have also been reported to suppress angiogenic process which leads to development of new blood vessels from already established vascular network. For example, AGE was shown to suppress proliferation and invasiveness of transformed human and rat endothelial cell line (Matsuura et al. 2006). DAS, DADS, and DATS were shown to inhibit the proliferation of human umbilical vein endothelial cells, and DATS was found to be most potent among them (Xiao et al. 2006). Alliin was also shown to inhibit the angiogenesis, and VEGF and FGF-2 induced capillary tube formation in human umbilical vein endothelial cells (Mousa and Mousa 2005). DAS and DADS were also shown to reduce proliferation and migration of endothelial cells along with the downregulation of MMP2 and MMP9 expressions (Thejass and Kuttan 2007a, b). *In vitro* studies, supporting the antimetastatic potential of OSCs of garlic, have also been confirmed by *in vivo* studies. For example, SAMC has been shown to reduce the pulmonary metastasis by 85% in mice with prostate cancer when administered orally (Howard et al. 2007). Oral administration of DATS also reduced the pulmonary metastasis by around 50% in TRAMP mice (Singh et al. 1995). Thus, it can be concluded from the above evidences that garlic and its OSCs have great potential as inhibitors of metastasis and angiogenesis.

9.3.10 Reversal of the Multidrug Resistance

One of the major hurdles in cancer chemotherapy is the multidrug resistance caused due to overexpression of P-glycoprotein (P-gp), a membrane transporter to efflux xenobiotic compounds from a cell. OSCs have been shown to modulate the P-gp-mediated multidrug resistance (Arora et al. 2004; Demeule et al. 2004). DAS has been shown to enhance the cytotoxic action of vinblastine and other *Vinca* alkaloids in K562 leukemic cells resistant to vinblastine. Interestingly, in this study, DAS treatment lead to reduced P-gp protein expression in K562 cells up to a normal level (Arora et al. 2004).

9.4 Clinical Trials Conducted with Garlic

Clinical evidences from cell culture and animal models have been reported in population-based studies and suggested the protective effects of garlic intake. To date, most of the clinical trials were conducted to observe the chemopreventive property of garlic on various types of cancer. A double-blind randomized clinical trial was conducted on 37 patients to study the effect of high-dose aged garlic extract (AGE) on colorectal adenomas by Tanaka et al. (2006). They found a significant reduction in size and number of adenomas in patients who received high dose of AGE for 1 year. Similarly, a factorial double-blind, placebo-controlled clinical trial, conducted with aged garlic extract and oil and/or vitamin supplement, showed a nonstatistically significant reduction in gastric cancer incidence and mortality (Ma et al. 2012). However, in another clinical trial, a borderline significant negative association between garlic and onion intake and gastric cancer risk was also observed (Millen et al. 2007). Similarly, the clinical trial in women having high consumption of onion and garlic showed a significant reduction in primary invasive epithelial ovarian cancer risk (Schulz et al. 2005), while a clinical study on Chinese men and women showed high intake of garlic significantly reduced the risk of liver cancer (Zhang et al. 2013). An Italian case-control study showed a significant association between high levels of garlic consumption and reduced risk of endometrial cancer (Galeone et al. 2009). Moreover, the hospital-based study, conducted by Wang et al. (2012a, b) showed that high amount of shallot and garlic intake was significantly associated with decreased risk of multiple myeloma, while Salem et al. (2011) found that the high consumption of garlic was linked with reduction of prostate cancer (Salem et al. 2011).

9.5 Conclusions and Future Prospects

During the past few decades, extensive research has been done on garlic and its organosulfur derivatives for its potential to control the growth and proliferation of various cancer cells under *in vitro* and *in vivo* conditions. The outcome of these researches suggested that the organosulfur compounds target many signaling pathways such as apoptosis, cell cycle, metastasis, angiogenesis, and drug resistance. This may be attributed to the anticancer property of garlic and its organosulfur compound derivatives. Moreover, the clinical trials conducted so far have mainly used aged garlic extract or garlic oil or total garlic, rather than the isolated OSCs in animal studies. To facilitate confident and meaningful decisions in the future, this gap of knowledge should be filled by conducting clinical trials with OSCs. Despite the availability of extensive information on OSCs and their mechanism of action on different cancers, more in-depth future investigations on garlic and its organosulfur derivatives are desired to unravel many unsolved issues related to their target genes regulating various cellular signaling pathways involved in cell survival, growth, proliferation, and maintenance of tissue architecture.

References

Abdullah TH, Kandil O, Elkadi A, Carter J (1988) Garlic revisited: therapeutic for the major diseases of our time. J Natl Med Assoc 80:439–445

Amagase H, Milner JA (1993) Impact of various sources of garlic and their constituents on 7,12-dimethylbenz[a]anthracene binding to mammary cell DNA. Carcinogenesis 14:1627–1631

Andorfer JH, Tchaikovskaya T, Listowsky I (2004) Selective expression of glutathione S-transferase genes in the murine gastrointestinal tract in response to dietary organosulfur compounds. Carcinogenesis 25:359–367

Arora A, Seth K, Shukla Y (2004) Reversal of P-glycoprotein mediated multidrug resistance by diallyl sulfide in K562 leukemic cells and in mouse liver. Carcinogenesis 25:941–949

Arunkumar A, Vijayababu MR, Srinivasan N, Aruldhas MM, Arunakaran J (2006) Garlic compound, Diallyl disulfideInduces cell cycle arrest in prostate cancer cell line PC-3. Mol Cell Biochem 288:107–113

Augusti KT, Sheela CG (1996) Antiperoxide effect of S-allyl cysteine sulfoxide, an insulin secretagogue, in diabetic rats. Experientia 52:115–120

Avato P, Tursil E, Vitali C, Miccolis V, Candido V (2000) Allylsulfide constituents of garlic volatile oil as antimicrobial agents. Phytomedicine 7:239–243

Block E (1985) The chemistry of garlic and onions. Sci Am 252:114–119

Borek C (2001) Antioxidant health effects of aged garlic extract. J Nutr 131:S1010–S1015

Bose C, Guo J, Zimniak L, Srivastava SK, Singh SP, Zimniak P, Singh SV (2002) Critical role of allyl groups and disulfide chain in induction of Pi class glutathione transferase in mouse tissues *in vivo* by diallyl disulfide, a naturally occurring chemopreventive agent in garlic. Carcinogenesis 23:1661–1665

Brady JF, Li DC, Ishizaki H, Yang CS (1988) Effect of diallyl sulfide on rat liver microsomal nitrosamine metabolism and other monooxygenase activities. Cancer Res 48:5937–5940

Brady JF, Ishizaki H, Fukuto JM, Lin MC, Fadel A, Gapac JM, Yang CS (1991) Inhibition of cytochrome P-450 2E1 by diallyl sulfide and its metabolites. Chem Res Toxicol 4:642–647

Caro AA, Adlong LW, Crocker SJ, Gardner MW, Luikart EF, Gron LU (2012) Effect of garlic-derived organosulfur compounds on mitochondrial function and integrity in isolated mouse liver mitochondria. Toxicol Lett 214:166–174

Cellini L, Di Campli E, Masulli M, Di Bartolomeo S, Allocati N (1996) Inhibition of *Helicobacter pylori* by garlic extract (*Allium sativum*). FEMS Immunol Med Microbiol 13:273–277

Chao DT, Korsmeyer SJ (1998) BCL-2 family: regulators of cell death. Annu Rev Immunol 16:395–419

Chung LY (2006) The antioxidant properties of garlic compounds:allyl cysteine, alliin, allicin, and allyl disulfide. J Med Food 9:205–213

Chung JG, Yeh CK, Chung KC, Lin SS, Lee JH (2004) Inhibition of N-acetyltransferase activity and gene expression in human colon cancer cell lines by diallyl sulphide. Food Chem Toxicol 42:195–202

Corzo-Martinez MCN, Villamiel M (2007) Biological properties of onions and garlic. Trends Food Sci Technol 18:609–625

Davenport DM, Wargovich MJ (2005) Modulation of cytochrome P450 enzymes by organosulfur compounds from garlic. Food Chem Toxicol 43:1753–1762

Demeule M, Brossard M, Turcotte S, Regina A, Jodoin J, Beliveau R (2004) Diallyl disulfide, a chemopreventive agent in garlic, induces multidrug resistance-associated protein 2 expression. Biochem Biophys Res Commun 324:937–945

Dion ME, Agler M, Milner JA (1997) S-allyl cysteine inhibits nitrosomorphine formation and bioactivation. Nutr Cancer 28:1–6

Dirsch VM, Gerbes AL, Vollmar AM (1998) Ajoene, a compound of garlic, induces apoptosis in human promyeloleukemic cells, accompanied by generation of reactive oxygen species and activation of nuclear factor-κB. Mol Pharmacol 53:402–407

Druesne-Pecollo N, Pagniez A, Mayeur C, Thomas M, Cherbuy C, Duee PH, Martel P, Chaumontet C (2004a) Repetitive treatments of colon HT-29 cells with diallyl disulfide induce a prolonged hyperacetylation of histone H3 K14. Ann N Y Acad Sci 1030:612–621

Druesne-Pecollo N, Pagniez A, Mayeur C, Thomas M, Cherbuy C, Duee PH, Martel P, Chaumontet C (2004b) Diallyl disulfide (DADS) increases histone acetylation and p21waf1/cip1 expression in human colon tumor cell lines. Carcinogenesis 25:1227–1236

Druesne-Pecollo N, Pagniez A, Thomas M, Cherbuy C, Duée PH, Martel P, Chaumontet C (2006) Diallyl disulfide increases CDKN1A promoter-associated histone acetylation in human colon tumor cell lines. J Agric Food Chem 54:7503–7507

Ejaz S, Woong LC, Ejaz A (2003) Extract of garlic (*Allium sativum*) in cancer chemoprevention. Exp Oncol 25:93–97

Fanelli SL, Castro GD, De Toranzo EG, Castro JA (1998) Mechanisms of the preventive properties of some garlic components in the carbon tetrachloride-promoted oxidative stress. Diallyl sulfide; diallyl disulfide; allyl mercaptan and allyl methyl sulfide. Res Commun Mol Pathol Pharmacol 102:163–174

Filomeni G, Aquilano K, Rotilio G, Ciriolo MR (2003) Reactive oxygen species-dependent c-Jun NH2-terminal kinase/c-Jun signaling cascade mediates neuroblastoma cell death induced by diallyl disulfide. Cancer Res 63:5940–5949

Folkman J (2003) Fundamental concepts of the angiogenic process. Curr Mol Med 3:643–651

Fukao T, Hosono T, Misawa S, Seki T, Ariga T (2004) The effects of allyl sulfides on the induction of phase II detoxification enzymes and liver injury by carbon tetrachloride. Food Chem Toxicol 42:743–749

Galeone C, Pelucchi C, Dal Maso L, Negri E, Montella M, Zucchetto A, Talamini R, La Vecchia C (2009) *Allium* vegetables intake and endometrial cancer risk. Public Health Nutr 12:1576–1579

Ghobrial IM, Witzig TE, Adjei AA (2005) Targeting apoptosis pathways in cancer therapy. CA Cancer J Clin 55:178–194

Goncagul G, Ayaz E (2010) Antimicrobial effect of garlic (*Allium sativum*). Recent Patent Anti-Infect Drug Discov 5:91–93

Gudi VA, Singh SV (1991) Effect of diallyl sulfide, a naturally occurring anti-carcinogen on glutathione dependent detoxification enzymes of female CD-1 mouse tissues. Biochem Pharmacol 42:1261–1265

Haber D, Siess MH, Canivenc-Lavier MC, Le Bon AM, Suschetet M (1995) Differential effects of dietary diallyl sulfide and diallyl disulfide on rat intestinal and hepatic drug metabolizing enzymes. J Toxicol Environ Health 44:423–434

Hageman GJ, van Herwijnen MH, Schilderman PA, Rhijnsburger EH, Moonen EJ (1997) Reducing effects of garlic constituents on DNA adduct formation in human lymphocytes *in vitro*. Nutr Cancer 27:177–185

Hengartner MO (2000) The biochemistry of apoptosis. Nature 407:770–776

Herman-Antosiewicz A, Singh SV (2005) Checkpoint kinase 1 regulates diallyl trisulfide-induced mitotic arrest in human prostate cancer cells. J Biol Chem 280:28519–28528

Herman-Antosiewicz A, Powolny AA, Singh SV (2007) Molecular targets of cancer chemoprevention by garlic-derived organosulfides. Acta Pharmacol Sin 28:1355–1364

Ho BE, Shen DD, Mccune JS, Bui T, Risler L, Yang Z, Ho RJY (2010) Effects of garlic on cytochromes p450 2c9 and 3a4-mediated drug metabolism in human hepatocytes. Sci Pharm 78:473–481

Hong YS, Ham YA, Choi JH, Kim J (2000) Effects of allyl sulfur compounds and garlic extract on the expression of Bcl-2, Bax, and p53 in non small cell lung cancer cell lines. Exp Mol Med 32:127–134

Hosono T, Fukao T, Ogihara J, Ito Y, Shiba H, Seki T, Ariga T (2005) Diallyl trisulfide suppresses the proliferation and induces apoptosis of human colon cancer cells through oxidative modification of β-tubulin. J Biol Chem 280:41487–41493

Hosono T, Hosono-Fukao T, Inada K, Tanaka R, Yamada H, Iitsuka Y, Seki T, Hasegawa I, Ariga T (2008) Alkenyl group is responsible for the disruption of microtubule network formation inhuman colon cancer cell line HT-29 cells. Carcinogenesis 29:1400–1406

Howard EW, Ling MT, Chua CW, Cheung HW, Wang XH, Wong YC (2007) Garlic-derived S-allylmercaptocysteine is a novel *in vivo* antimetastatic agent for androgen-independent prostate cancer. Clin Cancer Res 13(6):1847–1856

Howard EW, Lee DT, Chiu YT, Chua CW, Wang X, Wong YC (2008) Evidence of a novel docetaxel sensitizer, garlic-derived S-allylmercaptocysteine, as a treatment option for hormone refractory prostate cancer. J Int Cancer 122:1941–1948

Hsing AW, Chokkalingam AP, Gao YT, Madigan MP, Deng J, Gridley G, Fraumeni JF Jr (2002) *Allium* vegetables and risk of prostate cancer: a population-based study. J Natl Cancer Inst 94:1648–1651

Hu X, Singh SV (1997) Glutathione S-transferases of female A/J mouse lung and their induction by anticarcinogenic organosulfides from garlic. Arch Biochem Biophys 340:279–286

Hu X, Benson PJ, Srivastava SK, Xia H, Bleicher RJ, Zaren HA, Awasthi S, Awasthi YC, Singh SV (1997) Induction of glutathione S-transferase pi as a bioassay for the evaluation of potency of inhibitors of benzo(a)pyrene-induced cancer in a murine model. Int J Cancer 73:897–902

Hu X, Cao BN, Hu G, He J, Yang D, Wan Y (2002) Attenuation of cell migration and induction of cell death by aged garlic extract in rat sarcoma cells. Int J Mol Med 9(6):641–643

Iciek M, Kwiecien I, Wlodek L (2009) Biological properties of garlic and garlic-derived organosulfur compounds. Environ Mol Mutagen 50:247–265

Imai J, Ide N, Nagae S, Moriguchi T, Matsuura H, Itakura Y (1994) Antioxidant and radical scavenging effects of aged garlic extract and its constituents. Planta Med 60:417–420

Jin L, Baillie TA (1997) Metabolism of the chemoprotective agent diallyl sulfide to glutathione conjugates in rats. Chem Res Toxicol 10:318–327

Kalra N, Arora A, Shukla Y (2006) Involvement of multiple signaling pathways in diallyl sulfide mediated apoptosis in mouse skin tumors. Asian Pac J Cancer Prev 7:556–562

Karmakar S, Banik NL, Patel SJ, Ray SK (2007) Garlic compounds induced calpain and intrinsic caspase cascade for apoptosis in human malignant neuroblastoma SH-SY5Y cells. Apoptosis 12:671–684

Kaufmann SH, Gores GJ (2000) Apoptosis in cancer: cause and cure. BioEssays 22:1007–1017

Kim JY, Kwon O (2009) Garlic intake and cancer risk: an analysis using the food and drug administrations evidence-based review system for the scientific evaluation of health claims. Am J Clin Nutr 89:257–264

Kim Y, Xiao D, Xiao H, Powolny AA, Lew KL, Reilly ML (2007) Mitochondria-mediated apoptosis by diallyl trisulfide in human prostate cancer cells is associated with generation of reactive oxygen species and regulated by Bax/Bak. Mol Cancer Ther 6:1599–1609

Knowles LM, Milner JA (1998) Depressed p34 kinase activity and G_2/M phase arrest induced by diallyl disulfide in HCT-15 cells. Nutr Cancer 30:169–174

Knowles LM, Milner JA (2000) Diallyl disulfide inhibits p34(cdc2) kinase activity through changes in complex formation and phosphorylation. Carcinogenesis 21:1129–1134

Kumaraguruparan R, Chandra Mohan KV, Abraham SK, Nagini S (2005) Attenuation of N-methyl-N0-nitro-N-nitrosoguanidine induced genotoxicity and oxidative stress by tomato and garlic combination. Life Sci 76:2247–2255

Kwon KB, Yoo SJ, Ryu DG, Yang JY, Rho HW, Kim JS, Park JW, Kim HR, Park BH (2002) Induction of apoptosis by diallyl disulfide through activation of caspase-3 in human leukemia HL-60 cells. Biochem Pharmacol 63:41–47

Lai KC, Hsu SC, Kuo CL, Yang JS, Ma CY, Lu HF, Tang NY, Hsia TC, Ho HC, Chung JG (2011) Diallyl sulfide, diallyl disulfide, and diallyl trisulfide inhibit migration and invasion in human colon cancer colo 205 cells through the inhibition of matrix metalloproteinase-2, -7, and -9 expressions. Environ Toxicol 28:479–488

Lamm DL, Riggs DR (2000) The potential application of *Allium sativum* (garlic) for the treatment of bladder cancer. Urol Clin North Am 27:157–162

Lan H, Lu YY (2003) Effect of allitridi on cyclin D1 and p27(Kip1) protein expression in gastric carcinoma BGC823 cells. Ai Zheng 22:1268–1271

Lanzotti V (2006) The analysis of onion and garlic. J Chromatogr A 1112:3–22

Lau BH, Yamasaki T, Gridley DS (1991) Garlic compounds modulate macrophage and T-lymphocyte functions. Mol Biother 3:103–107

Lea MA, Randolph VM (2001) Induction of histone acetylation in rat liver and hepatoma by organosulfur compounds including diallyl disulfide. Anticancer Res 21:2841–2845

Lea MA, Randolph VM, Patel M (1999) Increased acetylation of histones induced by diallyl disulfide and structurally related molecules. Int J Oncol 15:347–352

Lea MA, Rasheed M, Randolph VM, Khan F, Shareef A, Des Bordes C (2002) Induction of histone acetylation and inhibition of growth of mouse erythroleukemia cells by S-allylmercaptocysteine. Nutr Cancer 43:90–102

Li TSC (2000) Medicinal plants. Technomic Publ Co, Pennsylvania, p 4

Li M, Min JM, Cui JR, Zhang LH, Wang K, Valette A (2002) Z-ajoene induces apoptosis of HL-60 cells: involvement of Bcl-2 cleavage. Nutr Cancer 42:241–247

Lin XY, Liu JZ, Milner JA (1994) Dietary garlic suppresses DNA adducts caused by N-nitroso compounds. Carcinogenesis 15:349–352

Lin JG, Chen GW, Su CC, Hung CF, Yang CC, Lee JH, Chung JG (2002) Effects of garlic components Diallyl sulfide and Diallyl disulfide on arylamine N-acetyltransferase activity and 2-aminofluorene-DNA adducts in human promyelocytic leukemia cells. The Am J Chinese Med 30:315–325

Liu JZ, Lin RI, Milner JA (1992) Inhibition of 7,12-dimethylbenz(a) anthracene induced mammary tumor and DNA adducts by garlic powder. Carcinogenesis 13:1847–1851

Lu X, Rasco BA, Jabal JMD, Aston E, Lin M, Konkel ME (2011) Investigating antibacterial effects of garlic (*Allium sativum*) concentrate and garlic-derived organosulfur compounds on campylobacter jejuni by using fourier transform infrared spectroscopy, raman spectroscopy, and electron microscopy. Appl Environ Microbiol 77:5257–5269

Ma JL, Zhang L, Brown LM, Li JY, Shen L, Pan KF, Liu WD, Hu Y, Han ZX, Crystal-Mansour S, Pee D, Blot WJ, Fraumeni JF Jr, You WC, Gail MH (2012) Fifteen-year effects of *Helicobacter pylori*, garlic, and vitamin treatments on gastric cancer incidence and mortality. J Natl Cancer Inst 104:488–492

Matsuura N, Miyamae Y, Yamane K, Nagao Y, Hamada Y, Kawaguchi N, Katsuki T, Hirata K, Sumi S, Ishikawa H (2006) Aged garlic extract inhibits angiogenesis and proliferation of colorectal carcinoma cells. J Nutr 136:S842–S846

Melino S, Sabelli R, Paci M (2011) Allyl sulfur compounds and cellular detoxification system: effects and perspectives in cancer therapy. Amino Acids 41:103–112

Mikaili P, Maadirad S, Moloudizargari M, Aghajanshakeri S, Sarahroodi S (2013) Therapeutic uses and pharmacological properties of garlic, shallot, and their biologically active compounds. Iran J Basic Med Sci 16:1031–1048

Molinari M (2000) Cell cycle checkpoints and their inactivation in human cancer. Cell Prolif 33:261–274

Mousa AS, Mousa SA (2005) Anti-angiogenesis efficacy of the garlic ingredient Alliin and antioxidants: role of nitric oxide and p53. Nutr Cancer 53:104–110

Munchberg U, Anwar A, Mecklenburg S, Jacob C (2007) Polysulfides as biologically active ingredients of garlic. Org Biomol Chem 5:1505–1518

Munday R, Munday CM (1999) Low doses of diallyl disulfide a compound derived from garlic increase tissue activities of quinone reductase and glutathione transferase in the gastrointestinal tract of the rat. Nutr Cancer 34:42–48

Munoz N, Plummer M, Vivas J, Moreno V, De Sanjose S, Lopez G, Oliver W (2001) A case-control study of gastric cancer in Venezuela. Int J Cancer 93:417–423

Murray AW (2004) Recycling the cell cycle: cyclins revisited. Cell 116:221–334

Nakagawa H, Tsuta K, Kiuchi K, Senzaki H, Tanaka K, Hioki A, Tsubura A (2001) Growth inhibitory effects of diallyl disulfide on human breast cancer cell lines. Carcinogenesis 22:891–897

Ng KT, Guo DY, Cheng Q, Geng W, Ling CC, Li CX, Liu XB, Ma YY, Lo CM, Poon RT, Fan ST, Man K (2012) A garlic derivative, S-allylcysteine (SAC), suppresses proliferation and metastasis of hepatocellular carcinoma. PLoS One 7:e31655

Nian H, Delage B, Pinto JT, Dashwood RH (2008) Allyl mercaptan, a garlic-derived organosulfur compound, inhibits histone deacetylase and enhances Sp3 binding on the P21WAF1 promoter. Carcinogenesis 29:1816–1824

Nian H, Delage B, Ho E, Dashwood RH (2009) Modulation of histone deacetylase activity by dietary isothiocyanates and allyl sulfides: studies with sulforaphane and garlic organosulfur compounds. Environ Mol Mutagen 50:213–221

Omar SH, Al-Wabel NA (2010) Organosulfur compounds and possible mechanism of garlic in cancer. Saudi Pharm J 18:51–58

Park EK, Kwon KB, Park KI, Park BH, Jhee EC (2002a) Role of Ca^{2+} in diallyl disulfide-induced apoptotic cell death of HCT-15 cells. Exp Mol Med 34:250–257

Park KA, Kweon S, Choi H (2002b) Anticarcinogenic effect and modification of cytochrome P450 2E1 by dietary garlic powder in diethylnitrosamine-initiated rat hepatocarcinogenesis. J Biochem Mol Biol 35:615–622

Pelucchi C, Bosetti C, Rossi M, Negri E, La Vecchia C (2009) Selected aspects of Mediterranean diet and cancer risk. Nutr Cancer 61:756–766

Perchellet JP, Perchellet EM, Abney NL, Zirnstein JA, Belman S (1986) Effects of garlic and onion oils on glutathione peroxidase activity, the ratio of reduced/oxidized glutathione and ornithine decarboxylase induction in isolated mouse epidermal cells treated with tumor promoters. Cancer Biochem Biophys 8:299–312

Pinto JT, Qiao C, Xing J, Rivlin RS, Protomastro ML (1997) Effects of garlic thioallyl derivatives on growth, glutathione concentration, and polyamine formation of human prostate carcinoma cells in culture. Am J Clin Nutr 66:398–405

Powolny AA, Singh SV (2008) Multitargeted prevention and therapy of cancer by diallyl trisulfide and related *Allium* vegetable-derived organosulfur compounds. Cancer Lett 269:305–314

Prasad K, Laxdal VA, Yu M, Raney BL (1995) Antioxidant activity of allicin, an active principle in garlic. Mol Cell Biochem 148:183–189

Prasad S, Kalra N, Srivastava S, Shukla Y (2008) Regulation of oxidative stressmediated apoptosis by diallyl sulfide in DMBA-exposed Swiss mice. Hum Exp Toxicol 27:55–63

Reddy BS, Rao CV, Rivenson A, Kelloff G (1993) Chemoprevention of colon carcinogenesis by organosulfur compounds. Cancer Res 53:3493–3498

Reicks MM, Crankshaw DL (1996) Modulation of rat hepatic cytochrome P-450 activity by garlic organosulfur compounds. Nutr Cancer 25:241–248

Rivlin RS (2001) Historical perspective on the use of garlic. J Nutr 131:S951–S954

Ross IA (1999) Medicinal plants of the world. Humana Press, Totowa, pp 25–63

Sakamoto K, Lawson LD, Milner JA (1997) Allyl sulfides from garlic suppress the *in vitro* proliferation of human A549 lung tumor cells. Nutr Cancer 29:152–156

Salem S, Salahi M, Mohseni M, Ahmadi H, Mehrsai A, Jahani Y, Pourmand G (2011) Major dietary factors and prostate cancer risk: a prospective multicenter case-control study. Nutr Cancer 63:21–27

Satia JA, Littman A, Slatore CG, Galanko JA, White E (2009) Associations of herbal and specialty supplements with lung and colorectal cancer risk in the vitamins and lifestyle study. Cancer Epidemiol Biomark Prev 18:1419–1428

Scherer C, Jacob C, Dicato M, Diederich M (2009) Potential role of organic sulfur compounds from *Allium* species in cancer prevention and therapy. Phytochem Rev 8:349–368

Schulz M, Lahmann PH, Boeing H, Hoffmann K, Allen N, Key TJ, Bingham S, Wirfält E, Berglund G, Lundin E, Hallmans G, Lukanova A, Martínez Garcia C, González CA, Tormo MJ, Quirós JR, Ardanaz E, Larrañaga N, Lund E, Gram IT, Skeie G, Peeters PH, van Gils CH, Bueno-de-Mesquita HB, Büchner FL, Pasanisi P, Galasso R, Palli D, Tumino R, Vineis P, Trichopoulou A, Kalapothaki V, Trichopoulos D, Chang-Claude J, Linseisen J, Boutron-Ruault MC, Touillaud M, Clavel-Chapelon F, Olsen A, Tjønneland A, Overvad K, Tetsche M, Jenab M, Norat T, Kaaks R, Riboli E (2005) Fruit and vegetable consumption and risk of epithelial ovarian cancer: the European prospective investigation into cancer and nutrition. Cancer Epidemiol Biomark Prev 14:2531–2535

Sengupta A, Ghosh S, Bhattacharjee S, Das S (2004a) Indian food ingredients and cancer prevention-an experimental evaluation of anticarcinogenic effects of garlic in rat colon. Asian Pac J Cancer Prev 5:126–132

Sengupta A, Ghosh S, Das S (2004b) Modulatory influence of garlic and tomato on cyclooxygenase-2 activity, cell proliferation and apoptosis during azoxymethane induced colon carcinogenesis in rat. Cancer Lett 208:127–136

Sheela CG, Kumud K, Augusti KT (1995) Anti-diabetic effects of onion and garlic sulfoxide amino acids in rats. Planta Med 61:356–357

Sheen LY, Li CK, Shen SF, Meng RH, Tsai SJ (1996) Effect of the active principle of garlic-diallyl sulphide on cell viability, detoxification capability and antioxidation system of primary rat hepatocytes. Food Chem Toxicol 34:971–978

Sheen LY, Chen HW, Kung YL, Liu CT, Li CK (1999) Effects of garlic oil and its organosulfur compounds on the activities of hepatic drug-metabolizing and antioxidant enzymes in rats fed high and low fat diets. Nutr Cancer 35:160–166

Shenoy NR, Choughuley AS (1992) Inhibitory effects of diet related sulfhydryl compounds on the formation of carcinogenic nitrosamines. Cancer Lett 65:227–232

Shin DY, Kim GY, Kim JI, Yoon MK, Kwon TK, Lee SJ, Choi YW, Kang HS, Yoo YH, Choi YH (2010) Anti-invasive activity of diallyl disulfide through tightening of tight junctions and inhibition of matrix metalloproteinase activities in LNCaP prostate cancer cells. Toxicol In Vitro 24:1569–1576

Shobana S, Naidu KA (2000) Antioxidant activity of selected Indian spices. Prost Leuk Essen Fatty Acids 62:107–110

Shojai TM, Langeroudi AG, Karimi V, Barin A, Sadri N (2016) The effect of *Allium sativum* (garlic) extract on infectious bronchitis virus in specific pathogen free embryonic egg. Avicen J Phytomed 6:458–467

Siegers CP, Steffen B, Robke A, Pentz R (1999) The effect of garlic reparation against human tumor cell proliferation. Phytomedicine 6:7–11

Siess MH, Le Bon AM, Canivenc-Lavier MC, Suschetet M (1997) Modification of hepatic drug metabolizing enzymes in rats treated with alkyl sulfides. Cancer Lett 120:195–201

Singh SP, Abraham SK, Kesavan PC (1995) *In vivo* radioprotection with garlic extract. Mutat Res 345:147–153

Singh SV, Pan SS, Srivastava SK, Xia H, Hu X, Zaren HA, Orchard JL (1998) Differential induction of NAD(P)H:quinone oxidoreductase by anticarcinogenic organosulfides from garlic. Biochem Biophys Res Commun 244:917–920

Sparnins VL, Barany G, Wattenberg LW (1988) Effects of organosulfur compounds from garlic and onions on benzo[a]pyrene-induced neoplasia and glutathione *S*-transferase activity in the mouse. Carcinogenesis 9:131–134

Steinmetz KA, Kushi LH, Bostick RM, Folsom AR, Potter JD (1994) Vegetables, fruit, and colon cancer in the Iowa women's health study. Am J Epidemiol 139:1–15

Sundaram SG, Milner JA (1996a) Diallyl disulfide inhibits the proliferation of human tumor cells in culture. Biochim Biophys Acta 1315:15–20

Sundaram SG, Milner JA (1996b) Diallyl disulfide induces apoptosis of human colon tumor cells. Carcinogenesis 17:669–673

Surh YJ, Lee RC, Park KK, Mayne ST, Liem A, Miller JA (1995) Chemoprotective effects of capsaicin and diallyl sulfide against mutagenesis or tumorigenesis by vinyl carbamate and N-nitrosodimethylamine. Carcinogenesis 10:2467–2471

Suzui N, Sugie S, Rahman KM, Ohnishi M, Yoshimi N, Wakabayashi K, Mori M (1997) Inhibitory effects of diallyl disulfide or aspirin on 2-amino-1-methyl-6-phenylimidazo[4,5-b]pyridine-induced mammary carcinogenesis in rats. Jpn J Cancer Res 88:705–711

Tan LM, Zhang MX, Luo HM, Zeng Y, Li J, Cui ZW, Zhang XH, Tang SS (2004) The initiation of G_2/M checkpoint by diallyl disulfide requires the activation of p38 MAP kinase in HL-60 cells. Zhonghua Xue Ye Xue Za Zhi 25:273–276

Tanaka S, Haruma K, Kunihiro M, Nagata S, Kitadai Y, Manabe N, Sumii M, Yoshihara M, Kajiyama G, Chayama K (2004) Effects of aged garlic extract (AGE) on colorectal adenomas: a double-blinded study. Hiroshima J Med Sci 53:39–45

Tanaka S, Haruma K, Yoshihara M, Kajiyama G, Kira K, Amagase H, Chayama K (2006) Aged garlic extract has potential suppressive effect on colorectal adenomas in humans. J Nutr 136:S821–S826

Thejass P, Kuttan G (2007a) Inhibition of angiogenic differentiation of human umbilical vein endothelial cells by diallyl disulfide (DADS). Life Sci 80:515–521

Thejass P, Kuttan G (2007b) Antiangiogenic activity of diallyl sulfide (DAS). Int Immunopharmacol 7:295–305

Thomson M, Ali M (2003) Garlic (*Allium sativum*): a review of its potential use as an anti-cancer agent. Curr Cancer Drug Targets 3:67–81

Thornberry N, Lazebnick Y (1998) Caspases: enemies within. Science 281:1312–1316

Velmurugan B, Mani A, Nagini S (2005) Combination of S-allylcysteine and lycopene induces apoptosis by modulating Bcl-2, Bax, Bim and caspases during experimental gastric carcinogenesis. Eur J Cancer Prev 14:387–393

Verma SK, Jain V, Verma D (2008) Garlic-the spice of life: composition, cooking, chemistry and preparations. J Herbal Med Toxicol 2:21–28

Wang HC, Pao J, Lin SY, Sheen LY (2012) Molecular mechanisms of garlic-derived allyl sulfides in the inhibition of skin cancer progression. Ann N Y Acad Sci 1271:44–52

Wang HC, Hsieh SC, Yang JH, Lin SY, Sheen LY (2012a) Diallyl trisulfide induces apoptosis of human basal cell carcinoma cells via endoplasmic reticulum stress and the mitochondrial pathway. Nutr Cancer 64:770–780

Wang Q, Wang Y, Ji Z, Chen X, Pan Y, Gao G, Gu H, Yang Y, Choi BC, Yan Y (2012b) Risk factors for multiple myeloma: a hospital-based case-control study in Northwest China. Cancer Epidemiol 36:439–444

Wanwimolruk S, Prachayasittikul V (2014) Cytochrome P450 enzyme mediated herbal drug interactions. EXCLI J 13:347–391

Wargovich MJ (2006) Diallylsulfide and allylmethylsulfide are uniquely effective among organosulfur compounds in inhibiting CYP2E1 protein in animal models. The J Nutr 136:S832–S834

Wu CC, Sheen LY, Chen HW, Kuo WW, Tsai SJ, Li CK (2002) Differential effects of garlic oil and its three major organosulfur components on the hepatic detoxification system in rats. J Agric Food Chem 50:378–383

Wu CC, Chung JG, Tsai SJ, Yang JH, Sheen LY (2004) Differential effects of allyl sulfides from garlic essential oil on cell cycle regulation in human liver tumor cells. Food Chem Toxicol 42:1937–1947

Xiao D, Pinto JT, Soh JW, Deguchi A, Gundersen GG, Palazzo AF, Yoon JT, Shirin H, Weinstein IB (2003) Induction of apoptosis by the garlic-derived compound S-allylmercaptocysteine (SAMC) is associated with microtubule depolymerization and c-Jun NH(2)-terminal kinase 1 activation. Cancer Res 63:6825–6837

Xiao D, Choi S, Johnson DE, Vogel VG, Johnson CS, Trump DL, Lee YJ, Singh SV (2004) Diallyl trisulfide-induced apoptosis in human prostate cancer cells involves c-Jun N-terminal kinase and extracellular-signal regulated kinase-mediated phosphorylation of Bcl-2. Oncogene 23:5594–5606

Xiao D, Herman-Antosiewicz A, Antosiewicz J, Xiao H, Brisson M, Lazo JS, Singh SV (2005) Diallyl trisulfide-induced G_2/M phase cell cycle arrest in human prostate cancer cells is caused by reactive oxygen species-dependent destruction and hyperphosphorylation of Cdc25C. Oncogene 24:6256–6268

Xiao D, Li M, Herman-Antosiewicz A, Antosiewicz J, Xiao H, Lew KL, Zeng Y, Marynowski SW, Singh SV (2006) Diallyl trisulfide inhibits angiogenic features of human umbilical vein endothelial cells by causing Akt inactivation and down-regulation of VEGF and VEGF-R2. Nutr Cancer 55:94–107

Yeh YY, Liu L (2001) Cholesterol lowering effect of garlic extracts and organosulfur compounds: human and animal studies. J Nutr 131:S989–S993

You WC, Blot WJ, Chang YS, Ershow A, Yang ZT, An Q, Henderson BE, Fraumeni JF Jr, Wang TG (1989) *Allium* vegetables and reduced risk of stomach cancer. J Natl Cancer Inst 81:162–164

Zeng T, Zhang CL, Song FY, Han XY, Xie KQ (2009) The modulatory effects of garlic oil on hepatic cytochrome P450s in mice. Hum Exp Toxicol 28:777–783

Zhang YW, Wen J, Xiao JB, Talbot SG, Li GC, Xu M (2006) Induction of apoptosis and transient increase of phosphorylated MAPKs by diallyl disulfide treatment in human nasopharyngeal carcinoma CNE2 cells. Arch Pharm Res 29:1125–1131

Zhang W, Xiang YB, Li HL, Yang G, Cai H, Ji BT, Gao YT, Zheng W, Shu XO (2013) Vegetable-based dietary pattern and liver cancer risk: results from the Shanghai women's and men's health studies. Cancer Sci 104:1353–1361

Chapter 10
Relevance of Traditional Unani (Greco-Arab) System of Medicine in Cancer: An Update

Mohd Kashif Husain, Mohd Khalid, Goli Penchala Pratap, and Munnawar Husain Kazmi

Abstract Cancer is one of the most dreadful diseases worldwide. The incidence of cancer has drastically increased in low- and middle-income countries and is a major cause of morbidity and mortality in the developed countries. Use of alternative medicines along with conventional ones against cancer managed the symptoms, control side effects, and improve the quality of life and mental state. Indian systems of medicine (ISM) including Unani (Greco-Arab medicine) have been used for healing and preventative health care all over the world since decades. Cancer (Arabic-*Sartan*) has been well described in Unani classical texts as a hard inflammation (*Warm-e-Salabat*) with lesion. The cancer management in Unani system includes regimenal therapy (*Ilaj bi'l-Tadbir*), dietotherapy (*Ilaj bi'l-Ghidhā*), pharmacotherapy (*Ilāj bi'l-Dawā*), and surgery (*Ilāj bi'l-Yad*). In pharmacotherapy, many Unani drugs have been evaluated preclinically (*in vitro* and *in vivo* studies) for its anticancer activities and found to be very effective, but unfortunately, no attention has been paid to explicate the efficacy of compound formulations of these Unani medicines in cancer management, till date. The aim of this chapter is to exploit and bring together the Unani system of medicine with conventional cancer therapies at interdisciplinary and integrative levels and also to discuss the future strategies for the prevention and management of cancer.

M. K. Husain (✉)
Survey of Medicinal Plants Unit, Central Research Institute of Unani Medicine (CRIUM, Under CCRUM), Hyderabad, Telangana, India

Department of Ilmul Adviya (Unani Pharmacology), Central Research Institute of Unani Medicine (CRIUM, Under CCRUM), Hyderabad, Telangana, India

M. Khalid
Swasthya Rakshan Parikshan Programme, Central Research Institute of Unani Medicine (CRIUM, Under CCRUM), Hyderabad, Telangana, India

G. Penchala Pratap
Survey of Medicinal Plants Unit, Central Research Institute of Unani Medicine (CRIUM, Under CCRUM), Hyderabad, Telangana, India

M. H. Kazmi
Department of Ilmul Adviya (Unani Pharmacology), Central Research Institute of Unani Medicine (CRIUM, Under CCRUM), Hyderabad, Telangana, India

M. S. Akhtar, M. K. Swamy (eds.), *Anticancer Plants: Clinical Trials and Nanotechnology*, https://doi.org/10.1007/978-981-10-8216-0_10

Keywords Sartan · Cancer · Unani medicine · Integrative oncology

10.1 Introduction

Cancer, a fatal noncommunicable disease (NCD), opened the eyes of the world. It is a cluster of disorders involving transformation, dysregulated apoptosis, cell proliferation, angiogenesis, and metastasis. An extensive research during the past century has unfolded various facts about the biomechanics of the cancer, and most of the anticancer drugs are employed to withhold signaling of cell cycle and growth factors, inflammation, and angiogenesis. The pathways linked to this process lead to chronic inflammation, a major mediator of tumor progression (Gupta et al. 2010). Cancer continues as a scourge of humankind and is considered as a burden on the populations of low- and middle-income countries. The occurrence of cancer has increased from 12.7 million in 2008 to 14.1 million in 2012, and this trend is projected to persist, with the number of new cases expected to go up further by 75%. This will take the number of cancer-related cases up to 25 million over the next two decades. Cancer is a major cause of mortality across the world with deaths projected to rise over 13.1 million in 2030. Cancer is the second largest NCD after ischemic heart disease with only 20% is getting cured and 20% get a prolonged life after treatment. The findings of GLOBOCAN showed that high-resource countries have the highest incidence of cancer (Ferlay et al. 2015) and also provide the finest services for detection, diagnosis, and treatment, as may be deduced from mortality and survival data. The results of the above finding showed approximately 14.1 million new cases of cancer were diagnosed worldwide (excluding non-melanoma skin cancer) and 8.2 million estimated deaths from cancer, affecting the entire global populations. These estimates correspond to age-standardized incidence and mortality rates of 182 and 102 per 100,000, respectively, but there were slightly more incident cases (53% of the total) and deaths (57%) among men than among women.

According to World Cancer Report (2014), more than 60% of the world's cancer cases occur in Africa, Asia, and Central and South America, which may account for about 70% of the cancer deaths. Among all types of cancers (excluding non-melanoma skin cancer), the highest incidence rates were reported from high-income countries such as North America and Western Europe (together with Japan, the Republic of Korea, Australia, and New Zealand). In case of men, the five most prevalent sites of cancer were the lung (16.7% of the total), prostate (15.0%), colorectum (10.0%), stomach (8.5%), and liver (7.5%), while, in case of women, the incident sites were the breast (25.2% of the total), colorectum (9.2%), lung (8.7%), cervix (7.9%), and stomach (4.8%). Among men, lung cancer had the highest incidence rate (34.2 per 100,000) followed by prostate cancer (31.1 per 100,000), while in women the breast cancer had a higher incidence rate (43.3 per 100,000) followed by colorectal cancer (14.3 per 100,000). In India, during the year 2011, it was estimated that 0.44 million died owing to cancer, and it is predicted to increase up to 0.60 and 0.70 million by year 2021–2026, respectively, as a result of change in size and composition of population (D'Souza et al. 2013). Murthy et al. (2008) reported that lung, esophagus, prostate, stomach, oral, and pharyngeal cancers are predominant in men, while cervix and breast cancers are common in women followed by oral cavity, stomach, and esophagus. The new cases of cancer were more prone in women's compared to men's, and most of the deaths occur at home (Dikshit et al. 2012). In India, the breast cancer has ranked first among Indian females with age-adjusted rate as high as 25.8 per 100,000 women and mortality 12.7 per 100,000 women (Malvia et al. 2017).

The Unani system of medicine originated in Greece, with its more than twenty-five centuries of legacy established by Hippocrates (Father of Medicine) and put forward by other eminent Unani physicians of well-known civilizations (Roman Egypt, Arabia, and India). The Unani stalwarts such as Galen, Rhazes (Razi), and Avicenna (Ibn Sina) have noticed and mentioned different types of cancer prevalence known to them at that period of time. They recommended myriad of treatment modalities in management, prevention, and prognosis of cancer. The earliest "humoral theory" given by Hippocrates on the basis of four humors was the first theory which explained the cause of cancer. This theory of cancer was passed on by the Romans and was embraced by the influential doctor Galen's medical teaching, which remained the unchallenged standard through the Middle Ages for over 1300 years (ACS 2014). The aim of this chapter is to exploit and bring together the Unani system of medicine with conventional cancer therapies at interdisciplinary and integrative levels and also to discuss the future strategies for the prevention and management of cancer.

10.2 Traditional Medicines and Their Significance

Traditional medicines have been used since thousands of years to cure health-related issues. The World Health Organization (WHO) defined the traditional medicine as "the sum total of all the knowledge and practices, whether explicable or not, used in diagnosis, prevention and elimination of physical, mental or social imbalance and relying exclusively on practical experience and observations handed down from generation to generation, whether verbally or in writing." Access to "modern" health-care services and medicine may be limited in developing countries. TM becomes the only affordable treatment available to poor people and in remote communities. Health-care providers worldwide including major pharmaceutical giants are turning to incorporate many of these into their mainstream activities.

Across the world, traditional medicine is either the mainstay of health-care delivery or serves as a complement to it. In some countries, traditional medicine or non-conventional medicine may be termed complementary medicine. Traditional medicine and complementary medicine are an important and often underestimated part of health care in almost every country in the world, and the demand for its services is increasing. Traditional medicine, of proven quality, safety, and efficacy, contributes to the goal of ensuring that all people have access to care. Many countries now recognize the need to develop a cohesive and integrative approach to health care that allows governmental health-care practitioners and, most importantly, those who use health-care services to access traditional medicines and complementary medicine in a safe, respectful, cost-efficient, and effective manner (WHO 2013). There are two well-defined terms which are related to traditional medicines: complementary treatment and alternative treatment. Complementary treatments are used in combination with mainstream therapy (conventional therapy) and are usually intended to treat side effects. On the other hand, alternative treatments are used in place of conventional therapy and usually have their own therapeutic potential. Half of the world's cancer patients use complementary medicine for treatment. Approximately 50–60% cancer patients in the USA utilize agents derived from different parts of plants or nutrients, either as stand-alone or concurrently with conventional therapeutic regime of chemotherapy or radiation therapy (Gutheil et al. 2012). The systematic review and meta-analysis done worldwide to estimate the prevalence of use of complementary medicine in treatment of cancer was 40% that has increased from 25% in 1970 to 49% after 2000 (Horneber et al. 2012). Though many research works are under progress, no literature highlights that traditional medicine can be used as an alternative to the mainstream therapies in curing cancer (Deng and Cassileth 2014). But, they can be considered complimentary to the regular therapies in reducing the side effects and improving their mental well-being.

Indian systems of medicine (Ayurveda, Unani, and Siddha) and traditional Chinese medicine (TCM) have been used for hundreds of years for healing and preventative health care all over the world. Among Indian systems, Ayurveda is the oldest system with documented history of its practice since more than 5000 years, whereas the Unani (Greco-Arab) system of medicine originated in Greece and passed through many countries before establishing itself in India during the medieval period. It is based on well-established knowledge and practices relating to the promotion of positive health and prevention of disease. With the fusion of traditional knowledge of ancient civilizations like Egypt, Arabia, Iran, China, Syria, and India, these systems are being practiced in the country with diverse preferences of people and infrastructural facilities. TCM has increasingly become popular in the West including in cancer patients (Boon et al. 2007) and plays an important role in minimizing disability, protecting patients suffering from complications, and helping patients to live well (Yoder 2005). Chinese medicine assisted in supportive and palliative care by reducing side effects of conventional treatment of cancer or improving quality of life (Ernst 2009). It is estimated that the US National Cancer Institute (NCI) spends around $120 million each year on complementary and alternative medicine (CAM)-related research projects (Jia 2012).

10.3 Origin of Unani (Greco-Arab) System of Medicine

The Unani system of medicine (*Yūnānī Niẓām-i Ṭibb or Ṭibb*) owes its immediate origin to ancient Greece (*Yūnān*) and was further adopted by the Arabs into a specialized medical science based on the principles given by Buqrat (Hippocrates) and Jalinoos (Galen). Greco-Arab refers to medicine developed during the Golden Age of the Arab-Islamic Empire, which extended from Andalusia (Spain) and Maghreb states (North Africa) in the west to Central Asia and India in the east, with the central lands of Egypt, Bilad al-Sham (Greater Syria), and Iraq. It spanned a period of roughly nine centuries, from the middle of the seventh century to the end of the fifteenth century, by which time it had broken up into three distinct empires, the Ottoman, the Safavid, and the Mughal (Saad and Said 2011). The herbal basis of Unani therapy can be traced back to its earliest originators in ancient Egypt, which gave primary place to plants in disease treatment. They had also initiated the use of surgery as a method of treatment. The studies of Papyri show the ability of Egyptians in the field of medicine. Imhotep (2800 BC) and Amenhotep (1550 BC) are some noted physicians of Egyptian period. Due to their great contributions in the field of medicine, Mesopotamia also occupied an important place in history. They had used urine sample as a diagnostic tool. The Greek period of Unani medicine began with Asclepius (*Asqalībūs*-1200 BC), who was a great scholar of medicine. During Asclepian period, the Greeks developed the art of medicine in the light of medical knowledge of Egyptians and Babylonians.

The theoretical framework of Unani medicine is based on the teachings of Hippocrates (460–370 BC), the overarching physician of the classical period of Unani medical history. He emphasized the natural causes of disease and recorded the existing medical knowledge to set the grounds for medicine to develop as a systematic science. The three fundamentals of Hippocratic medicine were observation, experience, and rational principles, which still hold valid in the field of medicine and science. The closing years of the creative age of Greece were graced by the great Roman scholar Galen (129–200 AD). He made valuable addition to medicine by conducting experiments and elevated the status of an art to the rank of a scientific discipline of medicine. Galen, one of the most illustrious scholars in the history of medicine, gathered most of the medical knowledge of his time and arranged it systematically in a way that continued to be authoritative for the next centuries. Greek and Roman medicinal practices, as preserved in the writings of Hippocrates and Galen, formed the roots for later Greco-Arab and Islamic medicine and modern Western medicine. Theophrastus (about 300 BC), in his book *History of Plants*, dealt with the medicinal qualities of herbs and noted the ability to change their characteristics through cultivation. Dioscorides (100 AD) mentioned the collection, storage, and use of medicinal herbs, and Galen wrote 30 books on these subjects and is well known for his complex prescriptions and formulas used in compounding drugs, sometimes containing dozens of ingredients. Taken together, a vast medical literature explored the synthesis and practice of medicinal plants was developed.

The Arabs introduced the Unani system of medicine to India in the eighth century, and soon it caught the attention of the masses. During the period between the thirteenth and seventeenth century, the Unani system of medicine established its roots in India. The Delhi Sultans, the Khiljis, the Tughlaqs, and the Mughal emperors provided state patronage to the scholars. This was a golden period of the Unani system of medicine as it virtually spread all over the country finding immediate favor with the people at large. The Unani system suffered a temporary setback during the colonial rule but was revived soon after independence, with the government recognizing it as one of the Indian systems of medicine and taking initiative for its development and propagation. During its over 1200 years' history in India, the Unani system of medicine made major advancements by the fusion of traditional knowledge of ancient civilizations like Egypt, Arabia, Iran, China, Syria, and India to get leadership at global level and successfully applied its principles to the Indian geo-human environment to emerge as one of the effective and commonly used systems of medicine. The Unani system of medicine also accommodated itself to Indian climate, temperament, culture, and medicinal resources successfully and catered to the health needs of all strata of the country's population (Anonymous 2016).

10.4 Unani (Greco-Arab) System of Medicine

The Unani system of medicine gave great importance to the Hippocratic "humoral theory" given in the fourth century BC, which emphasizes the unique character of living things/organic matter and their distinction from nonliving things/inorganic matter by explaining that the elements are metabolized and converted into humors, which are organic and possess the additional properties of living things. The theory describes four humors (*Akhlāṭ*, singular: *Khilt*) and elaborates that the most important of these principles is temperament (*Mizāj*) which classifies living human body into four (hot, cold, wet, and dry) qualitative types of states: (1) hot and dry, (2) hot and wet, (3) cold and wet, and (4) cold and dry. These are represented as earth, water, fire, and air. Temperament of human beings depends upon the dominant humor. Since each humor possesses a particular quality, it is possible to convert humoral temperament into qualitative temperament and then can easily be correlated with human states, drugs, and diets. The basic principles comprise of (a) the seven natural factors (*Umūr Ṭabī'iyya*), essential factors for the constitution of human body, (b) the basics of pathology (*Kulliyāt-i 'Ilm al-Amrāḍ*), (c) the principles of diagnosis (*Uṣūl-i Tashkhīṣ*), and (d) the principles of treatment (*Uṣūl-i 'Ilāj*). The Greek ideas were put forward by Arabian physicians as seven natural factors (*Umūr Ṭabī'iyya)*, which are responsible for the body constitution, its health, and diseased conditions (Anonymous 2016). The brief descriptions of these natural factors are as given below:

1. **Basic elements (*Arkān*)**
   Basic elements in Unani medicine are simple indivisible matters which provide the primary components for the human body. They can't be further resolved into simpler entities. The various substances (compounds) in nature depend for their existence on their chemical compositions (*Imtizāj*). Everything in the universe is composed of four basic elements in varying amount and proportion. Each element has two sets of basic qualities (*Kayfiyāt*): hot or cold and dry or wet (Fig. 10.1). The four elements are fire (*Nār*), air (*Hawā*), water (*Mā'*), and earth (*Arḍ*). Their basic qualities are fire (hot and dry), air (hot and wet), water (cold and wet), and earth (cold and dry).

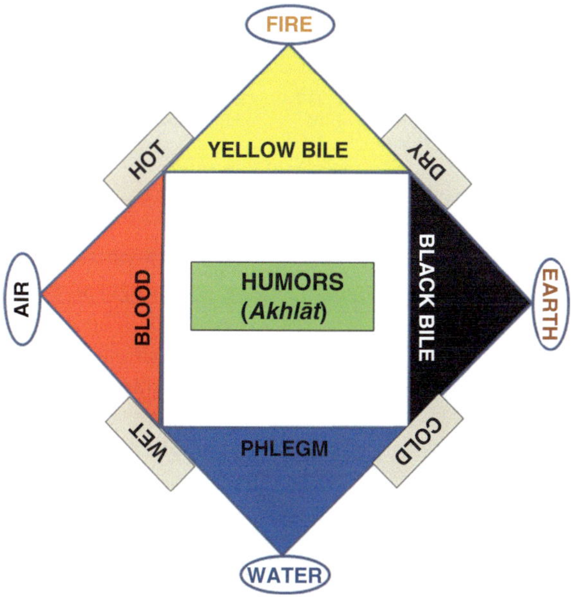

Fig. 10.1 Schematic diagram of four humors

Table 10.1 Humors (*Akhlāṭ*) and their basic qualities

Humors	Basic qualities
Blood (*Dam*)	Hot and wet
Phlegm (*Balgham*)	Cold and wet
Yellow bile (*Ṣafrā*')	Hot and dry
Black bile (*Sawdā*')	Cold and dry

2. **Humors (*Akhlāṭ*)**
   Humors (*Akhlāṭ*) are the wet and liquid parts of the body which are produced after normal metabolism. Their right proportion in terms of quality and quantity (homeostasis) is vital in maintaining health and any derangement results in disease. Their main function is to support biological processes related to nutrition, growth, repair, and production of energy for the preservation of individual's health. The four humors are characterized by the dominant basic qualities, and each humor has its own temperament (Fig. 10.1 and Table 10.1). The humors are formed within various organs, primarily in the liver, and run together in the blood vessels.
3. **Temperament (*Mizāj*)**
   Every individual has a unique *humoral* composition which represents his specific temperament. Entire human beings are classified accordingly as sanguine, phlegmatic, choleric, and melancholic, on the basis of the amount of dominant humor prevalent in the human body (Fig. 10.1 and Table 10.2). The temperament of man can be diagnosed by ten categories of bodily and mental parameters (*Ajnās 'Ashara*), which includes complexion, built, texture, hair, dreams, and

Table 10.2 Elements and basic qualities of humors on temperament

Elements	Basic qualities	Temperament
Air	Hot and wet	Sanguine (*Damawī*)
Water	Cold and wet	Phlegmatic (*Balghamī*)
Fire	Hot and dry	Choleric (*Ṣafrāwī*)
Earth	Cold and dry	Melancholic (*Sawdāwī*)

others. The pathological change of temperament, in whole man or in particular body parts and organs, can also be diagnosed by changes in these parameters.

4. **Organs (*A'ḍā'*)**
   The physical body is made up of tissues and organs (*A'ḍā'*). Organs are classified into two types: simple (*Basīṭ*) and compound (*Murakkab*). The simple organs are tissues like fat, bone, and so on, while compound organs are structures made up of poly-tissue such as the heart, brain, liver, and others.
5. **Pneumas (*Arwāḥ*)**
   The *Rūḥ* (p. *Arwāḥ*) is subtle, physical substance made from the subtle components of the humors. It possesses vitality and provides life to the organism. It also acts as the seat and carrier of the physiological powers or faculties (*Quwā*) which are responsible for producing the corresponding physiological processes including nutrition, growth, sensation, and movement. Humans possess three pneumas, each one generated in vital organs: vital pneuma (*Rūḥ Ḥaywānī*) in the heart, psychic pneuma (*Rūḥ Nafsānī*) in the brain, and natural pneuma (*Rūḥ Ṭabī'ī*) in the liver.
6. **Faculties (*Quwā*)**
   Faculties are the physiological power that performs the corresponding physiological functions of nutrition, growth, reproduction, propagation of species, respiration circulation, coordination, intelligence, sensation, and response. The primary faculties are Natural Faculty (*Quwwat Ṭabī'iyya*), Generative Faculty (*Quwwat Muwallida*), Vital Faculty (*Quwwat Ḥaywāniyya*), and Psychic Faculty (*Quwwat Nafsāniyya*) and their corresponding pneumas as Natural, Vital, and Psychic, respectively.
7. **Functions (*Af'āl*)**
   Af'āl refers to the functions and movement of all organs in the body due to the operation of faculties or physiological powers. In a healthy body, all the organs are in a proper shape or without any structural abnormality and perform their respective functions in an efficient manner. The humors themselves are assigned temperaments—blood is hot and moist, phlegm cold and moist, yellow bile hot and dry, and black bile cold and dry. There is a unique humoral constitution in every person representing his state of health. To maintain a healthy balance, there is a power of self-preservation or adjustment called *Quwwat-e-Mudabbira* (*medicatrix naturae*). A weakening of this power causes imbalance in the *humoral* composition and causes disease. What the Unani physician actually tries to do is to prescribe a medicine, which helps the body to regain this power and thereby restore the humoral balance. Correct diet and digestion is integral to this system.

10.4.1 Unani (Greco-Arab) System of Medicine and Cancer (Sartan)

The origin of the word cancer is credited to the Greek physician Hippocrates (460–370 BC). He has used the terms karkinos (carcinos), the Greek word (*καρκίνος*) for crab or crayfish, and carcinoma to connote ulcer- and non-ulcer-forming tumors. In Greek, these words used for a crab and most probably applied to the disease because the physical appearance of the cut surface of a solid malignant tumor, with the veins stretched on all sides (fingerlike spreading projections) as the crab has its feet. Celsus (ca. 25 BC–50 AD) translated karkinos into *cancer*, the Latin word for crab. Galen (130–200 AD), another Greek physician, for the first time used the term "*oncos*" (Greek word for swelling) to depict tumors. The crab analogy given by Hippocrates and Celsus is still in use to describe malignant tumors (ACS 2014). Hippocrates believed that a person is healthy, if all *humors* of the body are balanced, and if any *humor* is too much or little, then the disease will appear. An excess of black bile in body sites was considered to cause cancer. Hippocrates stated that cancer is a non-treatable disease. If treated, the patient may die; if not treated, patient may lead a longer life. It connotes that ablation and cauterization of cancer lesion may result in its infliction of vital organs (Tabri 2010). The Hippocratic theory of cancer was passed on and comprehends by influential Galen, who promulgated the concept in his medical education, which remained indisputable standard through the Middle Ages for over more than thousand years.

Arab-Muslim scholars in medicine, like Al Razi (Rhazes 860–930 AD), Al-Zahrawi (Abulcasis 936–1013 AD), and Ibn Sina (Avicenna 980–1037 AD), were well aware about the cancer. Al Razi explained most types of cancers known at his time and suggested several treatments for cancer on the basis of his knowledge on burnt black bile in the affected organs. He recommended the evacuation of black bile from the organ by emesis and purgation and using cold tempered foods and medications. Abulcasis was the first Andalusian scholar who conducted surgical removal of the cancerous breast and suggested that cancer can be managed surgically only in its initial stages, if complete removal is possible. Likewise, Avicenna also inferred that cancer is extremely difficult to treat caused by an excess of burnt black bile, which also produced excessive heat in the body. Avicenna believed that imbalance of any humor may lead to the inequilibrium of all the four humors. He differentiated benign tumor from a cancerous one by certain symptoms like pain, throbbing, and rapid growth and noted that cancerous tumors send out "crab like tracks" and predominantly occur in the hollow organs. Avicenna also stated that cancers often inflict muscles, tendons, and lymph nodes. Moreover, Abulcasis recognized that a cure was most likely if the cancer was detected at its earliest stage, and the ultimate goal should be to halt the cancerous growth (Said et al. 2011). Abulcasis for the first time clearly mentioned the cancer of kidney and was able to distinguish between acute kidney inflammation and cancer. All the Arab scholars including Rhazes, Abulcasis, and Avicenna realized that the possibilities of healing cancer are prognostically maximized if the cancer is diagnosed at an early stage

(Rhazes 925, Avi Senna 1037). According to Rhazes and Avicenna, cancer is a tumor that is complex to treat. They stated that a "cancerous tumor is destructive and progressively increases in size, spreading its roots which insinuate themselves in the tissue elements." Avicenna promulgated that if one of the body's humors goes out of balance, then all the four humors would also be disturbed. Excess of black bile generates excessive heat in the body which may cause all sorts of cancers. Avicenna further elaborated that cancers often involve muscles, tendons, and lymph nodes; cancerous lesions send out "*crab-like tracks*" and predominantly occur in the "*hollow*" organs. A benign tumor could be differentiated from a cancerous one on the basis of certain obvious symptoms such as throbbing pain and rapid growth. Diamandopoulos (1996) classified the tumors into three major types: (1) *onkoi* (lumps or masses), (2) *karkinomas* (non-ulcerating cancers), and (3) *karkinos* (malignant ulcers).

The first aim of the treatment strategy should be to arrest the growth of cancerous lesion. Avicenna advocated "When cancer starts, it may be possible to keep it as it is, so that it will not increase and keep it non-ulcerated. It may happen sometimes that the stating cancer may be cured. But when it is advanced, verily will not." The first description for treating the cancer was put forth by Avicenna in *Canon of Medicine*, wherein surgical removal of small and accessible tumor not closer to vital organs was described. It stated that the radical excision should be done to remove all diseased tissue, including amputation or the removal of veins extending in the direction of the tumor. Avicenna stated, "*can be arrested with anything, it can be so by vigorous excision.... including all the (blood) vessels supplying the tumor so that nothing of these will be left.*" The other citation by Avicenna states "*It was told by one of the predecessors that a physician had excised a cancerous breast radically then cancer developed in the other breast. My opinion is that the second breast might have been on its way to cancerization (a dormant cancer) which fits this case and it is possible to be a spread of the material (cancerous from the first breast) and this is more evident (opinion).*" Avicenna also recommended for purifying the body from excess of black bile by advising a nutritious and balanced diet to the patient in order to retain purity and strengthening the immune system before surgery. He has also attempted the earliest pharmacological treatments for cancer with a herbal drug, *Cichorium intybus* (Hindiba), which later was identified by Ibn al-Baitar as having "anticancer" property that could also be used for the treatment of other types of tumors and neoplastic diseases (Rhazes 925; Avi Senna 1037; Ibn AlBitar 1874, Avi Senna 1993; Saad et al. 2005, 2008; Khan 2006; Said et al. 2011).

10.4.2 Etiopathogenesis of Cancer (Sartan)

Sartan is a melancholic inflammation due to increased combustion of black bile (*Sawdā'*) that produces more intrinsic heat, thus becoming pathological. Avicenna noted that cancer may be caused by combustion of pure yellow bile, black bile mixed with that of yellow one, or pure black bile. It is characterized by pain, burning, and pulsation and increases in size very rapidly due to excess morbid matter and swelling. Cancer gives off newer vessels, blackish or greenish in color, which resemble those of crab's walking legs. Neovascularization in cancerous lesions simulates to the legs of crab which are penetrated deep into the skin as well as subcutaneous tissues and turn hard. The vessels of the cancerous organ filled the cancerous matter, get dilated, and solidified (Tabri 1995). It is primary in its origin, not occurring secondary to any other disease. It is a sensitive lesion that commonly occurs in the hollow organs; hence, it is more prevalent in women and is equally common in nervous organs. Initially, it remains dormant and then appears clinically in the form of tumor and swelling followed by signs and symptoms (Ibn Sina 1992). Rhazes noted that cancer in its initial stage is treatable. Chronic cancer may be treated with surgical removal of tumor. The three most significant complications of surgical ablation are severe hemorrhage, severe pain, and recurrence of cancer (Razi 2002).

Averroes (also known as Ibn Rushd) stated that the disease occurs either due to disturbance in quantity (*Kamiyat*) and quality (*Kaifiyat*) or both in melancholic bile. This derangement in normal constituent of black bile at times may be traced to those substances which are stimulated by excess and prolonged intake of cold (*Barid*) and dry (*Yabis*) edibles and predominance of cold and dry (*Sard Khushk*) or hot and dry (*Har Khushk*) dis-temperament on alimentary organs (Ibn Rushd 1980). The leading cause of *Sartan* may be any abnormal change occurring in the black bile (*Sawdā'*). There are five types of abnormal changes that occur in *Sawdā'*: (1) excess production of normal black bile, (2) burning of normal black bile into abnormal, (3) formation of black bile due to combustion of blood, (4) formation of black bile due to combustion of phlegm, and (5) formation of black bile due to combustion of (*Safra*) bile (Quraeishi 2002; Samarqandi 2010). When abnormal black bile is excessively produced in the human body, spleen succumbs to its complete absorption; later on, the same is spread to the blood circulation from which the organs get their nutrients resulting into a lethal disease. Cancer results due to these abnormal humors which may be of two kinds: ulcerative cancer due to combustion of abnormal black bile and non-ulcerative cancer due to combustion of normal black bile (Ibn Rushd 1980). The ulcerative lesion has intense burning sensation as though sparkle is put on it with its effect onto the internal organs. The margins of ulcer are lusterless, thick, dark red, and externally averted (Qamari 2008). External factors such as occupation and environment or both may result in derangement of black bile. At times, the causes to the abnormality in black bile either qualitative or quantitative may be related to hereditary factors such as leprosy (Ibn Rushd 1980). Nerves, veins, and arteries are commonly inflicted with the spread of cancerous lesions. It may also inflict the esophagus and larynx (Qamari 2008). As the morbid matter of cancer lesion is solid and thick, the tumor neither dissolves nor moves from its site. Foul smelling pale water oozes from the ulcerative lesions as some parts of the lesion are combusted, whereas some are infected (Samarqandi 2010). It is initially circular in shape, firm in consistency and small in size with earthy tinge, and later on develops into a big abscess (Ibn Qiff Maseehi 1995).The size ranges from gram to that of watermelon (Qamari 2008). It is a firm and solid tumor with its roots deeply seated into the organs. It is manifested by rough texture and distension with greenish vessels spread in its surroundings. The pain increases with the increase in its size. It starts with severe pain with no relief in pain from any drug. It is hot on touch. Some of its types may be without pain which is curable and has good prognosis (Baghdadi 2004).

Sartan is comparatively more common in women due to their softness of body which easily accepts the morbid thick matter; whereas the solid body as in case of men hardly accepts these matters, hence less prevalent in men (Qamari 2008). It is very common in wet organs such as women's breast, uterus, intestine, palate, face, and stomach; it may inflict any part and organ of the body, whereas cancer of intestine, genitals, and mouth is more common in men (Tabri 1997).

10.4.3 Management of Cancer (Sartan)

The pathological changes of temperament in particular body parts or organs amount to the emergence of a disease, which can be treated or cured by applying drugs and nondrug factors. In Unani system of medicine, there is a correlation in between the temperament and other nondrug factors (such as environmental, diet, mental state, and others), which directly or indirectly affect the temperament. The Unani system advocates to utilizing these nondrug factors very frequently and extensively. The principle modes of treatment in the Unani system of medicine include (1) regimenal therapy (*Ilāj bi'l-Tadbīr*), (2) dietotherapy (*Ilāj bi'l-Ghidhā*), (3) pharmacotherapy (*Ilāj bi'l-Dawā*), and (4) surgery (*Ilāj bi'l-Yad*).

The humoral theory of Unani system deciphered cancer as a result of diffusion of burnt black bile into the body organs which lead to "boiling" of the black bile in that cancerous site. The first step propounded in the management of cancer was to evacuate the black bile out of the body by the use of medications and regimental methods. According to Avicenna, medications for cancer have four purposes: (1) complete arrest of cancer lesion, (2) to arrest its progress, (3) to stop ulceration, and (4) healing of ulceration. Galen in his "methods of treatment" has hypothesized that cancer is a metabolic ailment associated with black bile which is difficult to be diagnosed at initial stages, and it should be treated systemically rather than locally. In order to treat cancer, Galen has proposed that removal of black bile from the body by administering a suitable purgative and then the production and buildup of black bile in the vessels should be prevented as far as possible. In case this method is not possible, then black bile should be evacuated from the body at regular time intervals (Said et al. 2011; Shamsi et al. 2016).

It is very important to know the stage of cancer before its treatment. The advanced stage of cancer is very difficult to treat. Avicenna and Galen clearly cautioned to treat the cancers which are developed inside the body and could not be observed by naked eyes. Initially, the evacuation of morbid substance (black bile) should be accomplished using steady purgation or venesection (*Fasd*) with highly nutritious, cold temperament balanced diets given to produce normal blood in patients, and then it can be treated safely (Razi 2002).

The patients of cancer (*Sartan*) can be treated on the principle of (1) correction of temperament (*Taadil-e-Mizaj*) and (2) cleansing of morbid substance (*Tanqia-e-Mawad*). In cancer, there is imbalance in black bile (*Sawdā'*) having cold and dry temperament, and therefore, by adopting the measures of *Ilaj-bil-zid*, the drugs and food having hot and wet temperament are given to antagonize the hotness of cancer. The diet in small quantity should be light (*Latif*), healthy (*Saaleh*), and easily digestible (*Jaiyyadul Kaimus*), having cold and wet temperament (Shamsi et al. 2016).

10.4.3.1 Correction of Temperament (*Taadil-e-Mizaj*)

Most of the ancient Unani physicians recommended the use of plants with cold temperaments in order to antagonize the cancer hotness. Rhazes and all the preceding Unani scholars had recommended cold-tempered plants in the category of third and fourth degrees such as *Brassica oleracea* (Karamkalla), *Bryonia syriaca*, *Lactuca sativa* (Kahu), *Lactuca serriola*, *Linum usitatissimum* (Alsi/Bazarul Kattan), *Malva sylvestris* (Khubbazi), and *Portulaca oleracea* (Khurfa). The recommended foods were also with cold temperament: milk, beans, *Hordeum vulgare*, *Portulaca oleracea* (Khurfa), *Cichorium intybus* (chicory), *Cucurbita maxima* (Kaddu), *Amaranthus polygamus* (Cholaee), *Chenopodium album* (Bathua), *Cucumis sativus* (Kheera), *Punica granatum* (anar), *Spinacia oleracea* (spinach), *Prunus amygdalus* (badaam), whey water (Ma-ul Jubn), barley water (Maus-Shaeer), Mufarreh, and exhilarant fruit juices (*Mashroobaat*) like apple juice (*Sharbat-e-saib*). *Arq-e-Neelofar* and *Arq-e-baid-Mushk* can effectively be used as a cardiac tonic (Samarqandi 2007a; Said et al. 2011; Shamsi et al. 2016).

10.4.3.2 Cleansing of Morbid Substance (*Tanqia-e-Mawad*)

Regimenal therapy (*Ilāj bi'l-Tadbīr*) is one of the most popular methods of treatment described by most of the ancient Unani physicians. Hippocrates, Galen, Rhazes, Avicenna, and Abulcasis described various regimens including *Munzij-o-Mus'hil* therapy for the management of diseases, either independently or in combination with other therapies. The regimen consists of two phases: (1) concoction (*Nuzj*) and (2) purgation (*Ishāl*).

Concoction is a method by which viscous *humors* (morbid materials) are modified into a form which could easily be expelled out of the body with the help of purgatives. Purgation (*Ishāl*) is the extraction of the morbid materials from the vessels and other deeper tissues. Concoction is essential in case of all chronic disorders as well as the diseases having a duration of more than 40 days soon after the signs of concoction (*Nuzj*) completion appear; the purgative (*Mus'hil*) drugs are added to the *Munzij* drugs. Purgative (Mus'hil) drugs bring out the purgation (*Ishāl*) of the morbid *humors* (*Phlegm*, *Sawdā'*, and *Safra*) that have been evacuated by the action of concoctive drugs (*Munzijat*). The *Munzij-o-Mus'hil* therapy can be used to evacuate the disease causing morbid materials out of the body (Kabiruddin 2006; Shamsi and Rehman 2012; Shamsi et al. 2016).

10.4.3.3 Concoctive Drugs (*Munzij-e-Sawdā'*)

Althaea officinalis (Khitmi), *Borago officinalis* (Gaozaban), *Cichorium intybus* (Tukhme Kasni), *Cordia dichotoma* (Sapistaan), *Foeniculum vulgare* (Badiyan), *Fumaria officinalis* (Barg-e-Shahatra), *Glycyrrhiza glabra* (Asl-ussoos), *Lavandula stoechas* (Ustukhuddus), *Rosa alba* (Gule Surkh), *Vitis vinifera* (Maweez Munaqqa), *Zizyphus vulgaris* (Unnab), and others.

10.4.3.4 Purgative Drugs (*Mus'hil-e-Sawdā'*)

Aloe barbadensis (Aelwa), *Cassia angustifolia* (Sana-e-Makki), *Cuscuta reflexa* (Afteemoon), *Emblica officinalis* (Amla), *Lavandula stoechas* (Ustukhuddus), *Polypodium vulgare* (Bisfayaj), *Terminalia bellerica* (Balela), *Terminalia chebula* (Halela Zard), and others.

10.5 Important Unani Formulations for *Munzij* and *Mus'hil* Therapy for Cancer

Different Unani formulations for *Munzij* and *Mus'hil* regimen are mentioned in classical text that can be used for the evacuation of morbid *Sawdā'* (Shamsi et al. 2016):

1. Unnab (*Zizyphus vulgaris*), 5 pieces; Gaozaban (*Borago officinalis*), 7 g; Shahatra (*Fumaria parviflora*), 7 g; Badranj boya (*Nepeta ruderalis*), 7 g; Badiyan (*Foeniculum vulgare*), 7 g; and Asl-us-Soos-muqashar (*Glycyrrhiza glabra*), 5 g. All the drugs are soaked in hot water overnight, and after serving add Gulqand, 20 g, or Turanjbeen (*Alhagi pseudalhagi*) 20 g (Azmi 2002).
2. Ayarjaat like Ayaraj-e-feyqra (*Aloe barbadensis*) or Ayaraj-e-Jalinoos or Ayaraj-e-Roofas to eliminate the morbid *Sawdā'* and for Taadil-e-Mizaj (Samarqandi 2007b).
3. Asl-us-Soos (*Glycyrrhiza glabra*); Gaozaban (*Borago officinalis*), 5 g; Gul-e-Gaozaban (flowers of *Borago officinalis*), 5 g; Badiyan (*Foeniculum vulgare)*; Beekh-e-Badyan (root of *Foeniculum vulgare*); and Tukhm and Bekh-e Kasni (seed and root of *Cichorium intybus*), 7 g each to be partially grounded. Gule Surkh (*Rosa alba*); Ustukhuddus (*Lavandula stoechas*), 7 g; Shahtara (*Fumaria officinalis*), 7 g; Tukhme Khatmi (*Althaea officinalis*), 7 g; Aloo Bukhara (*Prunus domestica*) 5 nos., medium size; and Maweez Munaqqa (*Vitis vinifera*) 9 nos., medium size. All the drugs to be soaked overnight in water followed by boiling

in the morning. Approximately 40 g Gulqand (sweet preserve of rose petals) to be added in filtrate. The preparation to be given as a drink (Kabiruddin 2006).

4. Kharbaq Siyah (*root of Picrorhiza kurroa*) 375 mg; Habbul Neel (*Ipomoea hederacea*), 256 mg; Shahtara (*Fumaria officinalis*) leaves and flowers, 375 mg; Habbul Ghar (*Laurus nobilis*), 375 mg; Hanzal (*Citrullus colocynthis*) fruit pulp, 512 mg; Afsanteen (*Artemisia absinthium*) whole plant, 1024 mg; Ghariqoon (*Agaricus alba*), mushroom, 1024 mg; Ayarij-e-Faiqra (Unani compound formulation), 1.75 g; Milhe Nafti (black salt), 630 mg; and Saqmonia (*Convolvulus scammonia*) resins, 630 mg; make powder of all drugs, and sieve it through silk cloth, and make pepper-sized tablet using cabbage water as a binder. Oral administration of 12 g tablet once in a week (Tabri 1995).

10.5.1 Dietotherapy (Ilāj bi'l-Ghidhā) in Unani System of Medicine

The Unani system of medicine lays rules on prevention of disease and promotion of health through a balanced lifestyle, which revolves around six essential factors (*Asbāb Sitta Ḍarūriyya*): air (*Hawā*), food and drinks (*Ma'kūl o Mashrūb*), bodily movement and repose (*Ḥarakat o Sukūn Badanī*), psychic movement and repose (*Ḥarakat o Sukūn Nafsānī*), sleep and wakefulness (*Nawm-o-Yaqẓa*), and evacuation and retention (*Istifrāgh-o-Iḥtibās*). Noncompliance with these principles leads to an errant lifestyle and ultimately leads to disease. The Unani system of medicine believes that a physician is not the healer but merely an assistant to physique (*Tabiyat*) of the body, which is the true healer; hence, the diet should be in accordance with it. Diet improves the health condition by preventing the complications. Since, *Sartan* is considered to be associated with black bile, hence, *Sawdā*' producing diet/foods may be harmful to cancer patients. The diet for cancer patient must be easily digestible (*Jaiyyadul Kaimoos*) and light (*Latif*) with cold and moist *temperament* including *Cucurbita maxima* (Kaddu), *Amaranthus polygamus* (Cholaee), *Chenopodium album* (Bathua), *Cucumis sativus* (Kheera), *Punica granatum* (Anar), whey water (*Ma-ul Jubn*), and barley water (*Maus-Shaeer*). Many items including wild animal beef, cow beef, red lentils, cabbage, some dry fruits, and concentrated liquor are possible contraindicated foods that should not be used in cancer patients (Tabri 1995; Alam et al. 2013; Anonymous 2016).

10.5.2 Pharmacotherapy ("Ilāj bi'l-Dawā") in Unani System of Medicine

The three primary sources of drugs (*Mawālīd Thalātha*) which are used for medication are plants, animals, and minerals. These drugs are either used singly or in compound formulations. Avicenna and other ancient scholars, like modern chemotherapy, used materials derived from mineral, mainly to stop and prevent cancer development. Minerals such as blue vitriol, zinc, and iron were applied onto the cancer lesion or immediately after surgical removal (Said et al. 2011). Avicenna advised to avoid irritant medications; however, he clarified to use good medications as washed (pure) minerals such as tutty (copper sulfate) mixed with rose grease oils and oil of yellow gillyflower. Medications for the prevention of ulceration were described as "smears that prevent cancer progress provided they will not be irritant; all of them are useful, particularly if mixed with the mixture of lead stone and stone pounder of aromatics."

Unani drugs/formulations for ulcerated and non-ulcerated lesions are described in classical Unani text. Sprinkling of Baid Saada (Kohna) powder on cancer ulcers and washing of cancerous wounds with decoction of Chunar leaves, liniment of *Curcuma longa* (Haldi), *Boswellia serrata* (Kundur), and *Aloe barbadensis* (Sibr) in equal quantity to be applied on ulcer mixed with rose oil (Roghan-e-Gulab) (Razi 2002). Ointment of Zoofa and Lu'ab Bahi Dana mixed with Roghan-e-Banafsha (Tabri 1995). Roghan-e-Gul, Kishneez Sabz, and Mako Sabz can be used to block cancerous growth. Pus formation can be prevented by the use of A'b-e-Kahu and Roghan-e-Zaitoon (Samarqandi 2010). Local application of suppository made up of Afyun, Nakhuna, and Za'fran may be used in eye and nasal cancers (Baghdadi 2004). Paste of pomegranate is highly efficacious in cancer ulcer, while bartang leaves (*Lahi'ya-tut-tees*) can be used for checking bleeding (Jurjani 2010). In case of uterine cancer, "sitz bath" can be taken by using the decoction of Khatmi, Methi, Alsi, and Ikleel-ul-Mulk. The paste of Methi, Alsi, Qinnab Seeds, dry Banafsha, Habbul Ghar, Babuna, Guggul, Sandal Surkh, Khatmi, and Soya can also be used (Majoosi 2010).

10.5.2.1 Plant Origin Unani Drugs with Anticancer Properties

Plants are the major source of Unani drugs. The plant origin drugs are prescribed with an intention to keep the tumor in its dormant stage, to stop its further growth, to dissolve it, and lastly to surgically remove it (Aslam et al. 1981). In the advancement of modern analytical tools and techniques, some of the single Unani drugs have been evaluated for anticancer activities (Table 10.3), but negligible attention has been paid to explicate the safety and efficacy of Unani compound formulations to treat the cancer patients through clinical trials.

Table 10.3 Important Unani medicinal plants (single drugs) with anticancer activity

Botanical name (family)	Unani name	Anticancer activity	References
Abelmoschus moschatus (Malvaceae)	Habbul-Mushk	Hydroalcoholic seed and leaf extracts exhibited antiproliferative activity against colorectal adenocarcinoma and retinoblastoma human cancer cell lines	Gul et al. (2011)
Abrus precatorius (Fabaceae)	Ghunchi	High antitumor activity of agglutinin protein purified extract from the seeds. *In vitro* anticancer activity of aqueous extract of leaves on the murine mastocytoma cancer cell line (P815)	Panneerselvam et al. (2000) and Lébri et al. (2015)
Acorus calamus (Araceae)	Waj-e-Turki, Bach	Methanolic and ethanolic extracts and essential oil of rhizome showed anticarcinogenic and antiangiogenic effects on AGS cells, and their components might act against the proliferation of AGS cells	Rahamooz Haghighi et al. (2017)
Aegle marmelos (Rutaceae)	Belgiri (Bel), Safarjale-Hindi	Leaf extracts inhibited the growth of leukemic K562, T-lymphoid Jurkat, B-lymphoid Raji, and breast cancer cell lines MCF-7 and MDA-MB-231 and increase *ERa* gene expression in MDA-MB-231 due to antineoplastic effects. The ethanolic fruit extract has cytotoxic effect on SKBR3 cells *in vitro*. Extracts also showed ability to inhibit the *in vitro* proliferation of human tumor cell lines, including the leukemic K562	Lampronti et al. (2003), Moongkarndi et al. (2004), and Jagetia et al. (2005), Subramaniam et al. (2008), Baliga et al. (2012), and Baskar et al. (2012)
Alpinia galanga (Zingiberaceae)	Khulanjaan	Antiproliferative activity and induction of apoptosis in human breast carcinoma cell line (MCF-7) and nonmalignant (MRC-5). ACA (1′S-1′-acetoxychavicol acetate) obtained from rhizome exhibited anti-colorectal adenocarcinoma activity in SW480 cell lines	Samarghandian et al. (2014) and Baradwaj et al. (2017)

(continued)

Table 10.3 (continued)

Botanical name (family)	Unani name	Anticancer activity	References
Alstonia scholaris (Apocynaceae)	Kashim	Demonstrated the chemopreventive potential of bark extract in DMBA-induced skin tumorigenesis in Swiss albino mice. Alkaloid fraction studied *in vitro* in cultured human neoplastic cell lines (HeLa, HepG(2), HL60, KB, and MCF-7) and in Ehrlich ascites carcinoma-bearing mice. Increase in the antineoplastic activity was observed in HeLa cells	Jagetia and Baliga (2006) and Jahan et al. (2009)
Ammi majus (Apiaceae)	Atrilal	8-MOP inhibited cell growth in several cancer cell lines. The SK-N-AS and SW620 cells were the most sensitive to the compound. 8-MOP impaired the PI3K/AKT signaling pathway and, independently of photoactivation, reported to inhibit the growth of neuroblastoma and colon cancer cells by induction of apoptosis via intrinsic and extrinsic pathways	Bartnik et al. (2017)
Argemone mexicana (Papaveraceae)	Satyanasi	Alkaloids strongly inhibited the cell proliferation in human colon cancer cell (SW480) line. Aqueous and methanolic extracts inhibited a cytotoxic effect on A549, SiHa, and KB immortalized cell lines	More and Kharat (2016) and Singh et al. (2015)
Artemisia absinthium (Asteraceae)	Afsanteen (Qaisoom)	Crude extracts of the aerial parts inhibited cell proliferation and promote apoptosis in a human breast carcinoma estrogenic-unresponsive cell line (MDA-MB 231) and an estrogenic-responsive cell line (MCF-7). Methanol extracts of flowers exhibited considerable cytotoxic effect on breast cancer cell line (MCF-7). Decoction of the plant reported to be potent inducer of *in vitro* apoptosis of B chronic lymphocytic leukemia (CLL) cells	Shafi et al. (2012), Gordanian et al. (2014), and Mirkin et al. (2017)

(continued)

Table 10.3 (continued)

Botanical name (family)	Unani name	Anticancer activity	References
Artemisia nilagirica (Asteraceae)	Nagdon (Branjasif)	Exhibited considerable cytotoxic activity against breast cancer cell line (MCF-7). Essential oil (EO) induced apoptosis in human promyelocytic leukemia (HL-60) cells. Inhibition of proliferation and apoptosis in a human monocytic leukemia (THP-1) cell line via mitochondria-dependent and death receptor-dependent apoptotic pathways	Gordanian et al. (2014), Saleh et al. (2014), and Gul et al. (2016)
Asparagus racemosus (Liliaceae)	Satawar	The isolated shatavarin showed potent cytotoxicity (tumor volume, packed cell volume, viable tumor cell count, and increased nonviable cell count) in tumor-bearing mice. Inhibited growth of renal cell carcinoma cell (RCC UOK 146) lines	Mitra et al. (2012) and Verma et al. (2014)
Averrhoa carambola (Oxalidaceae)	Kamrakh	Effective against hepatocellular carcinoma (HCC) in Swiss albino mice. Showed the cytotoxic effects of DMDD (2-dodecyl-6-methoxycyclohexa-2,5-diene-1,4-dione) against human breast, lung, and bone cancer cells *in vitro*; DMDD suppressed the growth of breast carcinoma cells	Singh et al. (2014) and Gao et al. (2015)
Bacopa monnieri (Scrophulariaceae)	Jal-Brahmi	Aqueous extract showed biological activity of apoptosis in Ehrlich ascites tumor (EAT) cell lines. *Bacopa monnieri* water extract treatment of EAT cells produced apoptotic morphological characteristics, and *in vivo* DNA fragmentation was able to induce apoptosis in EAT cells via Bax-related caspase-3 activation and also exhibited the cytotoxic activity of ethanolic extract of dichloromethane (DCM) fraction on two different cell lines MCF-7 and MDA-MB 231) under *in vitro* conditions	Kalyani et al. (2013), Mallick et al. (2015), and John et al. (2017)

(continued)

Table 10.3 (continued)

Botanical name (family)	Unani name	Anticancer activity	References
Butea monosperma (Fabaceae)	Tesu (Dhak), Darakht-e-Palasha	The chemopreventive and antiangiogenic effects evaluated in a HBV-related X15-myc mouse model of hepatocellular carcinoma (HCC). Butanol fraction (Bmbu) from bark showed cytotoxic and apoptosis inducing activity in MCF-7 breast cancer cell lines. Butrin, a novel compound isolated from *Butea monosperma* flowers on suppressing the expression of SIRT1 and Aurora B kinase-mediated apoptosis in colorectal cancer cells (SW480 CRC)	Choedon et al. (2010), Kaur et al. (2017), and Subramaniyan et al. (2017)
Cannabis sativa (Cannabinaceae)	Qinnab (Bhang)	Cannabinoids inhibited angiogenesis and decrease metastasis in various tumor types in laboratory animals. It showed specific cytotoxicity against tumor cells while protecting healthy tissue from apoptosis. It also displayed potent anticancer activity against tumor xenografts, including tumors that express high resistance to standard chemotherapeutics	Caffarel et al. (2012), Velasco et al. (2016), and Bogdanović et al. (2017)
Centella asiatica (Apiaceae)	Brahmi Buti	Anticancer potential of asiatic acid on human ovarian cancer cells (SKOV3 and OVCAR-3), through inactivation of the PI3K/Akt/mTOR pathway. MicroRNA (miR)-1290 promoted asiatic acid-induced apoptosis in non-small cell lung carcinoma cells (A549). An aqueous extract (AE) exhibited activity against mouse melanoma (B16F1), human breast cancer (MDA-MB-231), and rat glioma (C6) cell lines	Pittella et al. (2009), Kim et al. (2014), and Ren et al. (2016)
Cichorium intybus (Asteraceae)	Kasni	Exerted significant activity against amelanotic melanoma C32. Strong antiproliferative effect against HepG2 cell lines. Showed notable effects on the leukemia cell lines. Root extract exhibited tumor inhibitory activity against Ehrlich ascites carcinoma in mice	Hazra et al. (2002), Conforti et al. (2008), Yook et al. (2015), and Esmaeilbeig et al. (2015)

(continued)

Table 10.3 (continued)

Botanical name (family)	Unani name	Anticancer activity	References
Curcuma longa (Zingiberaceae)	Chob Zard (Haldi)	Extracts exhibited antiproliferative effects in cholangiocarcinoma (CCA) cells by inducing pro-apoptotic signals and modulating signal transduction molecules. Curcumin significantly induced apoptosis and cell cycle arrest in pancreatic cancer (PC) cell lines. Curcuma C20-dialdehyde exhibited anticancer effects against colon and cervical cancer cell	Chaithongyot et al. (2015), Zhou et al. (2016), Calaf and Roy (2017), and Leelawat and Leelawat (2016)
Cuscuta reflexa (Convolvulaceae)	Aftimun	Exhibited antiproliferation activities in HCT116 colorectal cell lines. Induced apoptosis in Hep3B cells through the upregulation of p53 and BAX and downregulation of Bcl-2 and survivin	Suresh et al. (2011) and Riaz et al. (2017)
Glycyrrhiza glabra (Fabaceae)	Asl-us-Soos	Licochalcone A (LCA) suppressed the oxidation of cancer cells and markedly inhibits the proliferation of cells. Aerial parts showed remarkable activity against human cancer cell lines (HeLa, MCF-7, MDA-MB-231, Caco-2, and PC3). Inhibited proliferation of the colon cancer cell line HT-29. Isoangustone-A (IAA), a novel flavonoid from licorice root, suppressed proliferation of human melanoma cell lines SK-MEL-28	Song et al. (2013), Nourazarian et al. (2015), Aiello et al. (2017), and Chen et al. (2017)
Justicia adhatoda (Acanthaceae)	Arusa	Isolated 2-acetyl-benzylamine screened for potent anticancer properties against leukemia cells: CEM, NB-4, MOLM-14, Jurkat, IM-9, K562, and HL-60. Significant cytotoxic properties reported against MOLM-14 and NB-4 cells. Vasicine acetate was obtained by acetylation of vasicine and showed prominent cytotoxic activity *in vitro* against A549 lung adenocarcinoma cancer cell line	Duraipandiyan et al. (2015) and Balachandran et al. (2017)

(continued)

Table 10.3 (continued)

Botanical name (family)	Unani name	Anticancer activity	References
Melia azedarach (Meliaceae)	Bakayin	Seed kernel extract exhibited highest cytotoxic activity and selectivity to cancer cell lines (HT-29, A549, MCF-7, and HepG-2 and MDBK). Hexane extract of the fruits exhibited cytotoxic activities against leukemia (HL60), lung (A549), stomach (AZ521), and breast (SK-BR-3) cancer cell lines	Jafari et al. (2013) and Pan et al. (2014)
Nigella sativa (Ranunculaceae)	Habbatus sauda (kalonji)	Thymoquinone (TQ), a major ingredient of black seed oil (*Nigella sativa*), exhibited antimetastatic capacity in prostate cancer DU145 and PC3 cells. TQ exhibited *in vitro* anticancer activity in animal models of cancer via numerous mechanisms of action likewise, antioxidant activity, interfering with DNA structure, affecting carcinogenic signaling molecules and immunomodulation activities. Moreover, the treatment of seed extract and seed oil significantly reduced the viability of human lung cancer cells (A549 cells)	Al-Sheddi et al. (2014), Asaduzzaman et al. (2017), Kou et al. (2017), and Mostofa et al. (2017)
Withania somnifera (Solanaceae)	Asgandh	The methanolic leaf extract showed a strong antiproliferative activity against MCF-7, HCT116, and HepH2 cell lines. Withaferin A (WFA) exhibited strong antiproliferative activity against osteosarcoma U2OS (a rare type of osteocancer), affecting human cell lines, via generation of ROS and disruption of mitochondrial membrane potential. Crude water extract of roots exhibited potent cytotoxic effect on human malignant melanoma A375 cells. Anticancer activity of withaferin A human and murine B cell lymphoma cell lines	Halder et al. (2015), McKenna et al. (2015), Alfaifi et al. (2016), and Li et al. (2017)

10.6 Current Status of Unani (Greco-Arab) System of Medicine in India

Nowadays, the pursuit for alternate to the conventional oncologic therapies has increased tremendously. The US National Cancer Institute (NCI) supports complementary and alternative medicine and has been instrumental in funding and collaborating with the institutions/group working on traditional Chinese medicine, Ayurveda, and other traditional therapies (Jia 2012). In India, the initiative on oncological research has been taken by Central Council for Research in Unani Medicine (CCRUM), New Delhi; so far EMR project "Evaluation of anti-proliferative activity of Unani drugs in cancer prevention- A mechanistic approach" has been completed, but its results are yet to come in public domain. Other projects that are in progress are Screening of Anticancer potential of Indian Medicinal Plants, Investigation of Anti-Cancer activities of some selected medicinal plants and their molecular targets on oral cancer cell lines, Analysis of efficacy of *in vitro* raised plant (*Catharanthus roseus*) extracts in protecting chemically induced carcinogenesis in model rat, and Screening and purification of plant extracts used in the Unani system of medicine against microbial flora of oral cavity: antimicrobial and carcinogenic activity (Anonymous 2017a).

Recently, the Ministry of AYUSH, Government of India, through CCRUM has started a program to integrate Unani Medicine with National Program for Prevention and Control of Cancer, Diabetes, Cardiovascular Diseases and Stroke (NPCDCS) in Lakhimpur district of Uttar Pradesh. The major objective of the program is to prevent the population from fatal diseases by early diagnosis and reducing the complications and drug dependency by the use of Unani medicine. In this program, patients are encouraged to make lifestyle changes and advised to take Unani medicines along with standard treatment for prevention of cancer and its complications (PIB 2016). Besides these measures, the Council is also making efforts to introduce Unani medicines as adjuvant therapy to improve the quality of life of patients with terminal cases of cancer in collaboration with Tata Memorial Cancer Hospital, Mumbai (Anonymous 2017b). Moreover, the Council has also started a collaborative study with Institute of Cytology and Preventive Oncology (ICPO), Noida (ICMR), to validate the efficacy of Unani pharmacopoeial drugs in the prevention of cancer of cervix (Anonymous 2017c).

At present, the research on cancer in Unani medicine is not much enthusiastic as the limited work has been done on the cancer. Thus, there is immense need to maximize the efforts to conduct the cancer trials with animal models on most common cancer types such as breast and colon cancers. There must be main concentration of Unani researchers on the evaluation and validation of Unani single as well as compound formulations in cancer on animal models as the work being done on cancer cell line is not much promising as the same is available across the scientific platforms in abundance. The focus should be on preclinical trial of Unani drugs on cancer animal models incorporating diagnostic and prognostic tools and carcinogenic biomarkers so that their efficacy can be elaborated more scientifically. Unani medicine should be studied as it incorporates herbs as whole or its part either singly or more than two in compound formulations. Each herb (drug) possesses various active principles that work synergistically and produce therapeutic effects and thereby lessen the prospects of adverse effects of formulations. The research work is already underway on single anticancer herbs. However, the question arises; does the findings give us any edge over other available conventional therapies. In current scenario, our main focus should be on compound formulations, which consists of two or more single drugs. The utilization of compound drugs in alleviation of cancer inflictions is still unexplored to a large extent. Hence, the clinical trial must be done with Unani compound formulation with minimum ingredients, i.e., two to three, so that their efficacy can be tested and proved scientifically.

10.7 Conclusions and Future Prospects

The thrust areas of research in oncologic therapies may encompass palliative pain management through Unani analgesics as an adjuvant therapy (immune-modulators such as *Khamīra Marwārīd* and *Ḥabb-e Jawāhar Mohra*) in management of hepato-toxicity (*Arqe Kasni*, *Majoon Dabidul Ward*), and nephrotoxicity (*Habb-e-Banadiqul Bazoor* and *Jawarish Zarooni*) develop due to anticancer agents, and also could be an alternate to the conventional therapies (Fig. 10.2). Interaction potential of some herbs with chemotherapeutic agents may be more critical. To avoid such

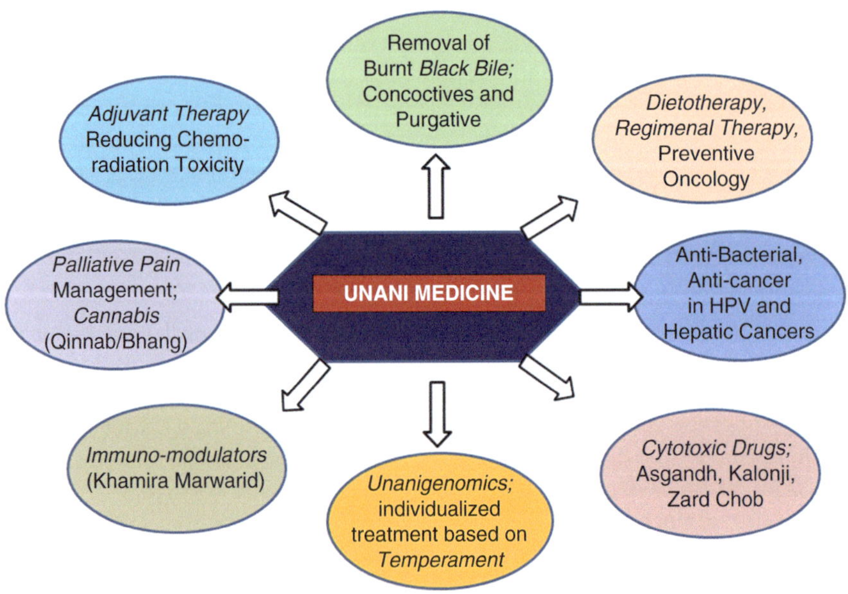

Fig. 10.2 Unani medicine and future strategies for research in cancer (*Sartan*)

circumstances, specific studies are warranted to optimize the dose requirement in cancer patients. The priority is to scientifically evaluate these valuable therapeutic options and generate evidence-based data for global acceptance of these age-old Unani medicines. But the path to these trends seems very difficult as the current status of Unani oncology is in its initial stage, but the same may be progressed on in a graded manner initiating with animal model trials on most prevalent cancer types in phases to the clinical trials. The future researches in Unani medicine may be undertaken to reduce the side effects of chemo- and radiotherapy in oncologic management as adjuvant to improve the quality of life of cancer patients. Moreover, Unani medicine is principally based on personalized medicine as the cure of every disease according to the humor involved, and patients are keenly monitored based on their specific condition. Thus, more clinical data are required to prove the effectiveness of the Unani system of medicine in the management of cancer.

Acknowledgment MK Husain is thankful to Central Council for Research in Unani Medicine (CCRUM), Ministry of AYUSH, Government of India, for the Young Scientist Award, 2017. The opinions expressed in this article are the author's own and do not reflect the view of either CCRUM or CRIUM.

References

ACS-American Cancer Society (2014) Early history of cancer. https://www.cancer.org/ cancer/cancer-basics/history-of-cancer/cancer-causes-theories-throughout-history.html. Accessed on 17 Oct 2017

Aiello F, Armentano B, Polerà N, Carullo G, Loizzo MR, Bonesi M, Cappello MS, Capobianco L, Tundis R (2017) From vegetable waste to new agents for potential health applications: antioxidant properties and effects of extracts, fractions and pinocembrin from *Glycyrrhiza glabra* L. aerial parts on viability of five human cancer cell lines. J Agric Food Chem 65:7944–7954

Alam A, Ahmed S, Alam T, Azeez A (2013) Cancer (Sartan) and its management in Unani (Greeco-Arab) medicine. Int J Pharm 1:612–630

Alfaifi MY, Saleh KA, El-Boushnak MA, Elbehairi SE, Alshehri MA, Shati AA (2016) Antiproliferative activity of the methanolic extract of *Withania somnifera* leaves from Faifa mountains, Southwest Saudi Arabia, against several human cancer cell lines. Asian Pac J Cancer Prev 17:2723–2726

Al-Sheddi ES, Farshori NN, Al-Oqail MM, Musarrat J, Al-Khedhairy AA, Siddiqui MA (2014) Cytotoxicity of *Nigella sativa* seed oil and extract against human lung cancer cell line. Asian Pac J Cancer Prev 15:983–987

Anonymous (2016) Unani system of medicine, the science of health and healing. Ministry of Ayurveda, Yoga and Naturopathy, Unani, Siddha and Homoeopathy (AYUSH), Government of India, New Delhi

Anonymous (2017a) CCRUM. http://ccrum.res.in/UserView/index?mid=1480. Accessed on 22 Oct 2017

Anonymous (2017b) CCRUM. http://ccrum.res.in/ViewData/Multiple?mid=1596. Accessed on 22 Oct 2017

Anonymous (2017c) Biovoice news. http://www.biovoicenews.com/clinical-trials-to-determine-efficacy-of-traditional-medicine-for-cancer. Accessed on 22 Oct 10.2017

Asaduzzaman Khan M, Tania M, Fu S, Fu J (2017) Thymoquinone, as an anticancer molecule: from basic research to clinical investigation. Oncotarget 8:51907–51919

Aslam M, Bano H, Vohora SB (1981) Sartan (cancer) and its treatment in Unani medicine. Am J Chin Med 9:95–107

Avi Senna AH (1993) Al Kanoon Fi Altib (The rules of medicine). IzAldin Publications, Beirut

Avi Senna AH (1037) Al Kanoon Fi Altib (The rules of medicine). Four volumes. Printed in 1993 by IzAldin Publications, Beirut (in Arabic)

Azmi AK (2002) Sartan aur Tibb Unani: Moalijatee imkaanat ka ijmali jaayezha. Pandhra Rozah-Tibb-E-Mashriq: Delhi, Cancer Special Issue. New Public Press, Gali Kasimjaan, Delhi, India, pp 36–57

Baghdadi IH (2004) Kitab *Al-Mokhtarat fit Tibb* (Urdu translation), vol-3 CCRUM. Ministry of Health and Family Welfare, Government of India, New Delhi, pp 90–128

Balachandran C, Arun Y, Sangeetha B, Duraipandiyan V, Awale S, Emi N, Ignacimuthu S, Perumal PT (2017) *In vitro* and *in vivo* anticancer activity of 2-acetyl-benzylamine isolated from *Adhatoda vasica* L. leaves. Biomed Pharmacother 93:796–806

Baliga MS, Thilakchand KR, Rai MP, Rao S, Venkatesh P (2012) *Aegle marmelos* (L.) Correa (Bael) and its phytochemicals in the treatment and prevention of cancer. Integr Cancer Ther 12:187–196

Baradwaj RG, Rao MV, Senthil Kumar T (2017) Novel purification of 1′S-1′-acetoxychavicol acetate from *Alpinia galanga* and its cytotoxic plus antiproliferative activity in colorectal adenocarcinoma cell line SW480. Biomed Pharmacother 91:485–493

Bartnik M, Sławińska-Brych A, Żurek A, Kandefer-Szerszeń M, Zdzisińska B (2017) 8-methoxypsoralen reduces AKT phosphorylation, induces intrinsic and extrinsic apoptotic pathways, and suppresses cell growth of SK-N-AS neuroblastoma and SW620 metastatic colon cancer cells. J Ethnopharmacol 207:19–29

Baskar AA, A Numair KS, Alsaif MA, Ignacimuthu S (2012) *In vitro* antioxidant and antiproliferative potential of medicinal plants used in traditional Indian medicine to treat cancer. Redox Rep 17:145–156

Bogdanović V, Mrdjanović J, Borišev (2017) A review of the therapeutic antitumor potential of cannabinoids. J Altern Compl Med (Online). https://doi.org/10.1089/acm.2017.0016

Boon HS, Olatunde F, Zick SM (2007) Trends in complementary/alternative medicine use by breast cancer survivors: comparing survey data from 1998 and 2005. BMC Women's Health 7:4. https://doi.org/10.1186/1472-6874-7-4

Caffarel MM, Andradas C, Pérez-Gómez E, Guzmán M, Sánchez C (2012) Cannabinoids, a new hope for breast cancer therapy. Cancer Treat Rev 38:911–918

Calaf GM, Roy D (2017) Metastatic genes targeted by an antioxidant in an established radiation – and estrogen-breast cancer model. Int J Oncol 51:1590–1600

Chaithongyot S, Asgar A, Senawong G, Yowapuy A, Lattmann E, Sattayasai N, Senawong T (2015) Anticancer effects of *Curcuma* C20-Dialdehyde against colon and cervical cancer cell lines. Asian Pac J Cancer Prev 16:6513–6519

Chen X, Liu Z, Meng R, Shi C, Guo N (2017) Antioxidative and anticancer properties of Licochalcone A from licorice. J Ethnopharmacol 198:331–337

Choedon T, Shukla SK, Kumar VJ (2010) Chemopreventive and anti-cancer properties of the aqueous extract of flowers of *Butea monosperma*. J Ethnopharmacol 129:208–213

Conforti F, Ioele G, Statti GA, Marrelli M, Ragno G, Menichini F (2008) Antiproliferative activity against human tumor cell lines and toxicity test on Mediterranean dietary plants. Food Chem Toxicol 46:3325–3332

D'Souza ND, Murthy NS, Aras RY (2013) Projection of cancer incident cases for India-till 2026. Asian Pac J Cancer Prev 14:4379–4386

Deng G, Cassileth B (2014) Integrative oncology: an overview. Am Soc Clin Oncol Educ Book:233–242

Diamandopoulos GT (1996) Cancer: an historical perspective. Anticancer Res 16:1595–1602

Dikshit R, Gupta PC, Ramasundarahettige C, Gajalakshmi V, Aleksandrowicz L, Badwe R, Kumar R, Roy S, Suraweera W, Bray F, Mallath M, Singh PK, Sinha DN, Shet AS, Gelband H, Jha P (2012) Cancer mortality in India: a nationally representative survey. Lancet 379:1807–1816

Duraipandiyan V, Al-Dhabi NA, Balachandran C, Ignacimuthu S, Sankar C, Balakrishna K (2015) Antimicrobial, antioxidant, and cytotoxic properties of vasicine acetate synthesized from vasicine isolated from *Adhatoda vasica* L. Biomed Res Int 2015:727304. https://doi.org/10.1155/2015/727304

Ernst E (2009) Complementary and alternative medicine (CAM) and cancer: the kind face of complementary medicine. Int J Surg 7:499–500

Esmaeilbeig M, Kouhpayeh SA, Amirghofran Z (2015) An investigation of the growth inhibitory capacity of several medicinal plants from Iran on tumor cell lines. Iran J Cancer Prev 8:e4032. https://doi.org/10.17795/ijcp-4032

Ferlay J, Soerjomataram I, Dikshit R, Eser S, Mathers C, Rebelo M, Parkin DM, Forman D, Bray F (2015) Cancer incidence and mortality worldwide: sources, methods and major patterns in GLOBOCAN 2012. Int J Cancer 136:E359–E386

Gao Y, Huang R, Gong Y, Park HS, Wen Q, Almosnid NM, Chippada-Venkata UD, Hosain NA, Vick E, Farone A, Altman E (2015) The antidiabetic compound 2-dodecyl-6-methoxycyclohexa-2,5-diene-1,4-dione, isolated from *Averrhoa carambola* L., demonstrates significant antitumor potential against human breast cancer cells. Oncotarget 6:24304–24319

Gordanian B, Behbahani M, Carapetian J, Fazilati M (2014) *In vitro* evaluation of cytotoxic activity of flower, leaf, stem and root extracts of five *Artemisia* species. Res Pharma Sci 9:91–96

Gul MZ, Bhakshu LM, Ahmad F, Kondapi AK, Qureshi IA, Ghazi IA (2011) Evaluation of *Abelmoschus moschatus* extracts for antioxidant, free radical scavenging, antimicrobial and antiproliferative activities using *in vitro* assays. BMC Complement Altern Med 11:64. https://doi.org/10.1186/1472-6882-11-64

Gul MZ, Chandrasekaran SKM, Bhat MY, Maurya R, Qureshi IA, Ghazi IA (2016) Bioassay-guided fractionation and in vitro anti-proliferative effects of fractions of *Artemisia nilagirica* on THP-1 cell line. Nutr Cancer 68:1210–1224

Gupta SC, Kim JH, Prasad S, Aggarwal BB (2010) Regulation of survival, proliferation, invasion, angiogenesis, and metastasis of tumor cells through modulation of inflammatory pathways by nutraceuticals. Cancer Metastasis Rev 29:405–434

Gutheil WG, Reed G, Ray A, Anant S, Dhar A (2012) Crocetin: an agent derived from saffron for prevention and therapy for cancer. Curr Pharm Biotechnol 13:173–179

Halder B, Singh S, Thakur SS (2015) *Withania somnifera* root extract has potent cytotoxic effect against human malignant melanoma cells. PLoS One 10:e0137498

Hazra B, Sarkar R, Bhattacharyya S, Roy P (2002) Tumour inhibitory activity of chicory root extract against Ehrlich ascites carcinoma in mice. Fitoterapia 73:730–733

Horneber M, Bueschel G, Dennert G, Less D, Ritter E, Zwahlen M (2012) How many cancer patients use complementary and alternative medicine: a systematic review and meta analysis. Integr Cancer Ther 11:187–203

Ibn AlBitar DAM (1874) AlJame Li-Mofradat al Adwiyah wal Aghthiyah (The collection of medical items and food). Two volumes. Dar Sader Publishing, Beirut (In Arabic)

Ibn Qiff Maseehi YNM (1995) Kitab Al-Umda fi Al-Jaraha (Urdu Translation), vol 1. Central Council for Research in Unani Medicine (CCRUM), Ministry of Health and Family Welfare, Government of India, New Delhi, pp 174–175

Ibn Rushd (1980) Kitab Al-Kulliyat (Urdu translation). 1st ed. Central Council for Research in Unani Medicine (CCRUM), Ministry of Health and Family Welfare, Government of India, New Delhi, pp 83–95

Ibn Sina (1992) Al-Qanoon fit Tibb (Urdu translation by Gulam Husain Kanturi), vol-4. Nigarishat Miyan Chambers, Lahore, pp 175–176

Jafari S, Saeidnia S, Hajimehdipoor H, Ardekani MR, Faramarzi MA, Hadjiakhoondi A, Khanavi M (2013) Cytotoxic evaluation of *Melia azedarach* in comparison with, *Azadirachta indica* and its phytochemical investigation. Daru 21:37. https://doi.org/10.1186/2008-2231-21-37

Jagetia GC, Baliga MS (2006) Evaluation of anticancer activity of the alkaloid fraction of *Alstonia scholaris* (Sapthaparna) *in vitro* and *in vivo*. Phytother Res 20:103–109

Jagetia GC, Venkatesh P, Baliga MS (2005) *Aegle marmelos* (L.) Correa inhibits the proliferation of transplanted Ehrlich ascites carcinoma in mice. Biol Pharm Bull 28:58–64

Jahan S, Chaudhary R, Goyal PK (2009) Anticancer activity of an Indian medicinal plant, *Alstonia scholaris*, on skin carcinogenesis in mice. Integr Cancer Ther 8:273–279

Jia L (2012) Cancer complementary and alternative medicine research at the US National Cancer Institute. Chin J Integr Med 18:325–332

John S, Sivakumar KC, Mishra R (2017) Bacoside-A induces tumor cell death in human glioblastoma cell lines through catastrophic macropinocytosis. Front Mol Neurosci 10:171. https://doi.org/10.3389/fnmol.2017.00171

Jurjani A (2010) Zakheera Khwarazm Shahi, vol 7. Idara Kitab AL-Shifa, New Delhi pp 30–32

Kabiruddin (2006) Alkarabadeen, 2nd edn. CCRUM, Ministry of Health and Family Welfare, Government of India, New Delhi, pp 1093–1094

Kalyani MI, Lingaraju SM, Salimath BP (2012) A pro-apoptotic 15-kDa protein from *Bacopa monnieri* activates caspase-3 and downregulates Bcl-2 gene expression in mouse mammary carcinoma cells. J Nat Med 67:123–136

Kaur V, Kumar M, Kaur P, Kaur S, Kaur S (2017) Inhibitory activities of butanol fraction from *Butea monosperma* (Lam.) Taub. bark against free radicals, genotoxins and cancer cells. Chem Biodivers 14:e1600484. https://doi.org/10.1002/cbdv.201600484

Khan A (2006) Avicenna (Ibn Sina): Muslim physician and philosopher of the eleventh century (Great Muslim Philosophers and Scientists of the Middle Ages). The Rosen Publishing Group, New York

Kim KB, Kim K, Bae S, Choi Y, Cha HJ, Kim SY, Lee JH, Jeon SH, Jung HJ, Ahn KJ, An IS, An S (2014) MicroRNA-1290 promotes asiatic acid-induced apoptosis by decreasing BCL2 protein level in A549 non-small cell lung carcinoma cells. Oncol Rep 32:1029–1036

Kou B, Liu W, Zhao W, Duan P, Yang Y, Yi Q, Guo F, Li J, Zhou J, Kou Q (2017) Thymoquinone inhibits epithelial-mesenchymal transition in prostate cancer cells by negatively regulating the TGF-β/Smad2/3 signaling pathway. Oncol Rep (Online). https://doi.org/10.3892/or.2017.6012

Lampronti I, Martello D, Bianchi N, Borgatti M, Lambertini E, Piva R, Jabbar S, Choudhuri MS, Khan MT, Gambari R (2003) In vitro antiproliferative effects on human tumor cell lines of extracts from the Bangladeshi medicinal plant *Aegle marmelos* Correa. Phytomedicine 10:300–308

Lébri M, Tilaoui M, Bahi C, Achibat H, Akhramez S, Fofié YBN, Gnahoué G, Lagou SM, Zirihi GN, Coulibaly A, Zyad A, Hafid A, Khouili M (2015) Phytochemical analysis and in vitro anticancer effect of aqueous extract of *Abrus precatorius* L. Der Pharma Chemica 7:112–117

Leelawat S, Leelawat K (2016) Molecular mechanisms of cholangiocarcinoma cell inhibition by medicinal plants. Oncol Lett 13:961–966

Li AX, Sun M, Li X (2017) Withaferin-A induces apoptosis in osteosarcoma U2OS cell line via generation of ROS and disruption of mitochondrial membrane potential. Eur Rev Med Pharmacol Sci 21:1368–1374

Majoosi AIA (2010) Kamil ul Sanaah (Urdu translation by Kantoori GH), vol 1. Idara Kitabus Shifa, New Delhi, pp 414–427

Mallick MN, Akhtar MS, Najm MZ, Tamboli ET, Ahmad S, Husain SA (2015) Evaluation of anticancer potential of *Bacopa monnieri* L. against MCF-7 and MDA-MB 231 cell line. J Pharm Bioallied Sci 7:325–328

Malvia S, Bagadi SA, Dubey US, Saxena S (2017) Epidemiology of breast cancer in Indian women. Asia Pac J Clin Oncol 13:289–295

McKenna MK, Gachuki BW, Alhakeem SS, Oben KN, Rangnekar VM, Gupta RC, Bondada S (2015) Anticancer activity of withaferin A in B-cell lymphoma. Cancer Biol Ther 16:1088–1098

Mirkin V, Berrebi A, Rakhman I, Haran M, Shvidel L (2017) The role of the plant *Artemisia* in survival and induction of apoptosis of b cells in Chronic Lymphocytic Leukemia (CLL). Harefuah 156:86–88

Mitra SK, Prakash NS, Sundaram R (2012) Shatavarins (containing Shatavarin IV) with anticancer activity from the roots of *Asparagus racemosus*. Indian J Pharm 44:732–736

Moongkarndi P, Kosem N, Luanratana O, Jongsomboonkusol S, Pongpan N (2004) Antiproliferative activity of Thai medicinal plant extracts on human breast adenocarcinoma cell line. Fitoterapia 75:375–377

More NV, Kharat AS (2016) Antifungal and anticancer potential of *Argemone mexicana* L. Medicines 3:28. https://doi.org/10.3390/medicines3040028

Mostofa AGM, Hossain MK, Basak D, Bin Sayeed MS (2017) Thymoquinone as a potential adjuvant therapy for cancer treatment: evidence from preclinical studies. Front Pharmacol 8:295. https://doi.org/10.3389/fphar.2017.00295

Murthy NS, Chaudhry K, Rath GK (2008) Burden of cancer and projections for 2016, Indian scenario: gaps in the availability of radiotherapy treatment facilities. Asian Pac J Cancer Prev 9:671–677

Nourazarian SM, Nourazarian A, Majidinia M, Roshaniasl E (2015) Effect of root extracts of medicinal herb *Glycyrrhiza glabra* on HSP90 gene expression and apoptosis in the HT-29 colon cancer cell line. Asian Pac J Cancer Prev 16:8563–8566

Pan X, Matsumoto M, Nakamura Y, Kikuchi T, Zhang J, Ukiya M, Suzuki T, Koike K, Akihisa R, Akihisa T (2014) Three new and other limonoids from the hexane extract of *Melia azedarach* fruits and their cytotoxic activities. Chem Biodivers 11:987–1000

Panneerselvam K, Lin SC, Liu CL, Liaw YC, Lin JY, Lu TH (2000) Crystallization of agglutinin from the seeds of *Abrus precatorius*. Acta Crystallogr D 56:898–899

Pittella F, Dutra RC, Junior DD, Lopes MT, Barbosa NR (2009) Antioxidant and cytotoxic activities of *Centella asiatica* (L.) Urb. Int J Mol Sci 10:3713–3721

PIB-Press Information Bureau (2016) Government of India, AYUSH, Government has Launched Special Programmes for Control and Prevention of Lifestyle Diseases. http://pib.nic.in/newsite/PrintRelease.aspx?relid=154809. Accessed on 22. Oct 2017

Qamari NBM (2008) Ghana Mina (Urdu translation), 1st ed. CCRUM, Ministry of Health and Family Welfare, Government of India, New Delhi, pp 508–510

Quraeishi AH (2002) Cancer ya *Sartan*: ek Ta'ruf. Pandhra Rozah-Tibb-e-Mashriq, Cancer Special Issue. New Public Press, Delhi, p 72

Rahamooz Haghighi S, Asadi MH, Akrami H, Baghizadeh A (2017) Anti-carcinogenic and anti-angiogenic properties of the extracts of *Acorus calamus* on gastric cancer cells. Avicenna J Phytomed 7:145–156

Razi AMBZ (2002) Kitab Al-Havi (Urdu translation by CCRUM), vol 12. CCRUM, Ministry of Health and Family Welfare, Government of India, New Delhi, pp 9–24

Ren L, Cao QX, Zhai FR, Yang SQ, Zhang HX (2016) Asiatic acid exerts anticancer potential in human ovarian cancer cells via suppression of PI3K/Akt/mTOR signalling. Pharm Biol 54:2377–2382

Rhazes (925) Al Hawy (The Comprehensive). Dar AlKalam Publishing Beirut (in Arabic)

Riaz M, Bilal A, Ali MS, Fatima I, Faisal A, Sherkheli MA, Asghar A (2017) Natural products from *Cuscuta reflexa* Roxb. with antiproliferation activities in HCT116 colorectal cell lines. Nat Prod Res 31:583–587

Saad B, Said O (2011) Greco-Arab and Islamic herbal medicine: traditional system, ethics, safety, efficacy, and regulatory issues. Wiley, USA

Saad B, Azaizeh H, Said O (2005) Tradition and perspectives of Arab herbal medicine: a review. Evid Based Compl Altern Med 2:475–479

Saad B, Azaizeh H, Said O (2008) Arab herbal medicine. Bot Med Clin Pract 4:31

Said O, Zaid H, Saad B (2011) Greco-Arab and Islamic herbal medicine and cancer treatment/prevention. In: Watson RR, Preedy RV (eds) Bioactive foods and extracts cancer treatment and prevention. CRC Press, Taylor and Francis, Boca Raton, pp 49–66

Saleh AM, Aljada A, Rizvi SA, Nasr A, Alaskar AS, Williams JD (2014) In vitro cytotoxicity of *Artemisia vulgaris* L. essential oil is mediated by a mitochondria-dependent apoptosis in HL-60 leukemic cell line. BMC Complement Altern Med 14:226

Samarghandian S, Hadjzadeh MA, Afshari JT, Hosseini M (2014) Antiproliferative activity and induction of apoptotic by ethanolic extract of *Alpinia galanga* rhizhome in human breast carcinoma cell line. BMC Complement Altern Med 14:192

Samarqandi AN (2007a) Sharah-e-Asbaab Tarjumah Kabeer: vol- III and IV (Urdu translation by Hakim Kabiruddin). Ejaj Publishing House, Darya Ganj, New Delhi, pp 223–226

Samarqandi AN (2007b) Sharah-e-AsbaabTarjumah Kabeer: vol-I (Urdu translation by HkmKabiruddin). Ejaj Publishing House, Darya Ganj, New Delhi

Samarqandi AN (2010) Moalajat Sharh-e-Asbab (Urdu translation by Khuwaja Rizwan), vol-3. CCRUM, Ministry of Health and Family Welfare, Government of India, New Delhi, pp 369–370

Shafi G, Hasan TN, Syed NA, Al-Hazzani AA, Alshatwi AA, Jyothi A, Munshi A (2012) *Artemisia absinthium* : a novel potential complementary and alternative medicine for breast cancer. Mol Biol Rep 39:7373–7379

Shamsi Y, Rehman RU (2012) Concept of cancer in Greco-Roman and Islamic medicine. Int J Biol Sci 3:63–67

Shamsi Y, Khan AA, Alam T, Jabeen A (2016) Management of cancer with Munzij and Mus'hil therapy: a regimen of Ilaj Bit Tadbeer (Regimenal therapy) in Unani system of medicine. Int J Curr Rev Res 8:4–9

Singh R, Sharma J, Goyal PK (2014) Prophylactic role of *Averrhoa carambola* (Star Fruit) extract against chemically induced hepatocellular carcinoma in swiss albino mice. Adv Pharmacol Sci 2014:158936. https://doi.org/10.1155/2014/158936

Singh S, Verma M, Malhotra M, Prakash S, Singh TD (2015) Cytotoxicity of alkaloids isolated from *Argemone mexicana* on SW480 human colon cancer cell line. Pharm Biol 54:740–745

Song NR, Lee E, Byun S, Kim JE, Mottamal M, Park JH, Lim SS, Bode AM, Lee HJ, Lee KW, Dong Z (2013) Isoangustone A, a novel licorice compound, inhibits cell proliferation by targeting PI3K, MKK4, and MKK7 in human melanoma. Cancer Prev Res 6:1293–1303

Subramaniam D, Giridharan P, Murmu N, Shankaranarayanan NP, May R, Houchen CW, Ramanujam RP, Balakrishnan A, Vishwakarma RA, Anant S (2008) Activation of apoptosis by 1-hydroxy-5,7-dimethoxy-2-naphthalene-carboxaldehyde, a novel compound from *Aegle marmelos*. Cancer Res 68:8573–8581

Subramaniyan B, Kumar V, Mathan G (2017) Effect of sodium salt of Butrin, a novel compound isolated from *Butea monosperma* flowers on suppressing the expression of SIRT1 and Aurora B kinase-mediated apoptosis in colorectal cancer cells. Biomed Pharmacother 90:402–413

Suresh V, Sruthi V, Padmaja B, Asha VV (2011) In vitro anti-inflammatory and anti-cancer activities of *Cuscuta reflexa* Roxb. J Ethnopharmacol 134:872–877

Tabri AAM (1995) Moalijat-e-Buqratia, vol 2. CCRUM, Ministry of Health and Family Welfare, Govt of India, New Delhi, pp 244–248

Tabri AAM (1997) Moalijat-e-Buqratia (Urdu translation) vol 2, CCRUM, Ministry of Health and Family Welfare, Govt of India, New Delhi

Tabri R (2010) Firdousul Hikmat, Idarae Kitabushifa, New Delhi

Velasco G, Sánchez C, Guzmán M (2016) Anticancer mechanisms of cannabinoids. Rev Curr Oncol 23:S23–S32

Verma SP, Tripathi VC, Das P (2014) *Asparagus racemosus* leaf extract inhibits growth of UOK 146 renal cell carcinoma cell line: simultaneous oncogenic PRCCTFE3 fusion transcript inhibition and apoptosis independent cell death. Asian Pac J Cancer Prev 15:1937–1941

World Cancer Report (2014) In: Stewart BW, Wild CP (eds) International Agency for Research on Cancer. World Health Organization (WHO), Lyon Cedex

World Health Organization (2013) World traditional medicine strategy (2014–2023). WHO Press, World Health Organization, Geneva

Yoder LH (2005) Let's talk cancer prevention. Med Surg Nursing 14:195–198

Yook JS, Kim M, Pichiah PB, Jung SJ, Chae SW, Cha YS (2015) The antioxidant properties and inhibitory effects on HepG2 cells of chicory cultivated using three different kinds of fertilizers in the absence and presence of pesticides. Molecules 20:12061–12075

Zhou X, Su J, Feng S, Wang L, Yin X, Yan J, Wang Z (2016) Antitumor activity of curcumin is involved in down-regulation of YAP/TAZ expression in pancreatic cancer cells. Onco Target 7:79076–79088

Chapter 11
Signaling Pathways of Anticancer Plants: Action and Reaction

Bilal Ahmad, Mohd Irfan Naikoo, Hassan Jaleel, Asfia Shabbir, Farha Rehman, Yawar Sadiq, and M. Masroor Akhtar Khan

Abstract Insights into the alterations of the mammalian genome in neoplastic diseases and the mechanism of action of the therapeutic anticancer drugs are one of the extremely important, diverse, and challenging areas of study currently. By the virtue of lingering toxicity of the reputable chemical drugs, plant-derived anticancer substances, viz., vinblastine, vincristine, Taxol, topotecan, camptothecin, and podophyllotoxin derivatives, are highly safe and efficient in the treatment and management of this monstrous disease. Among the list of accessible targets of the therapeutic drugs, DNA replication and mitosis, hormonal regulation of cell growth, aberrant signaling pathways, cell surface receptors, and second messengers are noteworthy. Nowadays, newer therapeutic approaches are being followed, and an increased understanding into the mechanism of action of the therapeutic anticancer agents is evolving due to continuous and relentless efforts of the researchers. The aim of the present chapter is to highlight the application of medicinal plants and their secondary metabolites as anticancer substances and also focus on the signaling aspects of potential anticancer compounds to find out their mechanisms of action against cancer cells.

Keywords Cancer cell · DNA methylation · Epigenetic factors · Secondary metabolites · Therapeutic potential

11.1 Introduction

Cancer, a neoplastic disease, exhibits the hallmark properties of metastasis and invasion. Despite over 110 years of exhausting cancer research, the reason behind this grave disease is still a subject of debate between diverse theories hypothesizing

B. Ahmad · M. I. Naikoo · H. Jaleel (✉) · A. Shabbir · Y. Sadiq · M. M. A. Khan
Department of Botany, Aligarh Muslim University, Aligarh, India

F. Rehman
Department of Botany, Mohammad Ali Jauhar University, Rampur, India

M. S. Akhtar, M. K. Swamy (eds.), *Anticancer Plants: Clinical Trials and Nanotechnology*, https://doi.org/10.1007/978-981-10-8216-0_11

either chromosomal or mutational alterations or epigenetic causes as reasons of cancer (Strauss 1992; Steinberg 2004). The cancer enigma is still mysterious as this debate has been monopolized by traditional theories of mutation, according to which the disease is a genetic one (Bishop 1987). The genetic reason of cancer development is the permanent mutations in the DNA of the cells resulting in the breakdown in the ability of the cell to control its own division. Normal cells repair most of the gene mutations while the cancer cells accumulate these mutations, and these become irreparable leading to abnormal proliferation of these cells (Krishnamurthi 2000). However, the reasons of cancer development are both genetic and epigenetic that result into the deregulated gene expression thereby leading to disruptive function (Jones and Baylin 2007). Moreover, epigenetic alterations unlike genetic ones are reversible hence the promising targets of the plant-derived drugs (Sharma et al. 2010; Best and Carey 2010).

Epigenetic mechanisms are indispensable for the maintenance of adult life. Deregulated gene expression is a consequence of disruption in the epigenetic processes which results in the development of grave diseases like cancer. Some of the important epigenetic components deregulated during the development of cancer include chromatin structure, DNA methylation, histone modifications, and noncoding RNAs. In view of the reversibility and reparability of epigenetic alterations, plant-derived epigenetic drugs or the drugs targeting epigenetic mechanisms have proven promising in cancer therapy (Mai and Altucci 2009; Sharma et al. 2010). Some of the most familiar epigenetic alterations that lead to the uncontrolled proliferation include methylation or deacetylation of histones and CpG (phosphorylated cytosine-guanine) islands methylation within the gene promoters (Esteller 2005; Jones and Baylin 2007; Best and Carey 2010). Epigenetic mechanisms pertaining to the alteration in the expression of certain important genes involved in cell proliferation and differentiation are often deregulated during the tumor development. Certain noncoding RNAs like miRNAs, siRNAs, etc. known for controlling the expression of plentiful cellular proteins by altering stability of the mRNA and/or translation constitute a vital part of the epigenome (Nelson and Weiss 2008; Guil and Esteller 2009). Evidences advocate that the epigenetic deregulation is only an introductory transforming event (Feinberg 2005). Epigenetic alterations commonly noticed in the preliminary-stage tumors include comprehensive DNA hypomethylation and hypermethylation of specific promoter sequences (Esteller 2007). This reveals that epigenetic changes comprise the early stages in the loss of cellular homeostasis and may pave way to the gene mutations and genomic instability. Deregulation of the epigenetic mechanisms lead to genetic instability which results in the inactivation of tumor suppressor genes leading to the activation of oncogenes (Ellis et al. 2009). Moreover, hypomethylation often takes place in the repetitive DNA sequences (heterochromatic DNA repeats and dispersed retrotransposons) (Ehrlich 2002).

11.2 Plant-Derived Anticancer Drugs

Even though a number of medicines are available to treat different types of cancers, no drug has been observed to be completely effective and harmless in anticancer therapy. The chief setback in the cancer chemotherapy is the lingering toxicity of the well-recognized chemical drugs. However, plant-derived products have proven safe and effective in the treatment as well as management of various types of cancers to an appreciable extent (Shukla and Mehta 2015). Herbal medicines are playing an irreplaceable role in the cancer prevention and treatment which implement their numerous medicinal effects and therapeutic benefits by restraining cancer, activating hormones and enzymes, promoting production of defensive enzymes including antioxidant enzymes, arousing DNA repair mechanism, and increasing immunity (Thakore et al. 2012). Vigorous investigations are in progress on a number of plant phytochemicals to gain insights into the antitumor mechanism of action against a variety of cancers, and a fair number of plants are under clinical trials for this purpose (Rao et al. 2008). A lot of natural products have been acknowledged as influential, effective, and intoxicating anticancer agents, and the anticancer properties of various phytochemicals or the plants as such are being identified. Plant-derived compounds have also proven promising in targeting these deregulated epigenetic processes as well as other genetic and cellular processes. The important anticancer drugs working through these different cellular and genetic as well as epigenetic mechanisms include polyphenols, terpenoids, alkaloids, and organosulfur compounds (Schneider-Stock et al. 2012). The aim of the present chapter is to highlight the application of medicinal plants and their secondary metabolites as anticancer substances and also focus on the signaling aspects of potential anticancer compounds to find out their mechanisms of action against cancer cells.

11.2.1 Alkaloids

Alkaloids are a remarkable assemblage of phytoprinciples, extracted from certain medicinal plants, having a diverse spectrum of effects ranging from analgesics and addictive to anticancer. These nitrogen-containing compounds do not follow into any proper classification, and no demarcated boundaries can distinguish them; however, most recent classifications rely on the carbon skeleton and the biosynthetic pathways (Meyers 2001). Owing to the substantial anticancer activity against a variety of cancers, these alkaloid structures are frequently mimicked and synthetic analogs in the form of chemical drugs are synthesized. Some of the important examples of anticancer alkaloids that are in wide clinical use throughout the world include vinblastine, vincristine, Taxol, camptothecin derivatives, topotecan, epipodophyllotoxins, etc. (Darwiche et al. 2007). The most commonly used anticancer mechanisms of the alkaloids involve cell cycle arrest at G2/M phase, microtubule destabilization leading to apoptosis, microtubule polymerization prevention, and inhibition of topoisomerase I activity (Li et al. 2012).

11.2.2 Polyphenols

Polyphenols, acknowledged for their anticancer properties, are ubiquitous in vegetables, fruits, and beverages and safe and well tolerated by humans except for a few cases where toxic effects have been reported (Thomasset et al. 2007). Polyphenols possess strong antioxidant effect which is the key property accredited for the chemopreventive and therapeutic effects against cancer. Among the other mechanisms of anticancer effects induced by polyphenols include antiproliferation, anti-inflammation, induction of cell cycle arrest as well as apoptosis, and inhibition of angiogenesis and metastasis (Lambert and Elias 2010; Araujo et al. 2011). Polyphenols have been categorized into flavonoids, stilbenes, tannins, coumarins, and phenolic acids. However, the flavonoids constitute the major and most copious group (Quideau et al. 2011). The best studied polyphenols for the anticancer activity include curcumin, resveratrol, and genistein.

11.2.3 Organosulfur Compounds

Organosulfurs like phenols are plant-derived drugs with substantial anticancer effects. Characterized by their sulfur group, these compounds occur naturally in a lot of plants, mainly garlic, coffee, nuts, and some tropical fruits. Isothiocyanates constitute the major subgroup of organosulfur compounds differentiated by the presence of an isothiocyanate group (-N=C=S); the central carbon renders it exceedingly electrophilic (Fahey and Talalay 1999). Sulforaphane, an important organosulfur compound belonging to isothiocyanate group, is found in vegetables belonging to cruciferous group. Present in fair amounts in broccoli, cauliflower, cabbage, and hoary weed, it is derived from glucoraphanin (a glucosinolate) by the activity of myrosinase enzyme upon damage while chewing or cooking. Sulforaphane exerts its antitumor effect by different mechanisms like cell cycle arrest at G2/M phase which has been observed to be associated with distorted microtubule dynamics (Azarenko et al. 2008). Altered cdc2 kinase activity, increased expression of proteins like cyclin B1 and p21, and histone H1 phosphorylation are among the other cell cycle arrest mechanisms induced by sulforaphane (Parnaud et al. 2004; Azarenko et al. 2008).

11.2.4 Terpenoids

Terpenoids are a very large and diverse class of plant secondary metabolites comprising a miscellaneous treasure of phytochemicals in drug discovery predominantly in the field of cancer (Huang et al. 2012). Terpenoids, derived from isoprene units by mevalonic acid and methylerythritol phosphate (MEP) pathways, are divided into monoterpenoids, diterpenoids, triterpenoids, tetraterpenoids, and sesquiterpenoids depending upon the number of isoprene units. The anticancer potential of this class of compounds have been a focus of vigorous and passionate research only recently. Several sesquiterpene lactones have been recently put to test during the clinical trials to investigate their anticancer potential, and promising results have been obtained (Ghantous et al. 2010). These studies have revealed that terpenoids prove as potent anticancer drugs at several tumor formation stages by inhibiting several processes (cell proliferation, angiogenesis, metastasis, and cell death). These terpenes induce apoptosis in the tumor cells by targeting specific molecular mechanisms, like inhibiting certain important signaling pathways like NF-κB, JAK-STAT, etc., blocking the activity of Ca^{2+} ATPase pump as well as the activities of DNA topoisomerases of both I and II classes, and modulating various epigenetic mechanisms, etc. (Huang et al. 2012; Ghantous et al. 2010). Furthermore, terpenoids are also known to induce cell death of malignant tumor cells by targeting extrinsic and intrinsic apoptotic pathways (Yang and Dou 2010). Parthenolide, betulinic acid, and triptolide are some of the important examples of terpenes inducing apoptotic pathways.

11.3 Plant-Derived Drugs as Epigenetic Modulators

Epigenetic variations and alterations are the hallmark of cancer. Insights into the cancer epigenetics have revealed the deregulation of epigenetic machinery components like chromatin structure, histone modifications, DNA methylation, and noncoding RNAs. Plant-derived compounds have proven promising in targeting epigenetic processes and are presently at the front of drug discovery. Besides targeting the proteins and enzymes at the genetic and cellular levels, plant-derived anticancer drugs also function by bringing about the alterations in the epigenetic mechanisms. Some of the important plant-derived anticancer products that function as epigenetic modulators include polyphenols, alkaloids, organosulfur compounds, and terpenoids.

11.3.1 Polyphenols as Epigenetic Modulators

Polyphenols include a very important class of secondary metabolites with strong anticancer properties (Scalbert et al. 2005; Zhao et al. 2010a). Flavonoids, the most important and best characterized class of polyphenols, include the flavonols, flavones, flavanones, anthocyanidins, catechins, and isoflavonoids (Jamison 2003). A number of polyphenolics intercede their anticancer and antitumor activities through the modulation of the acetylation pattern of some of the vital genes as well as by the inhibition of hypermethylation of several essential tumor suppressor genes, a landmark in the development of cancer (Hauser and Jung 2008). Noteworthy among these are anacardic acid and its derivatives, curcumin and its derivatives, epigallocatechin-3-gallate (EGCG), genistein, and resveratrol. Anacardic acid derivatives are major HATs inhibitors used in a variety of cancers including kidney, cervical, breast, and prostate cancers as well as in myeloid and lymphoid leukemias. Curcumin (diferuloylmethane), a natural flavonoid derived from the *Curcuma longa* rhizome, alters the functions of HATs and HDACs thus restricting the uncontrolled cell growth epigenetically through modulation of histone acetylation (Fu and Kurzrock 2010). Promotion of the deacetylation activity or the inhibition of the transferase activity leads to the decrease in the acetylation of H3 and H4 histones in particular as well as other histone proteins (Prives and Manley 2001; Chen et al. 2007). EGCG, the principle bioactive polyphenol found in green tea, is known for its antitumor properties like antiproliferative, anti-angiogenic, and apoptotic properties (Shankar et al. 2007; Shankar et al. 2008). Its anticancer functions are endorsed to the epigenetic modulation through histone modification and methylation of DNA (Mittal et al. 2003). Genistein (4′,5,7-trihydroxyisoflavone) is the most copious isoflavone found in *Glycine max* which has been in use as a chemopreventive drug against several cancers (Barnes 1995). Genistein displays estrogen agonist as well as antagonist effects and is known to suppress oxidative stress and inhibit angiogenesis and certain protein tyrosine kinases (Li et al. 2009). Similar in function to EGCG, genistein is known to decrease methylation of promoter sequences of many tumor suppressor genes thus reactivating them. Resveratrol, a phytoalexin found abundantly in nuts and red grapes, is well recognized for its diverse health benefits in neuro- and cardioprotection and prevention of breast, lung, kidney, and skin cancers (Wang et al. 2008; Lin et al. 2010). Resveratrol also reduces HAT activity of many important promoter sequences and deacetylates NF-kB leading to reduced transcriptional activity in many cancer types (Bourguignon et al. 2009; Yeung et al. 2004).

It has been recently shown that plant extracts that are rich in polyphenols possess epigenetic activities in a number of types of cancers by bringing about the modulation of some important promoter sequences of decisive tumor genes through methylation (Fini et al. 2007). Moreover, modulation of TBP-interacting protein (TIP60), p300, and PCAF in these cancers is also brought about by the anacardic acid and its analogues (Balasubramanyam et al. 2003; Chandregowda et al. 2009). Different polyphenols target different genes during the epigenetic modulation. An extract of chlorogenic and caffeic acids, catechin, and epicatechin inhibits the expression of two colon cancer genes (DNMT1 and DNMT3b) and reactivates the tumor suppressor genes by decreasing the methylation of important promoter sequences (*hMLH1*, *p14*, and *p16*) (Fini et al. 2007). Contrary to this, polyphenol extract (green tea rich) has been observed to decrease the methylation of CDX2 and p16 promoter sequences in the cells of primary gastric carcinoma as well as colon and gastric cancers (Yuasa et al. 2005, 2009). Green tea extract has further been found to be associated with the reduced promoter methylation of genes which are usually hyper-methylated in gastric cancers (Yuasa et al. 2009). On the other hand, polyphenol extracts (black tea) have been observed to inhibit HDAC1 expression, thereby modulating histone acetylation in the cells of hepatocarcinogenesis (Murugan et al. 2009).

11.3.2 Alkaloids as Epigenetic Modulators

Alkaloids, basic nitrogen-containing compounds, mostly comprise the well-recognized antitumor drugs, viz., the mitotic inhibitors, colchicines, vinblastine, and vincristine (Savel 1966). Procainamide is a non-nucleoside drug used against cardiac arrhythmias and shows potential anticancer (myeloid leukemia cells) activity through epigenetic modulation by affecting DNA methylation, specifically inhibiting the activity of DNMT1 in colon cancer cells (Lee et al. 2005). In addition, decreased methylation of RARβ, estrogen receptor (ER), and p16 promoter regions has also been endorsed to procainamide alkaloids which are proven effective against breast cancers (Segura-Pacheco et al. 2003). Mahanine, an alkaloid obtained from *Murraya koenigii* (curry leaf plant), not only reactivates the tumor suppressor gene (*RASSF1A*) which is epigenetically silenced during the cancer. In addition to the inhibition of the methylase activity, reactivation of *RASSFA1* has also been endorsed to the cell cycle arrest by specifically decreasing the expression of cyclin D1 in prostate cancer cells (Jagadeesh et al. 2007). Sanguinarine, a cell-permeable alkaloid purified from the roots of certain poppy and fumaria species like bloodroot or *Sanguinaria canadensis*, is a very effective cell growth inhibitor and inhibits the activity of NF-κB (Selvi et al. 2009; Chaturvedi et al. 1997). Unlike phenols and alkaloids like mahanine, sanguinarine has been observed to modulate histone expression other than DNA methylation. Further, it diffuses through cell membranes and has a strong affinity with the DNA. Its association with the core histones and chromatin and intercalation with the DNA consequently helps it in the alteration of the chromatin structure (Selvi et al. 2009).

11.3.3 Organosulfur Compounds as Epigenetic Modulators

Organosulfur compounds have been extensively applied in the avoidance of platelet aggregation as well as in cancer therapy (Moriarty et al. 2007). Sulforaphane (a derivative of glucoraphanin, extracted from the cruciferous vegetables like broccoli) is a vital organosulfur compound with miscellaneous biological effects comprising apoptosis, cell cycle arrest, and induction of heme oxygenase and phase II detoxifying enzyme, viz., NAD(P)H:quinone reductase (NQO1). Upregulation of phase II metabolism by sulforaphane helps in removing the genotoxins of the body helping to prevent the cancer at the initial phase (Ho et al. 2009). The anticancer effects of sulforaphane also mediate through epigenetic mechanisms, and it is a well-recognized histone deacetylase (HDAC) inhibitor (Ho et al. 2009). Diminished activity of HDAC leads to the enhanced promoter-specific acetylation of H3 and H4 in the cancer cells of the colon, kidney, and prostate (Myzak et al. 2004, 2007). Sulforaphane has recently been observed to inhibit HDAC6 specifically leading to the acetylation and subsequent degradation of Hsp90, an androgenic receptor (AR). In the absence of this AR chaperone, its target gene (*E-twenty-six-related gene, ERG*) expression is inhibited (Gibbs et al. 2009). ERG, an important transcription factor, is observed to be overexpressed in prostate cancers (Tomlins et al. 2005), which leads to increased growth and invasiveness (Tomlins et al. 2008).

11.3.4 Terpenoids as Epigenetic Modulators

Terpenes, hydrocarbon derivatives of isoprene units, have proven promising against cancer and inflammation and are in clinical trials of cancer at present (Ghantous et al. 2010). Parthenolide, a sesquiterpene lactone, is commonly obtained from *Tanacetum parthenium* commonly called as European feverfew herb. It is presently one of the most promising drugs against cancer in clinical development that is known to target cancer stem cells selectively without any effect on the normal ones (Guzman and Jordan 2005). Lycopene acts as an epigenetic modulator and brings about the GSTP1 (tumor suppressor gene) promoter demethylation, thereby expressing it in the breast cancer. It also demethylates HIN-1 and RARβ2 tumor suppressors partially but does not induce their expression in the breast carcinomas as in GSTP1 (King-Batoon et al. 2008). Certain terpenoids mediate a range of anticancer activities concurrently by modulating an array of epigenetic mechanisms. Triptolide obtained from *Tripterygium wilfordii* and thymoquinone purified from *Nigella sativa* inhibit polycomb group of proteins through modulation of chromatin modifications. It also inhibits histone methyltransferases, by decreasing the level of transcripts of catalytic subunits of polycomb recessive proteins (Zhao et al. 2010a). Triptolide has been recently observed to decrease the expression of HDAC8 mRNA and protein in myeloma cells (Zhao et al. 2010b). On the contrary, it enhances the methylation of *MMP*-9 promoter leading to the silencing of its expression in sarcoma cells (Yang et al. 2009). Thymoquinone upregulates p73, a regulator of cell cycle checkpoint, and stimulates apoptosis of lymphoma cells through the caspase-dependent mechanism (Alhosin et al. 2010).

11.4 Treatment of Cancer Involving Cellular Mechanisms

An orchestra of events is involved in cellular division and DNA replication and a set of protein phosphorylation cascades and many checkpoints which oversee these two crucial events and eventually lead to the completion of the cell cycle. Deregulation of the cell cycle is one among the ten cancer transformation hallmarks. Targeting one of its vital modulators (cyclins, cyclin-dependent kinases, and tumor suppressor proteins) is a very critical step in cell cycle arrest which consequently induces apoptosis of the cancer cell (Giacinti and Giordano 2006; Foster et al. 2010). A plethora of reports claim the pharmacological potential of biologically active compounds derived from the plants, especially with mounting exploration of fresh, innovative, and vastly successful candidates in the anticancer therapy. Phytochemicals have been observed to be implicated with various inflammatory phenomena as well as oncogenic transformation in a number of mammalian cells and tissues including cell cycle control alteration, angiogenesis, metastases, and evasion of apoptosis (Surh 2003). These plant-derived anticancer drugs are synergistic in action with those of the other pharmaceutical chemopreventive drugs which facilitate them in overcoming the resistance development in the uncontrolled proliferative cells, thus enabling the utilization of very small amounts of the anticancer drugs with enlarged anticancer efficiency (Liu 2004). The action of a phytochemical may include complete blockade of the metabolic alteration of a pro-carcinogen or suppression of the malignant transformation of a just commenced preneoplastic cell. Modulation of the various cellular networks and signaling events implicated in proliferation, invasion, and metastases is another important aspect and worth mentioning (Aggarwal and Shishodia 2006). Insights into the fundamental pharmacokinetic mechanisms through which these plant-derived drugs evoke the anticancer response involve a number of molecular targets. This panel comprises protein kinases of different classes like protein kinases A and C (PKA, PKC), mitogen-activated protein kinase (MAPK), and tyrosine kinase (TYK2), caspases and Bax apoptotic proteins, anti-apoptotic proteins (TRAF1, bcl2, survivin), transcription factors (p53, Ap1, Nrf2, NF-κB), growth factors (EGF, FGF, TNF, PDGF), cell cycle proteins (CDK1, CDK2, cyclin D, p21, p27), and cell adhesion molecules (ICAM-1, VCAM).

11.4.1 Phenols as Cellular Mechanism Modulators

Phenolics like flavanones, isoflavones, and lignans check the binding of the ligands (estrogen) to the cancer cell receptors, thereby reducing their proliferation. Apigenin is the flavone found in parsley, celery, and chamomile. Certain flavanones like apigenin extracted from chamomile and celery-like plants induce cell death in adenocarcinoma cells of the lungs by targeting leptin/leptin receptor pathway. This apoptosis of human EGF receptor 2 (HER-2) is extrinsic and caspase dependent which leads to the overexpression of BT-474 in the cells of breast cancer. Signal transducer and activator of transcription 3 (STAT3) signaling is inhibited in this overall process which leads to the caspase-dependent apoptosis (Seo et al. 2015). Curcumin, another important anticancer polyphenol obtained from the rhizomes of *Curcuma longa*, modulates a number of molecular components and halts the proliferation of the human glioblastoma cells. This modulation includes the upregulation of p21, p16, and p53; early growth response proteins (Egr-1); c-Jun N-terminal kinase (JNK); a member of ETS oncogenic family (ElK-1); extracellular signal-regulated kinase (Erk); caspase 3, 8, and 9 proteins; and Bcl-2-associated X protein (Bax) (Vallianou et al. 2015). It also downregulates the levels of cell division cycle protein 2 (cdc-2), NF-κB, cell division cycle protein 2 (cdc-2), mechanistic target of rapamycin (mTOR), p65, retinoblastoma protein (pRB), cyclin D1 proteins, B cell lymphoma protein (Bcl-xL), protein kinase B (Akt), cellular myelocytomatosis oncogenes (c-myc), etc. (Vallianou et al. 2015). Crocetin, a carotenoids obtained from *Crocus sativus* and *Gardenia jasminoides*, is very effective against cardiac hypertrophy. It has been observed to function through MEK-ERK1/2 pathway and GATA-binding protein 4 (Cai et al. 2009). Ellagic acid extracted from pomegranate restrains metastasis in many cancerous cells and stimulates apoptosis in cancer cells of the breast and prostate. Repression in the ornithine decarboxylase activity, enzyme signaling rapid proliferation of cells, and circumventing apoptosis by epigallocatechin gallate (EGCG) are an important cellular mechanism in a number of cancers. Obstruction of the epithelial mesenchymal transition is brought about by luteolin. Certain phenols (fisetin and hesperetin) alter an array of complex signaling networks like JAK-STAT, mTOR, MAPK, PI3K/Akt, NF-κB, and Wnt pathways leading to the cell cycle arrest in certain promyelocytic leukemias (Adan and Baran 2015). Genistein, an isoflavone obtained from soybean, inhibits the Akt and NF-κB signaling pathways (Li et al. 2012); gingerol inhibits the Erk1/2/JNK/AP-1 signaling and stimulates extrinsic caspase-dependent apoptosis of cancer cells (Radhakrishnan et al. 2014). Kaempferol acts via Erk1/2, Src, and Akt pathways and inhibits the growth as well as the migration of pancreatic cancer cells (Lee and Kim 2016). Cyanidin glycosides obtained from red berries execute anticancer and antioxidant functions through different mechanisms like suppression of the COX-2 expression in the cells of prostate carcinoma leading to their apoptosis and inhibition of matrix metallopeptidase-9 (MMP-9) expression in lung carcinoma cells (Singh et al. 2011).

11.4.2 Alkaloids as Cellular Mechanism Modulators

Alkaloids like vinca alkaloids and camptothecins have served as the well-reputed anticancer compounds (Darwiche et al. 2007). These compounds, mostly with ring structure containing basic nitrogen, provoke anticancer mechanisms mostly through arresting G2/M phase of the cell cycle and destabilize microtubule assembly leading to apoptosis, preventing polymerization of microtubules and topoisomerase I inhibition (Darwiche et al. 2007).The mechanisms of action are different for different alkaloids. In order to understand the diversity in the mechanism of action of the alkaloids, a handful of them have been discussed as follows.

Vinca alkaloids (antimicrotubule agents), alkaloids obtained from periwinkle, have been observed to be sedative, hypotensive, and anticancerous. These (vincristine, vinblastine vinorelbine, vindesine) alter the mitotic spindle dynamics by irreversibly binding to the receptor sites on β-tubulin obstructing microtubules depolymerization. Microtubules are very dynamic polymers (α,β-tubulin heterodimers). Two binding sites for the vinca alkaloids per mole of tubulin dimer have been observed. Vincristine prevents the chromosome separation at the metaphase, and vinblastine is known to arrest the mitotic phase of cell cycle by inappropriate kinetochore and mitotic spindle formation leading to apoptosis (Jordan 2002). Vinca alkaloids have been in use for more than 40 years in the treatment of different cancers, lymphomas, leukemias, Kaposi's sarcoma (KS), and breast, lung, and testicular cancers (Balunas and Kinghorn 2005). Likewise, taxanes (paclitaxel) also act as microtubule disruptors leading to cell cycle arrest and apoptosis (Jordan and Wilson 2004; Khazir et al. 2014). Taxanes inhibit this microtubule disassembly by stabilizing this dynamicity and alter the conformation of these microtubules (Prota et al. 2013). Taxanes have also been found to decrease the binding of proteins associated with the microtubules. Although binding of microtubule-associated proteins (MAP) induces conformational rearrangements of the tubulin proteins that lead to an overall stabilization, binding of taxanes to the MAP-microtubule complex further stabilizes this complex (Xiao et al. 2012). Such disturbances prevent the normal mitotic spindle formation leading to cell cycle arrest avoiding cell proliferation (Ganguly et al. 2010; Priyadarshini and Aparajitha 2012).

Colchicine, one among the most important bioactive nitrogen-containing alkaloids, is isolated from Liliaceae members like *Colchicum autumnale* and *Gloriosa superba* and has been in use as antimitotic drugs, since long (Dias et al. 2012). The antitumor properties have been endorsed to its irreversible binding to tubulin similar to those of vinca alkaloids, leading to hindered microtubule formation and in turn inhibiting the progression of the cell cycle as well as inducing apoptosis. Some of the colchicinoid drugs act as vascular disrupting agents (VDA) causing swift disruption and shutdown of the tumor vasculature which leads to tumor starvation and disproportionate cell death (Atkinson et al. 2010). Berberine, an isoquinoline alkaloid with potent apoptotic properties in cancer cells, is however different in action in comparison with the abovementioned alkaloids and interferes at a number of tumor progression stages and metastasis. Molecular targets of this alkaloid are diverse. Some of the crucial and noteworthy targets include DNA topoisomerases, telomerase, p53, COX-2, NF-jB, metalloproteinases, etc. (Tillhon et al. 2012). Additionally, berberine-induced antitumor effects have been observed in several hematological cancers involving G1 or G2/M phases of cell cycle arrest, intrinsic as well as extrinsic caspase-dependent apoptosis, involvement of Bcl-2 family proteins, and ROS production (Sun et al. 2009).

11.4.3 Organosulfur Compounds as Cellular Mechanism Modulators

Organosulfur compounds are reputed to modulate the activity of a group of related antioxidant enzymes imperative in carcinogen detoxification, namely, glutathione S-transferases (GST) (Sparnins et al. 1986). NRF2 (nuclear factor erythroid 2-related factor 2), an important transcription factor, is the chief regulator of the expression of antioxidant molecules within the cell (Sporn and Liby 2012). NRF2 arouses anti-stress signaling to repress electrophilic or oxidative stress and halts carcinogenesis thus leading to a protective response (Lee and Surh 2005). The phase II carcinogen detoxification enzymes (glutathione transferases, NAD(P)H:quinone oxidoreductase I, UDP-glucuronosyltransferase, and heme oxygenase-1) which function via ARE-NRF2 pathway are induced by sulforaphane, thereby helping to eliminate a diverse collection of oxidative and electrophilic toxicants prior to their devastation to vital cellular macromolecules (Misiewicz et al. 2004). Sulforaphane, a type of isothiocyanates, arrests cell cycle in G2/M phase which has been observed to be associated to the distorted microtubule dynamics (Jackson and Singletary 2004; Azarenko et al. 2008). This disturbed microtubule polarization leads to aneuploidy which is followed by apoptosis (Azarenko et al. 2008). Altered cdc2 kinase activity, enhanced expression of cyclin B1 and p21, and histone H1 phosphorylation are the other mechanisms of cell cycle arrest induced by sulforaphane (Azarenko et al. 2008; Mi et al. 2008). Additionally, improper activation of cdc2 which leads to cell cycle halt at G2/M phase is believed to be the reason for the discernible mitotic catastrophe (Jackson and Singletary 2004). Confirmation of the cell death in MCF-7 cells by mitotic catastrophe after exposure to sulforaphane has been observed by abnormal mitosis and micro-nucleation (Jackson and Singletary 2004).

11.4.4 Terpenoids as Cellular Mechanism Modulators

Several stages of tumor progression (cell proliferation, cell death, angiogenesis, and metastasis) are inhibited by the application of terpenoids. These target various precise molecular mechanisms like blockade of the calcium ATPase pump of sarcoendoplasmic reticulum; inhibition of JAK-STAT, NF-κB, AP-1, MMPs, and DNA polymerases; and activation of p53 leading to the programmed cell death of the cancer cells. Specific and alternative mechanisms of cell death have been endorsed to these alkaloids noteworthy among which are triptolide, parthenolide, and betulinic acid.

Parthenolide, a sesquiterpene lactone originally extracted and purified from *Tanacetum parthenium*, has grabbed substantial attention in the discovery of cancer drugs (Ghantous et al. 2013). It induces intrinsic as well as extrinsic apoptosis in a vast array of tumor cells mainly by disrupting NF-κB; regulating MAPK, JNK, and Bcl-2 family members; activating p53 signaling; interfering with STAT3; and generating ROS (Kreuger et al. 2012). Triptolide, a diterpenoid isolated from *Tripterygium wilfordii* roots, is a potent modulator of the transcriptional machinery and has been observed to act together with RNA polymerase leading to the modulation of transcription of some of the key transcription factors (NF-κB, p53, HSF-1, and NF-AT) (McCallum et al. 2007). It was established very recently that triptolide inhibits overall transcription in cancer cell genes due to the proteasome-mediated degradation of RNA polymerase II large subunit (Wang et al. 2011). Furthermore, it induces apoptosis in a wide spectrum of tumors containing wild-type as well as mutant forms of p53 (Yang et al. 2003). Moreover, this diterpenoid has also been observed to inhibit the expression of p53 regulator (MDM2) and also the downregulation of the apoptosis inhibitor (XIAP) in a p53-independent manner. It also inhibits the heat shock protein (HSP70) which has been found to be upregulated in tumor formation (Yang et al. 2003; McCallum et al. 2007; Phillips et al. 2007).

Betulinic acid, a triterpenoid, induces cell death by affecting the intrinsic apoptotic pathway leading to the increased permeability of mitochondrial membrane resulting in the increased liberation of cytochrome c. The other routes of inducing apoptosis by this terpenoid include inhibition of NF-κB activity and the regulation of members of Bcl-2 family proteins. Additionally, betulinic acid is known to have anti-metastatic and anti-angiogenic activities (Gheorgheosu et al. 2014; Csuk 2014). Furthermore, betulinic acid treatment has been shown to induce apoptosis via both ROS-mediated mitochondrial pathway and the ER pathway in HeLa cells (Xu et al. 2014).

11.5 Clinical Trials of Plant-Derived Drugs

Though a good number of plant-derived anticancer drugs have been put to clinical trials and their synthetic analogues have been prepared after analyzing their efficacy, we are restricting our discussion to only a few important ones. Curcumin and its synthetic analogues have been found significantly effective in inhibiting growth factors that mediated corneal neovascularization and angiogenesis *in vivo* as well as *in vitro* (Arbiser et al. 1998). In addition to these in vitro antitumor effects, curcumin is known for its effectiveness against gastric and colon cancers in rodents (Ikezaki et al. 2001). The protective effect of curcumin is endorsed to its ability to restrain the proliferation of several tumors and modulation in the activity of certain angiogenesis-associated genes (Arbiser et al. 1998; Toi et al. 2001). Additionally, curcumin has been put to trials and found effective in inhibiting the growth of about 19 diverse strains of *H. pylori* (Mahady et al. 2002).

Vinca alkaloids (vinblastine, vincristine) extracted from periwinkle plant opened new doors of anticipation in the use of plant-derived components as anticancer drugs. The initial clinical trial was carried out to make the possible use of these alkaloids as anticancer agents (Cragg and Newman 2005). These two vinca alkaloids have been put to trial along with other chemotherapeutic medicines for fighting a variety of tumors and cancers including leukemias, advanced testicular cancer, lymphomas, lung and breast cancers, and Kaposi's sarcoma (Cragg and Newman 2005). Similarly, paclitaxel extracted from *Taxus brevifolia* has revealed immense anticancer potential against advanced breast and ovarian cancer and lung cancer (Rowinsky et al. 1992). Camptothecin isolated from *Camptotheca acuminate* was also put to clinical trials; however, owing to a number of disputes on its severe bladder toxicity questioned its use as an anticancer drug (Potmeisel and Pinedo 1995). Its semisynthetic derivatives like irinotecan and topotecan have been exploited effectively in the treatment of ovarian and colorectal cancers as well as small-cell lung cancers (Creemers et al. 1996; Bertino 1997).

The chemoprotective effects of resveratrol have yet to be established in humans though these have been well standardized in animals. The shortcomings of resveratrol to establish its anticancer effect in humans are its meager availability and speedy metabolism limiting its occurrence in the bloodstream (Tome-Carneiro et al. 2013). Additionally, the actual metabolite(s) accountable for the resveratrol anticancer effects are still a mystery. In view of this, passionate research to enhance its bioavailability through encapsulating it alone or after combining it with standard chemotherapeutic drugs is being conducted (Mohan et al. 2014). In spite of these shortcomings, a good number of clinical trials are being conducted to analyze the effects of resveratrol on human health. Two inimitable alkaloids (schischkinnin and montamine) isolated from the seeds of *Centaurea schischkinii* and *Centaurea montana* when put to clinical trials showed remarkable anticancer potential in the cell lines of human colon cancer (Shoeb 2006). The exceptional structural features of these two alkaloids can be used as a template to a considerable level to develop drugs with improved anticancer activity.

11.6 Conclusions and Future Prospects

Presently the interpretation and understanding into the mechanism of cross talk between epigenetics and genetics are still illusive; however, pharmaco-epigenomics is continuously evolving and the approaches to target cancer cells using epigenetic mechanisms are becoming more comprehensive. Both types of alternative strategies (inhibition as well as activation) to counteract processes and enzymes (DNA methylation/demethylation; HATs/HDACs) are being followed. A perfect drug will alter histones like enhancing their degree of acetylation in the region of passive tumor suppressor genes of the chromatin to reactivate them. Such an ideal epi-drug will also lead to modification of the chromatin around the oncogenes to inhibit their activity. Equally, a drug of such kind can also encourage global DNA hypomethylation in tumor cells but will avoid such a change in normal cells. Further, such a drug will reduce promoter DNA methylation of tumor suppressor genes inactivated during the development of tumor to reactivate them by carrying out efficient promoter hypermethylation of the oncogenes leading to their reduced expression. Moreover, a number of medicinal plant products are being aimed at targeting several other genetic and cellular targets of destabilized and mutated cancer cells in addition to the epigenetic targets. Since the use of phytoproducts as anticancer agents is very promising and is increasing swiftly, the appropriate scientific study with miscellaneous and biologically important chemical structures of phytochemicals as well as their thorough anticancer role along with the clinical aspects can prove to be fascinating subject of cancer research in the future.

References

Adan A, Baran Y (2015) The pleiotropic effects of fisetin and hesperetin on human acute promyelocytic leukemia cells are mediated through apoptosis, cell cycle arrest, and alterations in signaling networks. Tumor Biol 36:8973–8984

Aggarwal BB, Shishodia S (2006) Guggulsterone inhibits NF-kappa and kappa B alpha kinase activation, suppresses expression of anti-apoptotic gene products, and enhances apoptosis. Biochem Pharmacol 71:1397–1421

Alhosin M, Abusnina A, Achour M, Sharif T, Muller C, Peluso J, Chataigneau T, Lugnier C, Schini-Kerth VB, Bronner C, Fuhrmann G (2010) Induction of apoptosis by thymoquinone in lymphoblastic leukemia Jurkat cells is mediated by a p73-dependent pathway which targets the epigenetic integrator UHRF1. Biochem Pharmacol 79:1251–1260

Araujo JR, Goncalves P, Martel F (2011) Chemopreventive effect of dietary polyphenols in colorectal cancer cell lines. Nutr Res 31:77–87

Arbiser JL, Klauber N, Rohan R, van Leeuwen R, Huang MT, Fisher C, Flynn E, Byers HR (1998) Curcumin is an *in vivo* inhibitor of angiogenesis. Mol Med 4:376–383

Atkinson JM, Falconer RA, Edwards DR, Pennington CJ, Siller CS, Shnyder SD, Bibby MC, Patterson LH, Loadman PM, Gill JH (2010) Development of a novel tumor-targeted vascular disrupting agent activated by membrane-type matrix metalloproteinases. Cancer Res 70:6902–6912

Azarenko O, Okouneva T, Singletary KW, Jordan MA, Wilson L (2008) Suppression of microtubule dynamic instability and turnover in MCF7 breast cancer cells by sulforaphane. Carcinogenesis 29:2360–2368

Balasubramanyam K, Swaminathan V, Ranganathan A, Kundu TK (2003) Small molecule modulators of histone acetyltransferase p300. J Biol Chem 278:19134–19140

Balunas MJ, Kinghorn AD (2005) Drug discovery from medicinal plants. Life Sci 78:431–441

Barnes S (1995) Effect of genistein on *in vitro* and *in vivo* models of cancer. J Nutr 125:S777–S783

Bertino JR (1997) Irinotecan for colorectal cancer. Semin Oncol 24:S18–S23

Best J, Carey N (2010) Epigenetic opportunities and challenges in cancer. Drug Discov Today 15:65–70

Bishop JM (1987) The molecular genetics of cancer. Science 235:305–311

Bourguignon L, Xia W, Wong G (2009) Hyaluronan mediated CD44 interaction with p300 and SIRT1 regulates β-catenin signaling and NFkB-specific transcription activity leading to MDR1 and Bcl-xL gene expression and chemoresistance in breast tumor cells. J Biol Chem 284:2657–2671

Cai J, Yi FF, Bian ZY, Shen DF, Yang L, Yan L et al (2009) Crocetin protects against cardiac hypertrophy by blocking MEK-ERK1/2 signaling pathway. J Cell Mol Med 13:909–925

Chandregowda V, Kush A, Reddy G (2009) Synthesis of benzamide derivatives of anacardic acid and their cytotoxic activity. Eur J Med Chem 44:2711–2719

Chaturvedi MM, Kumar A, Darnay BG, Chainy GB, Agarwal S, Aggarwal BB (1997) Sanguinarine (Pseudochelerythrine) is a potent inhibitor of NF-kB activation, Ik-Bα phosphorylation, and degradation. J Biol Chem 272:30129–30134

Chen Y, Shu W, Chen W, Wu Q, Liu H, Cui G (2007) Curcumin, both histone deacetylase and p300/CBP specific inhibitor, represses the activity of nuclear factor kappa B and Notch 1 in Raji cells. Basic Clin Pharmacol Toxicol 101:427–433
Cragg GM, Newman DJ (2005) Plants as a source of anticancer agents. J Ethanopharmacol 100:72–79
Creemers GJ, Bolis G, Gore M, Scarfone G, Lacave AJ, Guastalla JP, Despax R, Favalli G, Kreinberg R, Van Belle S, Hudson I, Verweij J, Ten Bokkel Huinink WW (1996) Topotecan, an active drug in the second-line treatment of epithelial ovarian cancer: results of a large European phase II study. J Clin Oncol 14:3056–3061
Csuk R (2014) Betulinic acid and its derivatives: a patent review (2008–2013). Exp Opin Ther Pat 24:913–923
Darwiche N, El-Banna S, Gali-Muhtasib H (2007) Cell cycle modulatory and apoptotic effects of plant-derived anticancer drugs in clinical use or development. Expert Opin Drug Discov 2:361–379
Dias DA, Urban S, Roessner U (2012) A historical overview of natural products in drug discovery. Metabolites 2:303–336
Ehrlich M (2002) DNA methylation in cancer: too much, but also too little. Oncogene 21:5400–5413
Ellis L, Atadja PW, Johnstone RW (2009) Epigenetics in cancer: targeting chromatin modifications. Mol Cancer Ther 8:1409–1420
Esteller M (2005) DNA methylation and cancer therapy: new developments and expectations. Curr Opin Oncol 17:55–60
Esteller M (2007) Cancer epigenomics: DNA methylomes and histone-modification maps. Nat Rev Genet 8:286–298
Fahey JW, Talalay P (1999) Antioxidant functions of sulforaphane: a potent inducer of phase II detoxication enzymes. Food Chem Toxicol 37:973–979
Feinberg AP (2005) Cancer epigenetics is no Mickey mouse. Cancer Cell 8:267–268
Fini L, Selgrad M, Fogliano V, Graziani G, Romano M, Hotchkiss E, Daoud YA, De Vol EB, Boland CR, Ricciardiello L (2007) Annurca apple polyphenols have potent demethylating activity and can reactivate silenced tumor suppressor genes in colorectal cancer cells. J Nutr 137:2622–2628
Foster DA, Yellen P, Xu L, Saqcena M (2010) Regulation of G1 cell cycle progression: distinguishing the restriction point from a nutrient-sensing cell growth checkpoint(s). Genes Cancer 1:1124–1131
Fu S, Kurzrock R (2010) Development of curcumin as an epigenetic agent. Cancer Ghantous A, Gali-Muhtasib H, Vuorela H, Saliba NA, Darwiche N (2010) What made sesquiterpene lactones reach cancer clinical trials. Drug Discov Today 15:668–678
Ganguly A, Yang H, Cabral F (2010) Paclitaxel-dependent cell lines reveal a novel drug activity. Mol Cancer Ther 9:2914–2923
Ghantous A, Gali-Muhtasib H, Vuorela H, Saliba NA, Darwiche N (2010) What made sesquiterpene lactones reach cancer clinical trials. Drug Discov Today 15:668–678
Ghantous A, Sinjab A, Herceg Z, Darwiche N (2013) Parthenolide: from plant shoots to cancer roots. Drug Discov Today 18:894–905
Gheorgheosu D, Duicu O, Dehelean C, Soica C, Muntean D (2014) Betulinic acid as a potent and complex antitumor phytochemical: a mini review. Anticancer Agents Med Chem 14:936–945
Giacinti C, Giordano A (2006) RB and cell cycle progression. Oncogene 25:5220–5227
Gibbs A, Schwartzman J, Deng V, Alumkal J (2009) Sulforaphane destabilizes the androgen receptor in prostate cancer cells by inactivating histone deacetylase 6. Proc Natl Acad Sci U S A 106:16663–16668
Guil S, Esteller M (2009) DNA methylomes, histone codes and miRNAs: tying it all together. Int J Biochem Cell Biol 41:87–95
Guzman M, Jordan C (2005) Feverfew: weeding out the root of leukaemia. Exp Opin Biol Ther 5:1147–1152

Hauser AT, Jung M (2008) Targeting epigenetic mechanisms: potential of natural products in cancer chemoprevention. Planta Med 74:1593–1601

Ho E, Clarke JD, Dashwood RH (2009) Dietary sulforaphane, a histone deacetylase inhibitor for cancer prevention. J Nutr 139:2393–2396

Huang M, Lu JJ, Huang MQ, Bao JL, Chen XP, Wang YT (2012) Terpenoids: natural products for cancer therapy. Exp Opin Invest Drugs 21:1801–1818

Ikezaki S, Nishikawa A, Furukawa F, Kudo K, Nakamura H, Tamura K, Mori H (2001) Chemopreventive effects of curcumin on glandular stomach carcinogenesis induced by N-methyl-N′-nitro-N-nitrosoguanidine and sodium chloride in rats. Anticancer Res 21:3407–3411

Jackson SJ, Singletary KW (2004) Sulforaphane inhibits human MCF-7 mammary cancer cell mitotic progression and tubulin polymerization. J Nutr 134:2229–2236

Jagadeesh S, Sinha S, Pal B, Bhattacharya S, Banerjee P (2007) Mahanine reverses an epigenetically silenced tumor suppressor gene RASSF1A in human prostate cancer cells. Biochem Biophys Res Commun 362:212–217

Jamison JR (2003) Clinical guide to nutrition and dietary supplements in disease management, 1st edn. Churchill Livingstone, London

Jones PA, Baylin SB (2007) The epigenomics of cancer. Cell 128:683–692

Jordan MA (2002) Mechanism of action of antitumor drugs that interact with microtubules and tubulin. Curr Med Chem Anticancer Agents 2:1–17

Jordan MA, Wilson L (2004) Microtubules as a target for anticancer drugs. Nat Rev Cancer 4:253–266

Khazir J, Mir BA, Pilcher L, Riley DL (2014) Role of plants in anticancer drug discovery. Phytochem Lett 7:173–181

King-Batoon A, Leszczynska JM, Klein CB (2008) Modulation of gene methylation by genistein or lycopene in breast cancer cells. Environ Mol Mutagen 49:36–45

Kreuger MR, Grootjans S, Biavatti MW, Vandenabeele P, D'Herde K (2012) Sesquiterpene lactones as drugs with multiple targets in cancer treatment: focus on parthenolide. Anticancer Drugs 23:883–896

Krishnamurthi K (2000) Screening of natural products for anticancer and antidiabetic properties. Health Adm 20:69–75

Lambert JD, Elias RJ (2010) The antioxidant and prooxidant activities of green tea polyphenols: a role in cancer prevention. Arch Biochem Biophys 501:65–72

Lee J, Kim JH (2016) Kaempferol inhibits pancreatic cancer cell growth and migration through the blockade of EGFR-related pathway in vitro. PLoS One 11:e0155264

Lee JS, Surh YJ (2005) Nrf2 as a novel molecular target for chemoprevention. Cancer Lett 224:171–184

Lee BH, Yegnasubramanian S, Lin X, Nelson WG (2005) Procainamide is a specific inhibitor of DNA methyltransferase 1. J Biol Chem 280:40749–40756

Li Y, Liu L, Andrews LG, Tollefsbol TO (2009) Genistein depletes telomerase activity through cross-talk between genetic and epigenetic mechanisms. Int J Cancer 125:286–296

Li XL, Hu YJ, Wang H, Yu BQ, Yue HL (2012) Molecular spectroscopy evidence of berberine binding to DNA: comparative binding and thermodynamic profile of intercalation. Biomacromolecules 13:873–880

Lin JN, Lin VC, Rau KM, Shieh PC, Kuo DH, Shieh JC, Chen WJ, Tsai SC, Way TD (2010) Resveratrol modulates tumor cell proliferation and protein translation via SIRT1-dependent AMPK activation. J Agric Food Chem 58:1584–1592

Liu RH (2004) Potential synergy of phytochemicals in cancer prevention: mechanism of action. J Nutr 134:3479–3485

Mahady GB, Pendland SL, Yun G, Lu ZZ (2002) Turmeric (*Curcuma longa*) and curcumin inhibit the growth of *Helicobacter pylori*, a group 1 carcinogen. Anticancer Res 22:4179–4181

Mai A, Altucci L (2009) Epi-drugs to fight cancer: from chemistry to cancer treatment, the road ahead. Int J Biochem Cell Biol 41:199–213

McCallum C, Kwon S, Leavitt P, Shen DM, Liu W, Gurnett A (2007) Triptolide binds covalently to a 90 kDa nuclear protein. Role of epoxides in binding and activity. Immunobiology 212:549–556
Meyers R (2001) Encyclopedia of physical science and technology, 3rd edn. Academic, San Diego
Mi L, Xiao Z, Hood BL, Dakshanamurthy S, Wang X, Govind S, Conrads TP, Veenstra TD, Chung FL (2008) Covalent binding to tubulin by isothiocyanates. A mechanism of cell growth arrest and apoptosis. J Biol Chem 283:22136–22146
Misiewicz I, Skupińska K, Kowalska E, Lubiński J, Kasprzycka-Guttman T (2004) Sulforaphane-mediated induction of a phase 2 detoxifying enzyme NAD(P)H:quinone reductase and apoptosis in human lymphoblastoid cells. Acta Biochim Pol 51:711–721
Mittal A, Piyathilake C, Hara Y, Katiyar SK (2003) Exceptionally high protection of photocarcinogenesis by topical application of (-)-epigallocatechin-3-gallate in hydrophilic cream in SKH-1 hairless mouse model: relationship to inhibition of UVB-induced global DNA hypomethylation. Neoplasia 5:555–565
Mohan A, Narayanan S, Sethuraman S, Krishnan UM (2014) Novel resveratrol and 5 fluorouracil coencapsulated in PEGylated nanoliposomes improve chemotherapeutic efficacy of combination against head and neck squamous cell carcinoma. Bio Med Res Intern 2014:424239
Moriarty R, Naithani R, Surve B (2007) Organosulfur compounds in cancer chemoprevention. Mini Rev Med Chem 7:827–838
Murugan R, Vinothini G, Hara Y, Nagini S (2009) Black tea polyphenols target matrix metalloproteinases, RECK, proangiogenic molecules and histone deacetylase in a rat hepatocarcinogenesis model. Anticancer Res 29:2301–2305
Myzak M, Karplus P, Chung F, Dashwood R (2004) A novel mechanism of chemoprotection by sulforaphane: inhibition of histone deacetylase. Cancer Res 64:5767–5774
Myzak M, Tong P, Dashwood W, Dashwood R, Ho E (2007) Sulforaphane retards the growth of human PC-3 xenografts and inhibits HDAC activity in human subjects. Exp Biol Med 232:227–234
Nelson KM, Weiss GJ (2008) MicroRNAs and cancer: past, present, and potential future. Mol Cancer Ther 7:3655–3660
Parnaud G, Li P, Cassar G, Rouimi P, Tulliez J, Combaret L, Gamet-Payrastre L (2004) Mechanism of sulforaphane-induced cell cycle arrest and apoptosis in human colon cancer cells. Nutr Cancer 48:198–206
Phillips PA, Dudeja V, McCarroll JA, Borja-Cacho D, Dawra RK, Grizzle WE, Vickers SM, Saluja AK (2007) Triptolide induces pancreatic cancer cell death via inhibition of heat shock protein 70. Cancer Res 67:9407–9416
Potmeisel M, Pinedo H (1995) Camptothecins: new anticancer agents. CRC Press, Boca Raton, pp 149–150
Prives C, Manley JL (2001) Why is p53 acetylated. Cell 107:815–818
Priyadarshini K, Aparajitha UK (2012) Paclitaxel against cancer: a short review. Med Chem 2:139–141
Prota AE, Bargsten K, Zurwerra D, Field JJ, Díaz JF, Altmann KH, Steinmetz MO (2013) Molecular mechanism of action of microtubule-stabilizing anticancer agents. Science 339:587–590
Quideau S, Deffieux D, Douat-Casassus C, Pouysegu L (2011) Plant polyphenols: chemical properties, biological activities, and synthesis. Angew Chem 50:586–621
Radhakrishnan EK, Bava SV, Narayanan SS, Nath LR, Thulasidasan AKT, Soniya EV et al (2014) 6-Gingerol induces caspase dependent apoptosis and prevents PMA- induced proliferation in colon cancer cells by inhibiting MAPK/AP-1 signaling. PLoS One 9:e104401
Rao GV, Kumar S, Islam M, Saber EM (2008) Folk medicines for anticancer therapy-a current status. Cancer Ther 6:913–922
Rowinsky EK, Onetto N, Canetta RM, Arbuck SG (1992) Taxol-the 1st of the texanes, an important new class of anti-tumor agents. Semin Oncol 19:646–662
Savel H (1966) The metaphase arresting plant alkaloids and cancer chemotherapy. Prog Exp Tumor Res 8:189–224

Scalbert A, Manach C, Morand C, Remesy Jimenez L (2005) Dietary polyphenols and the prevention of diseases. Crit Rev Food Sci Nutr 45:287–306
Schneider-Stock R, Ghantous A, Bajbouj K, Saikali M, Darwiche N (2012) Epigenetic mechanisms of plant-derived anticancer drugs. Front Biosci 17:129–173
Segura-Pacheco B, Trejo-Becerril C, Perez-Cardenas E, Taja-Chayeb L, Mariscal I, Chavez A, Acuña C, Salazar AM, Lizano M, Dueñas-Gonzalez A (2003) Reactivation of tumor suppressor genes by the cardiovascular drugs hydralazine and procainamide and their potential use in cancer therapy. Clin Cancer Res 9:1596–1603
Selvi R, Pradhan BSK, Shandilya J, Das C, Sailaja BS, Shankar N, Gadad GSS, Reddy A, Dasgupta D, Kundu TK (2009) Sanguinarine interacts with chromatin, modulates epigenetic modifications, and transcription in the context of chromatin. Chem Biol 16:203–216
Seo HS, Jo JK, Ku JM, Choi HS, Choi YK et al (2015) Induction of caspase dependent extrinsic apoptosis through inhibition of signal transducer and activator of transcription 3 (STAT3) signaling in HER2-overexpressing BT-474 breast cancer cells. Biosci Rep 35:e00276
Shankar S, Suthakar G, Srivastava RK (2007) Epigallocatechin-3-gallate inhibits cell cycle and induces apoptosis in pancreatic cancer. Front Biosci 12:5039–5051
Shankar S, Ganapathy S, Hingorani SR, Srivastava RK (2008) EGCG inhibits growth, invasion, angiogenesis and metastasis of pancreatic cancer. Front Biosci 13:440–452
Sharma S, Kelly TK, Jones PA (2010) Epigenetics in cancer. Carcinogenesis 31:27–36
Shoeb M (2006) Anticancer agents from medicinal plants. Bangladesh J Pharmacol 1:35–41
Shukla S, Mehta A (2015) Anticancer potential of medicinal plants and their phytochemicals: a review. Braz J Bot 38:199–210
Singh BN, Shankar S, Srivastava RK (2011) Green tea catechin, epigallocatechin-3-gallate (EGCG): mechanisms, perspectives and clinical applications. Biochem Pharmacol 82:1807–1821
Sparnins VL, Mott AW, Barany G, Wattenberg LW (1986) Effects of allyl methyl trisulfide on glutathione S-transferase activity and BP-induced neoplasia in the mouse. Nutr Cancer 8:211–215
Sporn MB, Liby KT (2012) NRF2 and cancer: the good, the bad and the importance of context. Nat Rev Cancer 12:564–571
Steinberg D (2004) Appraising aneuploidy as a cancer cause. The Sci 18:26–27
Strauss BS (1992) The origin of point mutations in human tumor cells. Cancer Res 52:249–253
Sun Y, Xun K, Wang Y, Chen X (2009) A systematic review of the anticancer properties of berberine, a natural product from Chinese herbs. Anti-Cancer Drugs 20:757–769
Surh YJ (2003) Cancer chemoprevention with dietary phytochemicals. Nat Rev Cancer 3:768–780
Thakore P, Mani RK, Singh J, Kavitha (2012) A brief review of plants having anticancer property. Int J Pharm Res Dev 3:129–136
Thomasset SC, Berry DP, Garcea G, Marczylo T, Steward WP, Gescher AJ (2007) Dietary polyphenolic phytochemicals promising cancer chemopreventive agents in humans. A review of their clinical properties. Int J Cancer 3:451–458
Tillhon M, Guaman Ortiz LM, Lombardi P, Scovassi AI (2012) Berberine: new perspectives for old remedies. Biochem Pharmacol 84:1260–1267
Toi M, Matsumoto T, Bando H (2001) Vascular endothelial growth factor: its prognostic, predictive, and therapeutic implications. Lancet Oncol 2:667–673
Tome-Carneiro J, Larrosa M, Gonzalez-Sarrias A, Tomas-Barberan FA, Garcia-Conesa MT, Espin JC (2013) Resveratrol and clinical trials: the crossroad from in vitro studies to human evidence. Curr Pharm Des 19:6064–6093
Tomlins SA, Rhodes DR, Perner S, Dhanasekaran SM, Mehra R, Sun XW, Varambally S, Cao X, Tchinda J, Kuefer R, Lee C, Montie JE, Shah RB, Pienta KJ, Rubin MA, Chinnaiyan AM (2005) Recurrent fusion of TMPRSS2 and ETS transcription factor genes in prostate cancer. Science 310:644–648
Tomlins SA, Laxman B, Varambally S, Cao X, Yu J, Helgeson BE, Cao Q, Prensner JR, Rubin MA, Shah RB, Mehra R, Chinnaiyan AM (2008) Role of the TMPRSS2-ERG gene fusion in prostate cancer. Neoplasia 10:177–188

Vallianou NG, Evangelopoulos A, Schizas N, Kazazis C (2015) Potential anticancer properties and mechanisms of action of curcumin. Anticancer Res 35:645–651

Wang R, Zheng Y, Kim H, Xu X, Cao L, Lahusen T, Lee M, Xiao C, Vassilopoulos A, Chen W (2008) Interplay among BRCA1, SIRT1, and Survivin during BRCA1-associated tumorigenesis. Mol Cell 32:11–20

Wang Y, Lu JJ, He L, Yu Q (2011) Triptolide (TPL) inhibits global transcription by inducing proteasome-dependent degradation of RNA polymerase II. PLoS One 6:e23993

Xiao H, Wang H, Zhang X, Tu Z, Bulinski C, Khrapunovich-Baine M, Angeletti RH, Horwitz SB (2012) Structural evidence for cooperative microtubule stabilization by Taxol and the endogenous dynamics regulator MAP 4. ACS Chem Biol 7:744–752

Xu T, Pang Q, Zhou D, Zhang A, Luo S, Wang Y, Yan X (2014) Proteomic investigation into betulinic acid-induced apoptosis of human cervical cancer HeLa cells. PLoS One 9:e105768

Yang H, Dou QP (2010) Targeting apoptosis pathway with natural terpenoids: implications for treatment of breast and prostate cancer. Curr Drug Targets 11:733–744

Yang S, Chen J, Guo Z, Xu XM, Wang L, Pei XF, Yang J, Underhill CB, Zhang L (2003) Triptolide inhibits the growth and metastasis of solid tumors. Mol Cancer Ther 2:65–72

Yang S, Wen H, Zhang G, Zhao S, Luo Y, Lu Q (2009) Triptolide evaluates DNA methylation level of matrix metalloproteinase 9 gene in human fibrosarcoma HT-1080 cells. China J Chin Mat Med 34:611

Yeung F, Hoberg JE, Ramsey CS, Keller MD, Jones DR, Frye RA, Mayo MW (2004) Modulation of NF-αB-dependent transcription and cell survival by the SIRT1 deacetylase. EMBO J 23:2369–2380

Yuasa Y, Nagasaki H, Akiyama Y, Sakai H, Nakajima T, Ohkura Y, Takizawa T, Koike M, Tani M, Iwai T (2005) Relationship between CDX2 gene methylation and dietary factors in gastric cancer patients. Carcinogenesis 26:193–200

Yuasa Y, Nagasaki H, Akiyama Y, Hashimoto Y, Takizawa T, Kojima K, Kawano T, Sugihara K, Imai K, Nakachi K (2009) DNA methylation status is inversely correlated with green tea intake and physical activity in gastric cancer patients. Int J Cancer 124:2677–2682

Zhao F, Chen Y, Li R, Liu Y, Wen L, Zhang C (2010a) Triptolide alters histone H3K9 and H3K27 methylation state and induces G0/G1 arrest and caspase dependent apoptosis in multiple myeloma in vitro. Toxicology 267:70–79

Zhao F, Zeng LL, Chen Y, Li R, Liu Y, Wen L, Cheng YQ, Zhang C (2010b) Effects of triptolide on histone acetylation and HDAC8 expression in multiple myeloma. Chin J Cancer Res 22:148–155

MIX
Papier aus verantwortungsvollen Quellen
Paper from responsible sources
FSC® C105338
FSC
www.fsc.org

If you have any concerns about our products,
you can contact us on
ProductSafety@springernature.com

In case Publisher is established outside the EU,
the EU authorized representative is:
Springer Nature Customer Service Center GmbH
Europaplatz 3, 69115 Heidelberg, Germany

Printed by Libri Plureos GmbH
in Hamburg, Germany